TEACHER'S PLANNING GUIDE

Project-Based Inquiry Science™

GOOD FRIENDS and GERMS

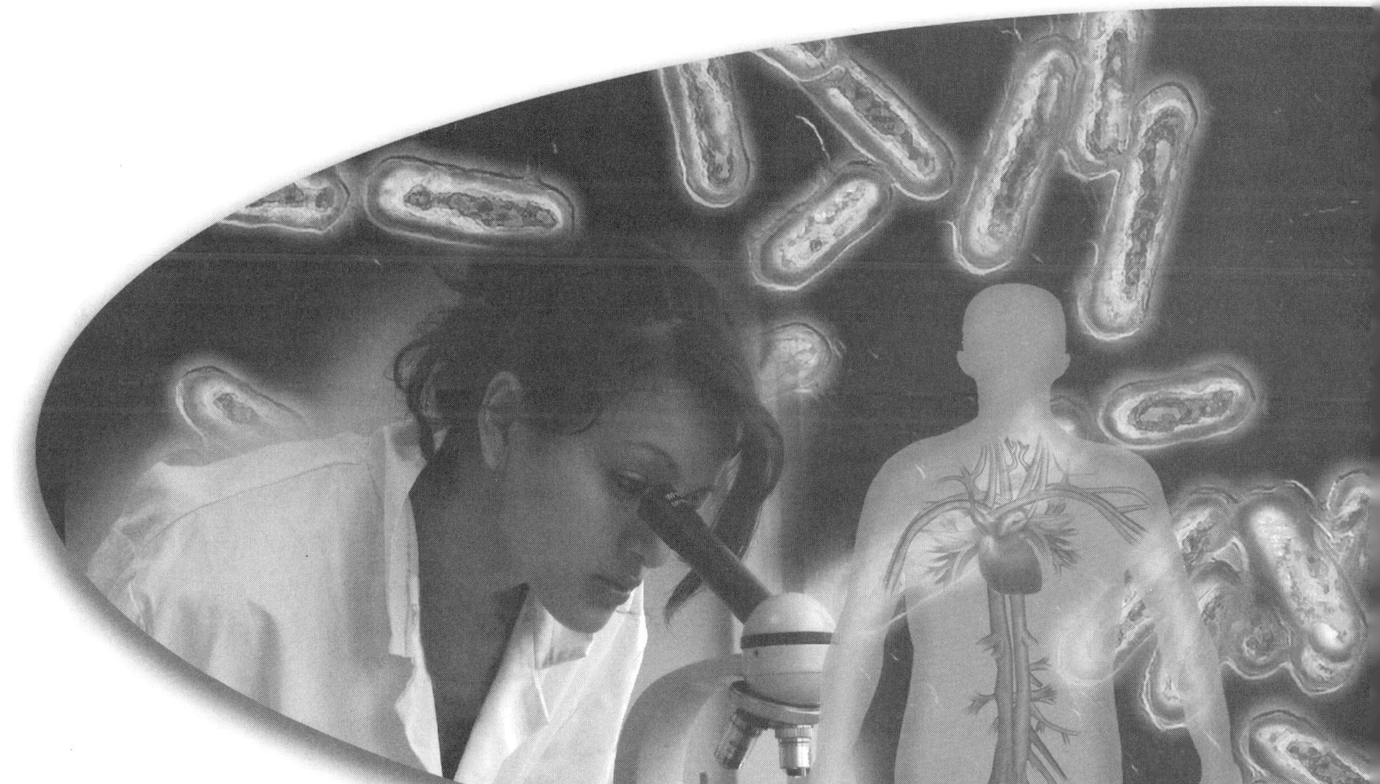

NSF

IT's ABOUT TIME®

HERFF JONES EDUCATION DIVISION

84 Business Park Drive, Armonk, NY 10504
Phone (914) 273-2233 Fax (914) 273-2227
www.its-about-time.com

Program Components

Student Edition
Teacher's Planning Guide
Teacher's Resources Guide
Durable Equipment Kit
Consumable Equipment Kit

Printed and bound in the United States of America.

ISBN-13: 978-1-58591-631-3
1 2 3 4 5 VH 12 11 10 09 08

This project was supported, in part, by the ***National Science Foundation*** under grant nos. 0137807, 0527341, 0639978.
Opinions expressed are those of the authors and not necessarily those of the National Science Foundation.

PBIS Principal Investigators

Janet L. Kolodner is a Regents' Professor in the School of Interactive Computing in the Georgia Institute of Technology's College of Computing. Since 1978, her research has focused on learning from experience, both in computers and in people. She pioneered the Artificial Intelligence method called *case-based reasoning*, providing a way for computers to solve new problems based on their past experiences. Her book, *Case-Based Reasoning*, synthesizes work across the case-based reasoning research community from its inception to 1993.

Since 1994, Dr. Kolodner has focused on the applications and implications of case-based reasoning for education. In her approach to science education, called Learning by Design™ (LBD), students learn science while pursuing design challenges. Dr. Kolodner has investigated how to create a culture of collaboration and rigorous science talk in classrooms, how to use a project challenge to promote focus on science content, and how students learn and develop when classrooms function as learning communities. Currently, Dr. Kolodner is investigating how to help young people come to think of themselves as scientific reasoners. Dr. Kolodner's research results have been widely published, including in *Cognitive Science, Design Studies,* and the *Journal of the Learning Sciences.*

Dr. Kolodner was founding Director of Georgia Tech's EduTech Institute, served as coordinator of Georgia Tech's Cognitive Science program for many years, and is founding Editor in Chief of the *Journal of the Learning Sciences*. She is a founder of the International Society for the Learning Sciences, and she served as its first Executive Officer. She is a fellow of the American Association of Artificial Intelligence.

Joseph S. Krajcik is a Professor of Science Education and Associate Dean for Research in the School of Education at the University of Michigan. He works with teachers in science classrooms to bring about sustained change by creating classroom environments in which students find solutions to important intellectual questions that subsume essential curriculum standards and use learning technologies as productivity tools. He seeks to discover what students learn in such environments, as well as to explore and find solutions to challenges that teachers face in enacting such complex instruction.

Dr. Krajcik has authored and co-authored over 100 manuscripts and makes frequent presentations at international, national, and regional conferences that focus on his research, as well as presentations that translate research findings into classroom practice. He is a fellow of the American Association for the Advancement of Science and served as president of the National Association for Research in Science Teaching. Dr. Krajcik co-directs the Center for Highly Interactive Classrooms, Curriculum and Computing in Education at the University of Michigan and is a co-principal investigator in the Center for Curriculum Materials in Science and The National Center for Learning and Teaching Nanoscale Science and Engineering. In 2002, Dr. Krajcik was honored to receive a Guest Professorship from Beijing Normal University in Beijing, China. In winter 2005, he was the Weston Visiting Professor of Science Education at the Weizmann Institute of Science in Rehovot, Israel.

Daniel C. Edelson is Vice President for Education and Children's Programs at the National Geographic Society. Previously, he was the director of the Geographic Data in Education (GEODE) Initiative at Northwestern University, where he led the development of Planetary Forecaster and Earth Systems and Processes. Since 1992, Dr. Edelson has directed a series of projects exploring the use of technology as a catalyst for reform in science education and has led the development of a number of software environments for education. These include My World GIS, a geographic information system for inquiry-based learning, and WorldWatcher, a data visualization and analysis system for gridded geographic data. Dr. Edelson is the author of the high school environmental science text, *Investigations in Environmental Science: A Case-Based Approach to the Study of Environmental Systems*. His research has been widely published, including in the *Journal of the Learning Sciences*, the *Journal of Research on Science Teaching*, *Science Educator*, and *Science Teacher*.

Brian J. Reiser is a Professor of Learning Sciences in the School of Education and Social Policy at Northwestern University. Professor Reiser served as chair of Northwestern's Learning Sciences Ph.D. program from 1993, shortly after its inception, until 2001. His research focuses on the design and enactment of learning environments that support students' inquiry in science, including both science curriculum materials and scaffolded software tools. His research investigates the design of learning environments that scaffold scientific practices, including investigation, argumentation, and explanation; design principles for technology-infused curricula that engage students in inquiry projects; and the teaching practices that support student inquiry. Professor Reiser also directed BGuILE (Biology Guided Inquiry Learning Environments) to develop software tools for supporting middle school and high school students in analyzing data and constructing explanations with biological data. Reiser is a co-principal investigator in the NSF Center for Curriculum Materials in Science. He served as a member of the NRC panel authoring the report Taking Science to School.

Mary L. Starr is a Research Specialist in Science Education in the School of Education at the University of Michigan. She collaborates with teachers and students in elementary and middle school science classrooms around the United States who are implementing *Project-Based Inquiry Science*. Before joining the PBIS team, Dr. Starr created professional learning experiences in science, math, and technology, designed to assist teachers in successfully changing their classroom practices to promote student learning from coherent inquiry experiences. She has developed instructional materials in several STEM areas, including nanoscale science education, has presented at national and regional teacher education and educational research meetings, and has served in a leadership role in the Michigan Science Education Leadership Association. Dr. Starr has authored articles and book chapters, and has worked to improve elementary science teacher preparation through teaching science courses for pre-service teachers and acting as a consultant in elementary science teacher preparation. As part of the PBIS team, Dr. Starr has played a lead role in making units cohere as a curriculum, in developing the framework for PBIS Teachers Planning Guides, and in developing teacher professional development experiences and materials.

Acknowledgements

Three research teams contributed to the development of *Project-Based Inquiry Science* (PBIS): a team at the Georgia Institute of Technology headed by Janet L. Kolodner, a team at Northwestern University headed by Daniel Edelson and Brian Reiser, and a team at the University of Michigan headed by Joseph Krajcik and Ron Marx. Each of the PBIS units was originally developed by one of these teams and then later revised and edited to be a part of the full three-year middle-school curriculum that became PBIS.

PBIS has its roots in two educational approaches, Project-Based Science and Learning by Design™. Project-Based Science suggests that students should learn science through engaging in the same kinds of inquiry practices scientists use, in the context of scientific problems relevant to their lives and using tools authentic to science. Project-Based Science was originally conceived in the hi-ce Center at the University of Michigan, with funding from the National Science Foundation. Learning by Design™ derives from Problem-Based Learning and suggests sequencing, social practices, and reflective activities for promoting learning. It engages students in design practices, including the use of iteration and deliberate reflection. LBD was conceived at the Georgia Institute of Technology, with funding from the National Science Foundation, DARPA, and the McDonnell Foundation.

The development of the integrated PBIS curriculum was supported by the National Science Foundation under grants no. 0137807, 0527341, and 0639978. Any opinions, findings and conclusions, or recommendations expressed in this material are those of the authors and do not necessarily reflect the views of the National Science Foundation.

PBIS Team

Principal Investigator
Janet L. Kolodner

Co-Principal Investigators
Daniel C. Edelson
Joseph S. Krajcik
Brian J. Reiser

NSF Program Officer
Gerhard Salinger

Curriculum Developers
Michael T. Ryan
Mary L. Starr

Teacher's Planning Guide Developers
Rebecca M. Schneider
Mary L. Starr

Literacy Specialist
LeeAnn M. Sutherland

NSF Program Reviewer
Arthur Eisenkraft

Project Coordinator
Juliana Lancaster

External Evaluators
The Learning Partnership
Steven M. McGee
Jennifer Witers

The Georgia Institute of Technology Team

Project Director:
Janet L. Kolodner

Development of PBIS units at the Georgia Institute of Technology was conducted in conjunction with the Learning by Design™ Research group (LBD), Janet L. Kolodner, PI.

Lead Developers, Physical Science:
David Crismond
Michael T. Ryan

Lead Developer, Earth Science:
Paul J. Camp

Assessment and Evaluation:
Barbara Fasse
Daniel Hickey
Jackie Gray
Laura Vandewiele
Jennifer Holbrook

Project Pioneers:
JoAnne Collins
David Crismond
Joanna Fox
Alice Gertzman
Mark Guzdial
Cindy Hmelo-Silver
Douglas Holton
Roland Hubscher
N. Hari Narayanan
Wendy Newstetter
Valery Petrushin
Kathy Politis
Sadhana Puntambekar
David Rector
Janice Young

The Northwestern University Team

Project Directors:
Daniel Edelson
Brian Reiser

Lead Developer, Biology:
David Kanter

Lead Developers, Earth Science:
Jennifer Mundt Leimberer
Darlene Slusher

Development of PBIS units at Northwestern was conducted in conjunction with:

The Center for Learning Technologies in Urban Schools (LeTUS) at Northwestern, and the Chicago Public Schools
Louis Gomez, PI;
Clifton Burgess, PI
for Chicago Public Schools.

The BioQ Collaborative
David Kanter, PI.

The Biology Guided Inquiry Learning Environments (BGuILE) Project
Brian Reiser, PI.

The Geographic Data in Education (GEODE) Initiative
Daniel Edelson, Director

The Center for Curriculum Materials in Science at Northwestern
Brian Reiser,
Daniel Edelson,
Bruce Sherin, PIs.

The University of Michigan Team

Project Directors:
Joseph Krajcik
Ron Marx

Literacy Specialist:
LeeAnn M. Sutherland

Project Coordinator:
Mary L. Starr

Development of PBIS units at the University of Michigan was conducted in conjunction with:

The Center for Learning Technologies in Urban Schools (LeTUS)
Ron Marx, Phyllis Blumenfeld,
Barry Fishman,
Joseph Krajcik,
Elliot Soloway, PIs.

The Detroit Public Schools
Juanita Clay-Chambers
Deborah Peek-Brown

The Center for Highly Interactive Computing in Education (hi-ce)
Ron Marx,
Phyllis Blumenfeld,
Barry Fishman,
Joseph Krajcik,
Elliot Soloway,
Elizabeth Moje,
LeeAnn Sutherland, PIs.

Field-Test Teachers

National Field Test
Tamica Andrew
Leslie Baker
Jeanne Bayer
Gretchen Bryant
Boris Consuegra
Daun D'Aversa
Candi DiMauro
Kristie L. Divinski
Donna M. Dowd
Jason Fiorito
Lara Fish
Christine Gleason
Christine Hallerman
Terri L. Hart-Parker
Jennifer Hunn
Rhonda K. Hunter
Jessica Jones
Dawn Kuppersmith
Anthony F. Lawrence
Ann Novak
Rise Orsini
Tracy E. Parham
Cheryl Sgro-Ellis
Debra Tenenbaum
Sarah B. Topper
Becky Watts
Debra A. Williams
Ingrid M. Woolfolk
Ping-Jade Yang

New York City Field Test
Several sequences of PBIS units have been field- tested in New York City under the leadership of Whitney Lukens, Staff Developer for Region 9, and Greg Borman, Science Instructional Specialist, New York City Department of Education

6th Grade
Norman Agard
Tazinmudin Ali
Heather Guthartz Aniba
Asher Arzonane
Asli Aydin
Shareese Blakely
John J. Blaylock
Joshua Blum
Tsedey Bogale
Filomena Borrero
Zachary Brachio
Thelma Brown
Alicia Browne-Jones
Scott Bullis
Maximo Cabral
Lionel Callender
Matthew Carpenter
Ana Maria Castro
Diane Castro
Anne Chan
Ligia Chiorean
Boris Consuegra
Careen Halton Cooper
Cinnamon Czarnecki
Kristin Decker
Nancy Dejean
Gina DiCicco
Donna Dowd
Lizanne Espina
Joan Ferrato
Matt Finnerty
Jacqueline Flicker
Helen Fludd
Leigh Summers Frey
Helene Friedman-Hager
Diana Gering
Matthew Giles
Lucy Gill
Steven Gladden
Greg Grambo
Carrie Grodin-Vehling
Stephan Joanides
Kathryn Kadei
Paraskevi Karangunis
Cynthia Kerns
Martine Lalanne
Erin Lalor
Jennifer Lerman
Sara Lugert
Whitney Lukens
Dana Martorella
Christine Mazurek
Janine McGeown
Chevelle McKeever
Kevin Meyer
Jennifer Miller
Nicholas Miller
Diana Neligan
Caitlin Van Ness
Marlyn Orque
Eloisa Gelo Ortiz
Gina Papadopoulos
Tim Perez
Albertha Petrochilos
Christopher Poli
Kristina Rodriguez
Nadiesta Sanchez
Annette Schavez
Hilary Sedgwitch
Elissa Seto
Laura Shectman
Audrey Shmuel
Katherine Silva
Ragini Singhal
C. Nicole Smith
Gitangali Sohit
Justin Stein
Thomas Tapia
Eilish Walsh-Lennon
Lisa Wong
Brian Yanek
Cesar Yarleque
David Zaretsky
Colleen Zarinsky

7th Grade
Mayra Amaro
Emmanuel Anastasiou
Cheryl Barnhill
Bryce Cahn
Ligia Chiorean
Ben Colella
Boris Consuegra
Careen Halton Cooper
Elizabeth Derse
Urmilla Dhanraj
Gina DiCicco
Lydia Doubleday
Lizanne Espina
Matt Finnerty
Steven Gladden
Stephanie Goldberg
Nicholas Graham
Robert Hunter
Charlene Joseph
Ketlynne Joseph
Kimberly Kavazanjian
Christine Kennedy
Bakwah Kotung
Lisa Kraker
Anthony Lett
Herb Lippe
Jennifer Lopez
Jill Mastromarino
Kerry McKie
Christie Morgado
Patrick O'Connor
Agnes Ochiagha
Tim Perez
Nadia Piltser
Chris Poli
Carmelo Ruiz
Kim Sanders
Leslie Schiavone
Ileana Solla
Jacqueline Taylor
Purvi Vora
Ester Wiltz
Carla Yuille
Marcy Sexauer Zacchea
Lidan Zhou

8th Grade
Emmanuel Anastasio
Jennifer Applebaum
Marsha Armstrong
Jenine Barunas
Vito Cipolla
Kathy Critharis
Patrecia Davis
Alison Earle
Lizanne Espina
Matt Finnerty
Ursula Fokine
Kirsis Genao
Steven Gladden
Stephanie Goldberg
Peter Gooding
Matthew Herschfeld
Mike Horowitz
Charlene Jenkins
Ruben Jimenez
Ketlynne Joseph
Kimberly Kavazanjian
Lisa Kraker
Dora Kravitz
Anthony Lett
Emilie Lubis
George McCarthy
David Mckinney
Michael McMahon
Paul Melhado
Jen Miller
Christie Morgado
Ms. Oporto
Maria Jenny Pineda
Anastasia Plaunova
Carmelo Ruiz
Riza Sanchez
Kim Sanders
Maureen Stefanides
Dave Thompson
Matthew Ulmann
Maria Verosa
Tony Yaskulski

Good Friends and Germs

Good Friends and Germs is adapted from the unit, *How can good friends make me sick?* developed at University of Michigan as part of the work of the Center for Learning Technologies in Urban Schools and as a project of University of Michigan's Center for Highly Interactive Computing in Education. *Good Friends and Germs* was developed by the Project-Based Inquiry Science team in conjunction with the development team at the University of Michigan.

Project Directors:

Joseph Krajcik

Ron Marx

Lead Developer:

Barbara Hug

Other developers:

Amy Wefel

University of Michigan School of Public Health collaborators:

Toby Citrin

Renee Bayer

Detroit Public Schools Urban Systemic Initiative collaborator:

Deborah Peek¬-Brown

Pilot teachers

Yulonda Hale

Barbara Case

Susan Clark

Alicia Merriweather

Joy Reynolds

PBIS Team

Lead Developer:

Mary L. Starr

Contributing Field-test Teachers

Asher Arzonane

Suzy Bachman

Greg Borman

Matthew Carpenter

Anne Chan

Lizanne Espina

Enrique Garcia

Steven Gladden

Greg Grambo

Lillian Arlia Grippo

Dani Horowitz

Nicole Shiu Horowitz

Stephan Joanides

Verneda Johnson

Sunny Kam

Crystal Marsh

Kristin McNichol

Melissa Nathan

Tim Perez

Christopher Poli

Nadiesta Sanchez

Caitlin Van Ness

Melanie Wenger

Cesar Yarleque

Renee Zalewitz

The development of *Good Friends and Germs* was supported by National Science Foundation under grants no. 9553583, 9818828, and 0208059. The development of *How can good friends make me sick?* was supported by a grant from the Centers for Disease Control and Prevention and monitored by the Michigan Department of Community Health under contract no. DHHS-CDC-20030755 and by the National Science Foundation under grant no. 0830 310 A605. Any opinions, findings, and conclusions or recommendations expressed in this material are those of the authors and do not necessarily reflect the views of the CDC or the National Science Foundation.

Good Friends and Germs Teacher's Planning Guide

Learning Set 2

What Kinds of Things Make You Sick?

Science Concepts: *Symptoms, infection, gastrointestinal disease, sanitation and hygiene, animal cells and plant cells, microscope use, one-celled organisms, metabolism and homeostasis, microbiology, growing bacteria, viruses, experiment planning, explanation, use of evidence, recommendation.*

Learning Set 4

How Do Scientists Identify and Stop Disease Outbreaks?

Science Concepts: *Disease outbreak, virus mutation, disease containment, tracking disease, epidemic, explanation, use of evidence, recommendation.*

Answer the Big Question

How Can You Prevent Your Good Friends from Getting Sick?

Welcome to Project-Based Inquiry Science!

Welcome to Project-Based Inquiry Science (PBIS): A Middle-School Science Curriculum!

This year, your students will be learning the way scientists learn, exploring interesting questions and challenges, reading about what other scientists have discovered, investigating, experimenting, gathering evidence, and forming explanations. They will learn to collaborate with others to find answers and to share their learning in a variety of ways. In the process, they will come to see science in a whole new, exciting way that will motivate them throughout their educational experiences and beyond.

What is PBIS?

In project-based inquiry learning, students investigate scientific content and learn science practices in the context of attempting to address challenges in or answer questions about the world around them. Early activities introducing students to a challenge help them to generate issues that need to be investigated, making inquiry a student-driven endeavor. Students investigate as scientists would, through observations, designing and running experiments, designing, building, and running models, reading written material, and so on, as appropriate. Throughout each project, students might make use of technology and computer tools that support their efforts in observation, experimentation, modeling, analysis, and reflection. Teachers support and guide the student inquiries by framing the guiding challenge or question, presenting crucial lessons, managing the sequencing of activities, and

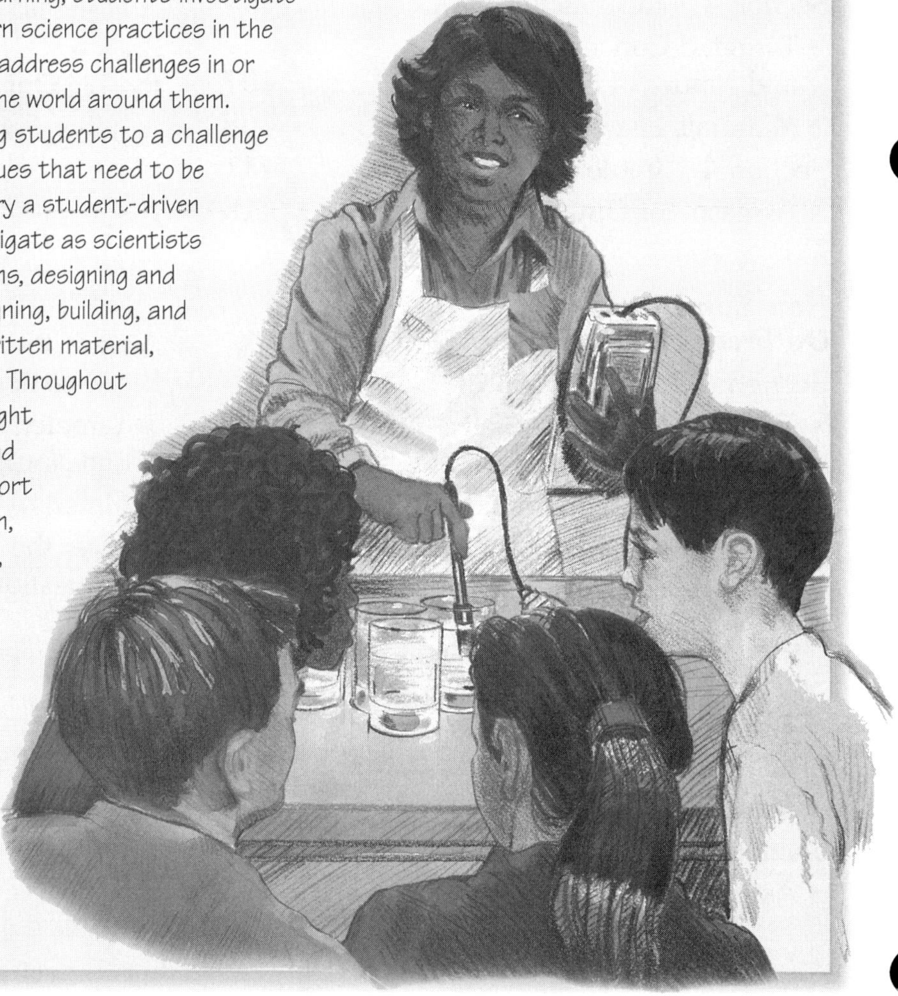

eliciting and steering discussion and collaboration among the students. At the completion of a project, students publicly exhibit what they have learned along with their solutions to the specific challenge. Personal reflection to help students learn from the experience is embedded in student activities, as are opportunities for assessment.

The curriculum will provide three years of piloted project-based inquiry materials for middle-school science. Individual curriculum units have been defined that cover the scope of the national content and process standards for the middle-school grades. Each Unit focuses on helping students acquire qualitative understanding of targeted science principles and move toward quantitative understanding, is infused with technology, and provides a foundation in reasoning skills, science content, and science process that will ready them for more advanced science. The curriculum as a whole introduces students to a wide range of investigative approaches in science (e.g., experimentation, modeling) and is designed to help them develop scientific reasoning skills that span those investigative approaches.

Technology can be used in project-based inquiry to make available to students some of the same kinds of tools and aids used by scientists in the field. These range from pencil-and-paper tools for organized data recording, collection, and management to software tools for analysis, simulation, modeling, and other tasks. Such infusion provides a platform for providing prompts, hints, examples, and other kinds of aids to students as they are engaging in scientific reasoning. The learning technologies and tools that are integrated into the curriculum offer essential scaffolding to students as they are developing their scientific reasoning skills, and are seamlessly infused into the overall completion of project activities and investigations.

Standards-Based Development

Development of each curriculum Unit begins by identifying the specific relevant national standards to be addressed. Each Unit has been designed to cover a specific portion of the national standards. This phase of development also includes an analysis of curriculum requirements across multiple states. Our intent is to deliver a product that will provide coverage of the content deemed essential on the widest practical scope and that will be easily adaptable to the needs of teachers across the country.

Once the appropriate standards have been identified, the development team works to define specific learning goals built from those standards, and takes into account conceptions and misunderstandings common among middle-school students. An orienting design challenge or driving question for investigation is chosen that motivates achieving those learning goals, and the team then sequences activities and the presentation of specific concepts so that students can construct an accurate understanding of the subject matter.

Inquiry-Based Design

The individual curriculum Units present two types of projects: engineering-design challenges and driving-question investigations. Design-challenge Units begin by presenting students with a scenario and problem and challenging them to design a device or plan that will solve the problem. Driving-question investigations begin by presenting students with a complex question with real-world implications. Students are challenged to develop answers to the questions. The scenario and problem in the design Units and the driving question in the investigation Units are carefully selected to lead the students into investigation of specific science concepts, and the solution processes are carefully structured to require use of specific scientific reasoning skills.

Pedagogical Rationale

Research shows that individual project-based learning units promote excitement and deep learning of the targeted concepts. However, achieving deep, flexible, transferable learning of cross-disciplinary content (e.g., the notion of a model, time scale, variable, experiment) and science practice requires a learning environment that consistently, persistently, and pervasively encourages the use of such content and practices over an extended period of time. By developing project-based inquiry materials that cover the spectrum of middle-school science content in a coherent framework, we provide this extended exposure to the type of learning environment most likely to produce competent scientific thinkers who are well grounded in their understanding of both basic science concepts and the standards and practices of science in general.

Evidence of Effectiveness

There is compelling evidence showing that a project-based inquiry approach meets this goal. Working at Georgia Tech, the University of Michigan, and Northwestern University, we have developed, piloted, and/or field-tested many individual project-based units. Our evaluation evidence shows that these materials engage students well and are manageable by teachers, and that students learn both content and process skills. In every summative evaluation, student performance on post-tests improved significantly from pretest performance (Krajcik, et al., 2000; Holbrook, et al., 2001; Gray et. al. 2001). For example, in the second year in a project-based classroom in Detroit, the average student at post-test scored at about the 95th percentile of the pre-test distribution. Further, we have repeatedly documented significant gains in content knowledge relative to other inquiry-based (but not project-based) instructional methods. In one set of results, performance by a project-based class

in Atlanta doubled on the content test while the matched comparison class (with an excellent teacher) experienced only a 20% gain (significance $p < .001$). Other comparisons have shown more modest differences, but project-based students consistently perform better than their comparisons. Most exciting about the Atlanta results is that results from performance assessments show that, within comparable student populations, project-based students score higher on all categories of problem-solving and analysis and are more sophisticated at science practice and managing a collaborative scientific investigation. Indeed, the performance of average-ability project-based students is often statistically indistinguishable from or better than performance of comparison honors students learning in an inquiry-oriented but not project-based classroom. The Chicago group also has documented significant change in process skills in project-based classrooms. Students become more effective in constructing and critiquing scientific arguments (Sandoval, 1998) and in constructing scientific explanations using discipline-specific knowledge, such as evolutionary explanations for animal behavior (Smith & Reiser, 1998).

Researchers at Northwestern have also investigated the change in classroom practices that are elicited by project-based units. Analyses of the artifacts students produce indicate that students are engaging in ambitious learning practices, requiring weighing and synthesizing many results from complex analyses of data, and constructing scientific arguments that require synthesizing results from multiple complex analyses of data (Edelson et al, 1998; Reiser et al, 2001). Students are engaged in planning, performing, monitoring and revising their investigations, and reporting on their investigation processes as well as their results (Loh et al. 1998). In general, the classrooms engaging in project-based activities reveal substantial moves toward a scientific discourse community in which students focus on arguing from evidence, critiquing ideas, and conjecturing, rather than simply reporting on what they have read or been told (Tabak & Reiser, 1997).

Introducing PBIS

What Do Scientists Do?

1) Scientists...address big challenges and big questions.

Students will find many different kinds of *Big Challenges* and *Questions* in PBIS Units. Some ask them to think about why something is a certain way. Some ask them to think about what causes something to change. Some challenge them to design a solution to a problem. Most are about things that can and do happen in the real world.

Understand the Big Challenge or Question

As students get started with each Unit, they will do activities that help them understand the *Big Question* or *Challenge* for that Unit. They will think about what they already know that might help them, and they will identify some of the new things they will need to learn.

Project Board

The *Project Board* helps you and your students keep track of their learning. For each challenge or question, they will use a *Project Board* to keep track of what they know, what they need to learn, and what they are learning. As they learn and gather evidence, they will record that on the *Project Board*. After they have answered each small question or challenge, they will return to the *Project Board* to record how what they have learned helps them answer the *Big Question* or *Challenge*.

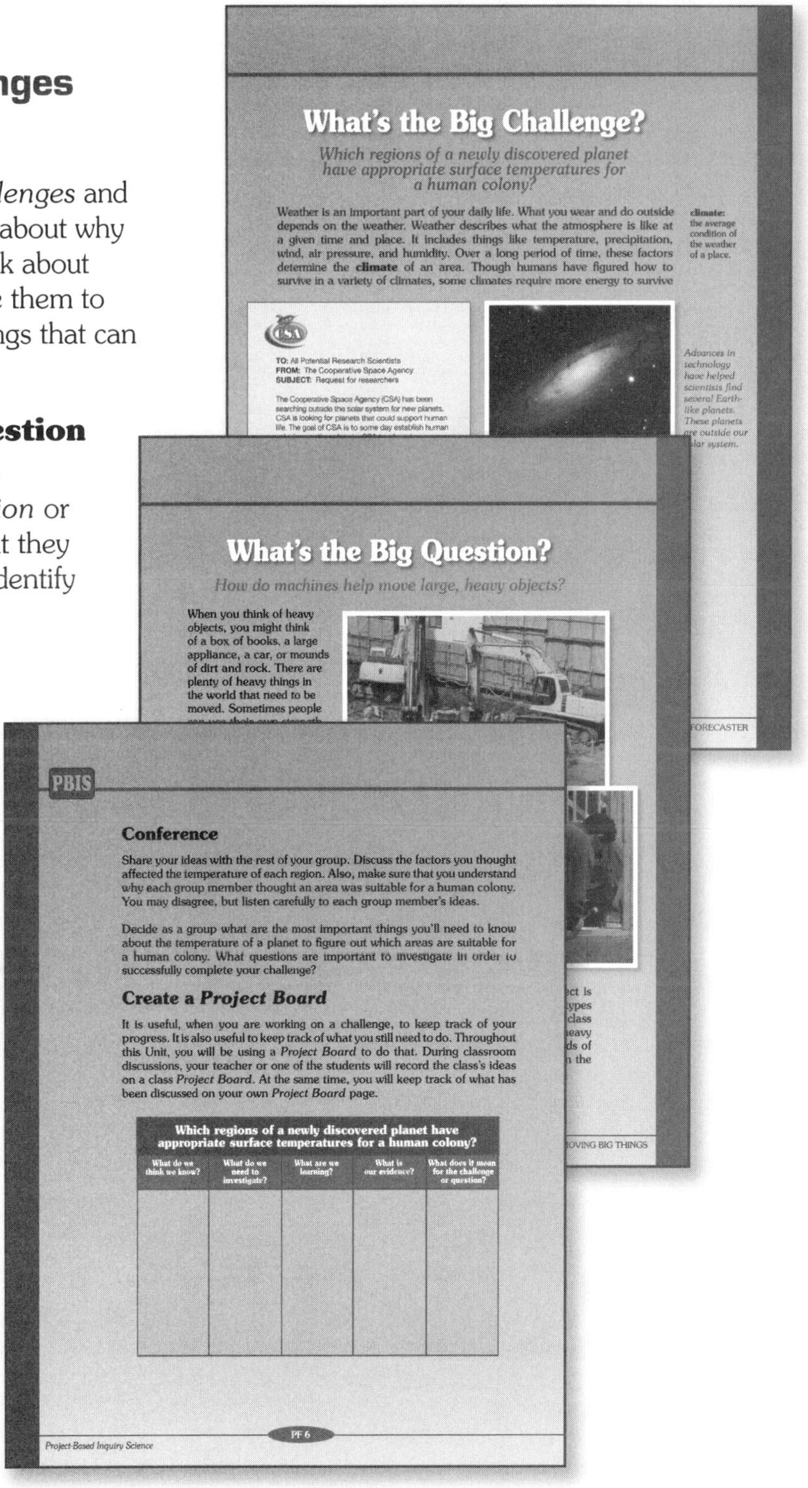

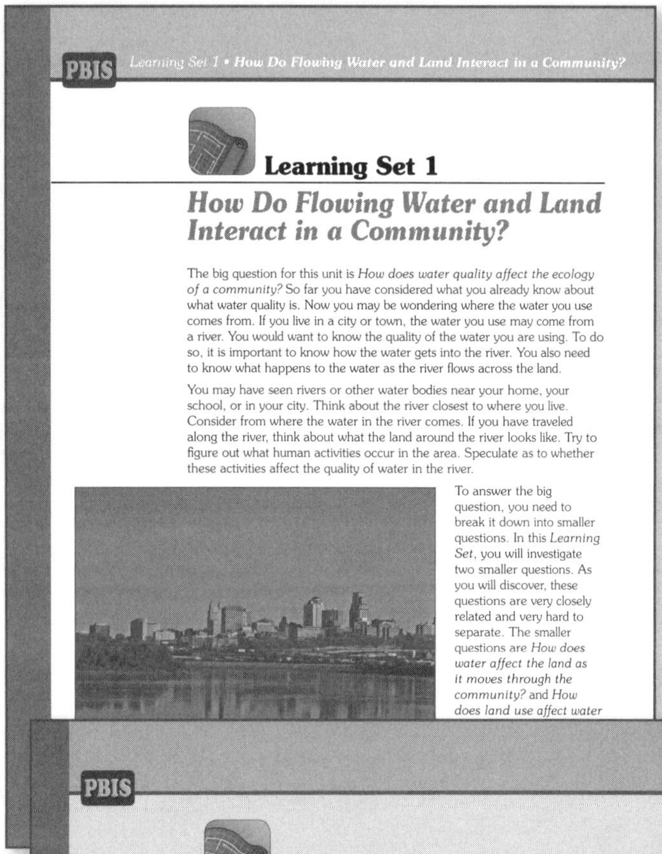
PBIS Learning Set 1 • How Do Flowing Water and Land Interact in a Community?

Learning Set 1

How Do Flowing Water and Land Interact in a Community?

The big question for this unit is *How does water quality affect the ecology of a community?* So far you have considered what you already know about what water quality is. Now you may be wondering where the water you use comes from. If you live in a city or town, the water you use may come from a river. You would want to know the quality of the water you are using. To do so, it is important to know how the water gets into the river. You also need to know what happens to the water as the river flows across the land.

You may have seen rivers or other water bodies near your home, your school, or in your city. Think about the river closest to where you live. Consider from where the water in the river comes. If you have traveled along the river, think about what the land around the river looks like. Try to figure out what human activities occur in the area. Speculate as to whether these activities affect the quality of water in the river.

To answer the big question, you need to break it down into smaller questions. In this *Learning Set*, you will investigate two smaller questions. As you will discover, these questions are very closely related and very hard to separate. The smaller questions are *How does water affect the land as it moves through the community?* and *How does land use affect water*

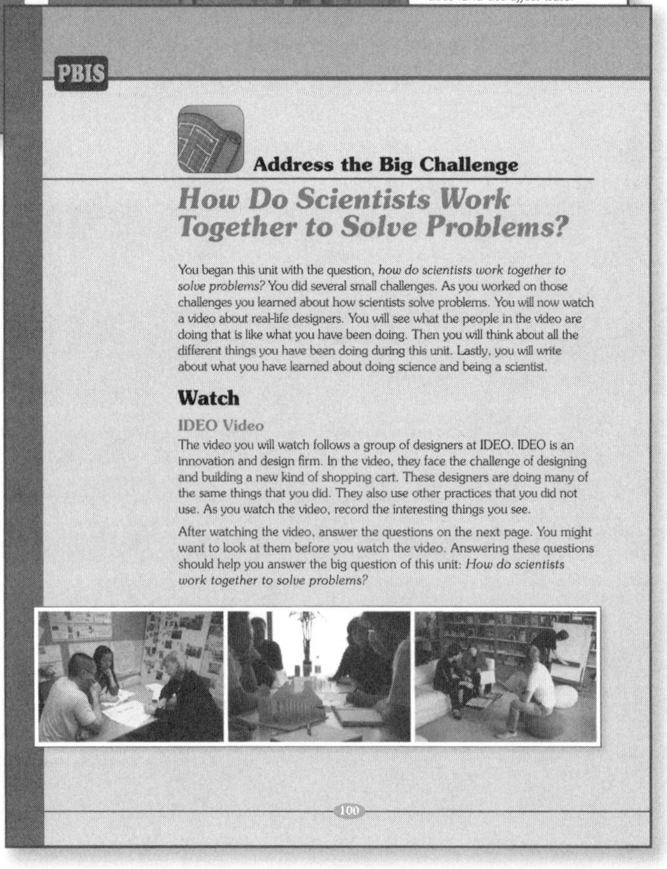
PBIS

Address the Big Challenge

How Do Scientists Work Together to Solve Problems?

You began this unit with the question, *how do scientists work together to solve problems?* You did several small challenges. As you worked on those challenges you learned about how scientists solve problems. You will now watch a video about real-life designers. You will see what the people in the video are doing that is like what you have been doing. Then you will think about all the different things you have been doing during this unit. Lastly, you will write about what you have learned about doing science and being a scientist.

Watch

IDEO Video

The video you will watch follows a group of designers at IDEO. IDEO is an innovation and design firm. In the video, they face the challenge of designing and building a new kind of shopping cart. These designers are doing many of the same things that you did. They also use other practices that you did not use. As you watch the video, record the interesting things you see.

After watching the video, answer the questions on the next page. You might want to look at them before you watch the video. Answering these questions should help you answer the big question of this unit: *How do scientists work together to solve problems?*

100

Learning Sets

Each Unit is composed of a group of *Learning Sets*, one for each of the smaller questions that needs to be answered to address the *Big Question* or *Challenge*. In each *Learning Set*, students will investigate and read to find answers to the *Learning Set's* question. They will also have a chance to share the results of their investigations with their classmates and work together to make sense of what they are learning. As students come to understand answers to the questions on the *Project Board*, you will record those answers and the evidence they collected. At the end of each *Learning Set*, they will apply their knowledgeto the *Big Question* or *Challenge*.

Answer the Big Question/ Address the Big Challenge

At the end of each Unit, students will put everything they have learned together to tackle the *Big Question* or *Challenge*.

2) Scientists...address smaller questions and challenges.

What Students Do in a Learning Set

Understanding the Question or Challenge

At the start of each *Learning Set*, students will usually do activities that will help them understand the *Learning Set's* question or challenge and recognize what they already know that can help them answer the question or achieve the challenge. Usually, they will visit the *Project Board* after these activities and record on it the even smaller questions that they need to investigate to answer a *Learning Set's* question.

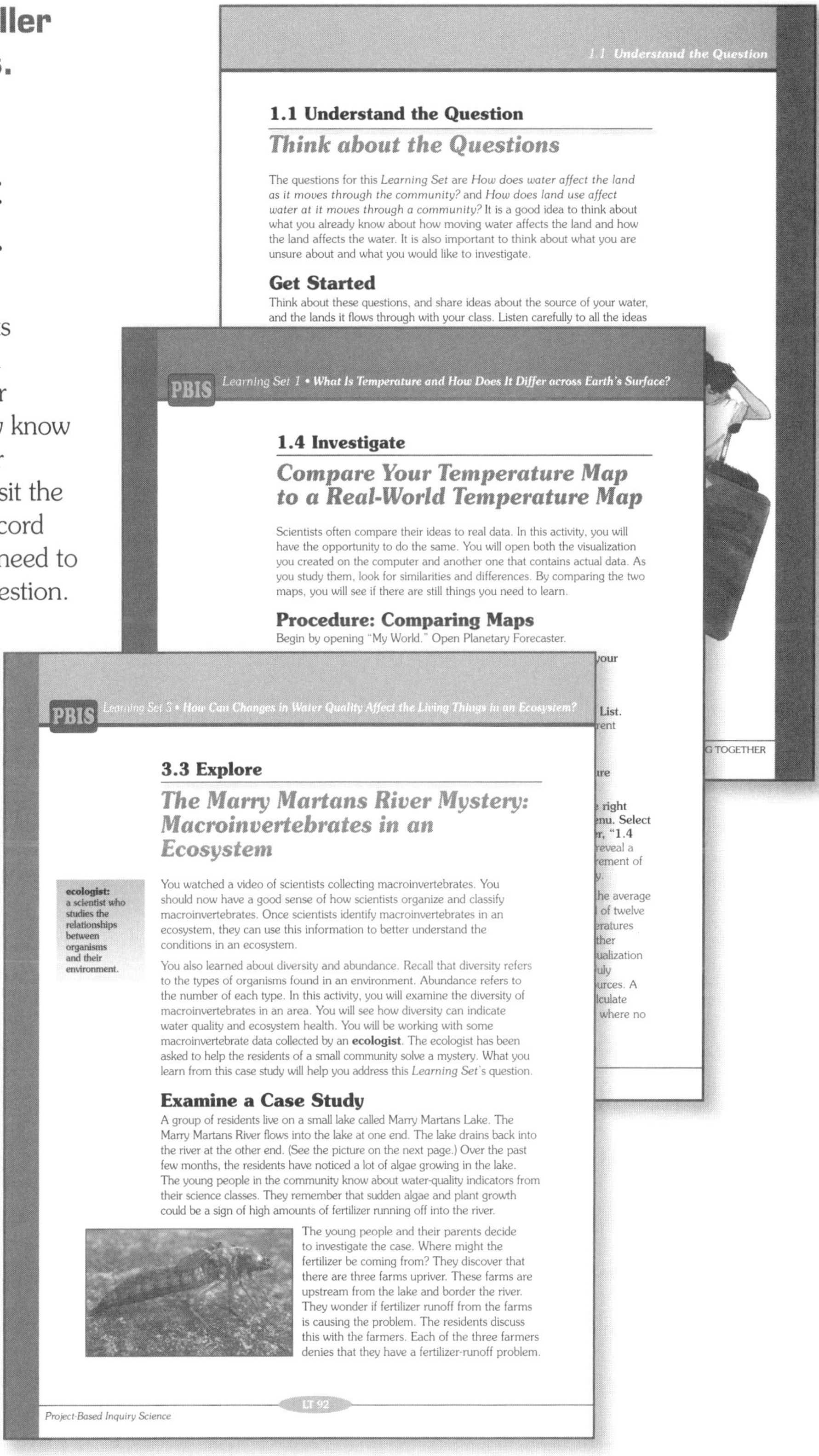
1.1 Understand the Question

1.1 Understand the Question

Think about the Questions

The questions for this *Learning Set* are *How does water affect the land as it moves through the community?* and *How does land use affect water at it moves through a community?* It is a good idea to think about what you already know about how moving water affects the land and how the land affects the water. It is also important to think about what you are unsure about and what you would like to investigate.

Get Started

Think about these questions, and share ideas about the source of your water, and the lands it flows through with your class. Listen carefully to all the ideas

PBIS Learning Set 1 • What Is Temperature and How Does It Differ across Earth's Surface?

1.4 Investigate

Compare Your Temperature Map to a Real-World Temperature Map

Scientists often compare their ideas to real data. In this activity, you will have the opportunity to do the same. You will open both the visualization you created on the computer and another one that contains actual data. As you study them, look for similarities and differences. By comparing the two maps, you will see if there are still things you need to learn.

Procedure: Comparing Maps

Begin by opening "My World." Open Planetary Forecaster.

PBIS Learning Set 3 • How Can Changes in Water Quality Affect the Living Things in an Ecosystem?

3.3 Explore

The Marry Martans River Mystery: Macroinvertebrates in an Ecosystem

ecologist: a scientist who studies the relationships between organisms and their environment.

You watched a video of scientists collecting macroinvertebrates. You should now have a good sense of how scientists organize and classify macroinvertebrates. Once scientists identify macroinvertebrates in an ecosystem, they can use this information to better understand the conditions in an ecosystem.

You also learned about diversity and abundance. Recall that diversity refers to the types of organisms found in an environment. Abundance refers to the number of each type. In this activity, you will examine the diversity of macroinvertebrates in an area. You will see how diversity can indicate water quality and ecosystem health. You will be working with some macroinvertebrate data collected by an **ecologist**. The ecologist has been asked to help the residents of a small community solve a mystery. What you learn from this case study will help you address this *Learning Set's* question.

Examine a Case Study

A group of residents live on a small lake called Marry Martans Lake. The Marry Martans River flows into the lake at one end. The lake drains back into the river at the other end. (See the picture on the next page.) Over the past few months, the residents have noticed a lot of algae growing in the lake. The young people in the community know about water-quality indicators from their science classes. They remember that sudden algae and plant growth could be a sign of high amounts of fertilizer running off into the river.

The young people and their parents decide to investigate the case. Where might the fertilizer be coming from? They discover that there are three farms upriver. These farms are upstream from the lake and border the river. They wonder if fertilizer runoff from the farms is causing the problem. The residents discuss this with the farmers. Each of the three farmers denies that they have a fertilizer-runoff problem.

Project-Based Inquiry Science LT 92

Investigate/Explore

There are many different kinds of investigations students might do to find answers to questions. In the *Learning Sets,* they might

- design and run experiments;
- design and run simulations;
- design and build models;
- examine large sets of data.

Don't worry if your students haven't done these things before. The text will provide them with lots of help in designing their investigations and in analyzing thier data.

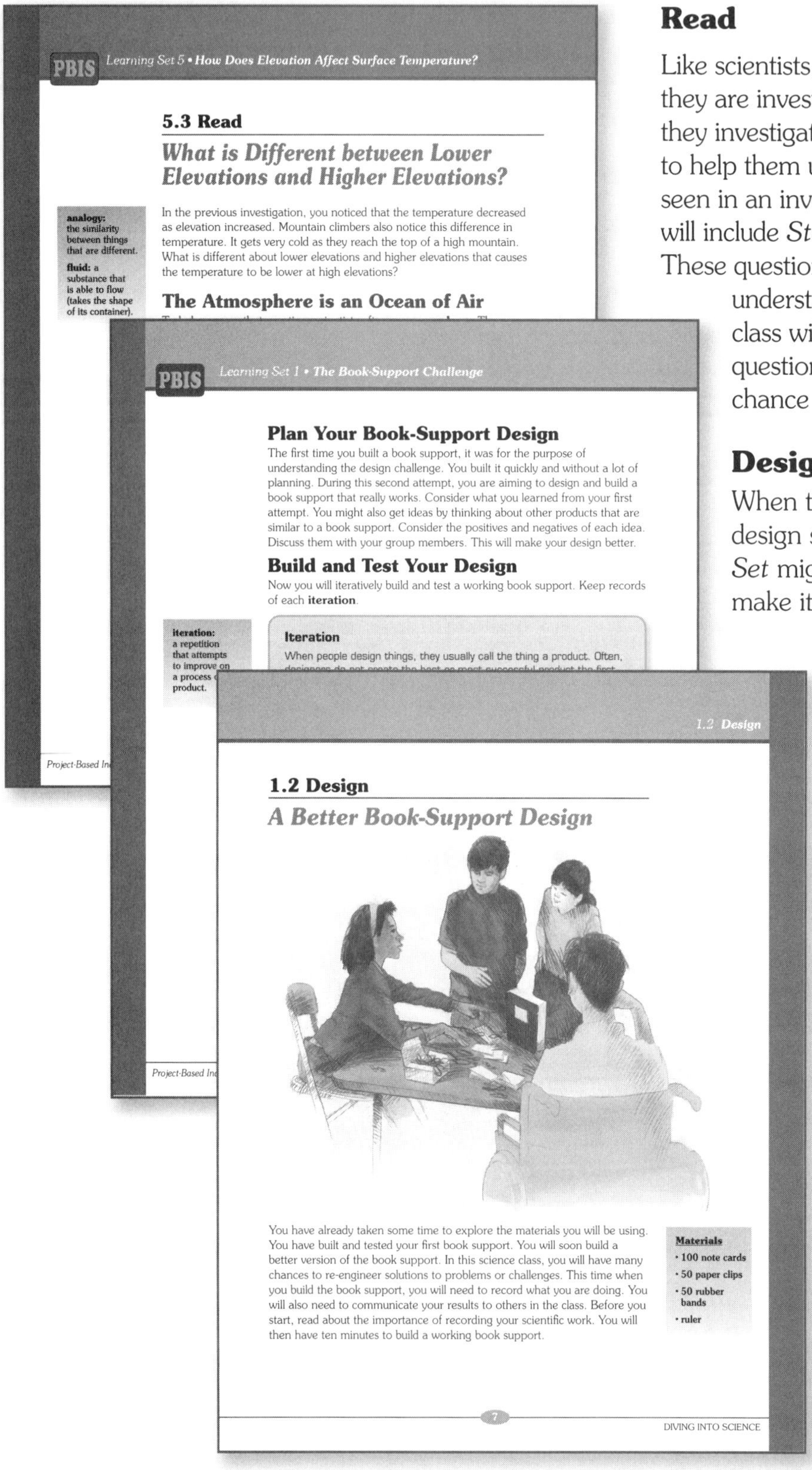
PBIS Learning Set 5 • How Does Elevation Affect Surface Temperature?

5.3 Read

What is Different between Lower Elevations and Higher Elevations?

analogy: the similarity between things that are different.

fluid: a substance that is able to flow (takes the shape of its container).

In the previous investigation, you noticed that the temperature decreased as elevation increased. Mountain climbers also notice this difference in temperature. It gets very cold as they reach the top of a high mountain. What is different about lower elevations and higher elevations that causes the temperature to be lower at high elevations?

The Atmosphere is an Ocean of Air

PBIS Learning Set 1 • The Book-Support Challenge

Plan Your Book-Support Design

The first time you built a book support, it was for the purpose of understanding the design challenge. You built it quickly and without a lot of planning. During this second attempt, you are aiming to design and build a book support that really works. Consider what you learned from your first attempt. You might also get ideas by thinking about other products that are similar to a book support. Consider the positives and negatives of each idea. Discuss them with your group members. This will make your design better.

Build and Test Your Design

Now you will iteratively build and test a working book support. Keep records of each **iteration**.

iteration: a repetition that attempts to improve on a process or product.

Iteration

When people design things, they usually call the thing a product. Often,

1.2 Design

1.2 Design

A Better Book-Support Design

You have already taken some time to explore the materials you will be using. You have built and tested your first book support. You will soon build a better version of the book support. In this science class, you will have many chances to re-engineer solutions to problems or challenges. This time when you build the book support, you will need to record what you are doing. You will also need to communicate your results to others in the class. Before you start, read about the importance of recording your scientific work. You will then have ten minutes to build a working book support.

Materials
- 100 note cards
- 50 paper clips
- 50 rubber bands
- ruler

7 DIVING INTO SCIENCE

Read

Like scientists, students will also read about the science they are investigating. They will read a little bit before they investigate, but most of the reading they do will be to help them understand what they have experienced or seen in an investigation. Each time they read, the text will include *Stop and Think* questions after the reading. These questions will help students gauge how well they understand what they have read. Usually, the class will discuss the answers to *Stop and Think* questions before going on so that everybody has a chance to make sense of the reading.

Design and Build

When the *Big Challenge* for a Unit asks them to design something, the challenge in a *Learning Set* might also ask them to design something and make it work. Often students will design a part of the thing they will design and build for the *Big Challenge*. When a *Learning Set* challenges students to design and build something, they will do several things:

- identify what questions they need to answer to be successful
- investigate to find answers to those questions
- use those answers to plan a good design solution
- build and test their design

Because designs don't always work the way one wants them to, students will usually do a design challenge more than once. Each time through, they will test their design. If their design doesn't work as well as they would like, they will determine why it is not working and identify other things they need to investigate to make it work better. Then, they will learn those things and try again.

Explain and Recommend

A big part of what scientists do is explain, or try to make sense of why things happen the way they do. An explanation describes why something is the way it is or behaves the way it does. An explanation is a statement one makes built from claims (what you think you know), evidence (from an investigation) that supports the claim, and science knowledge. As they learn, scientists get better at explaining. You will see that students get better, too, as they work through the *Learning Sets*.

A recommendation is a special kind of claim—one where you advise somebody about what to do. Students will make recommendations and support them with evidence, science knowledge, and explanations.

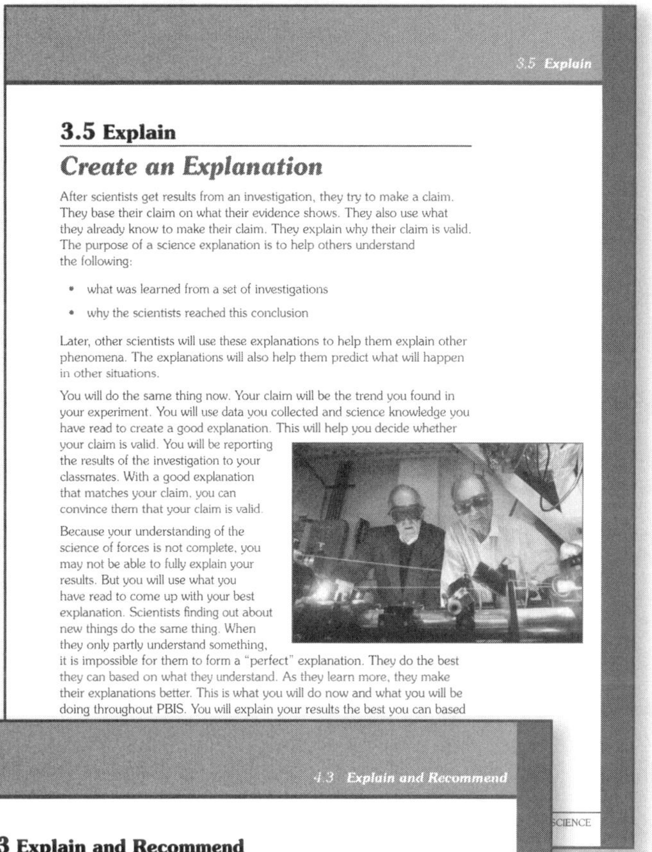
3.5 Explain

3.5 Explain

Create an Explanation

After scientists get results from an investigation, they try to make a claim. They base their claim on what their evidence shows. They also use what they already know to make their claim. They explain why their claim is valid. The purpose of a science explanation is to help others understand the following:

- what was learned from a set of investigations
- why the scientists reached this conclusion

Later, other scientists will use these explanations to help them explain other phenomena. The explanations will also help them predict what will happen in other situations.

You will do the same thing now. Your claim will be the trend you found in your experiment. You will use data you collected and science knowledge you have read to create a good explanation. This will help you decide whether your claim is valid. You will be reporting the results of the investigation to your classmates. With a good explanation that matches your claim, you can convince them that your claim is valid.

Because your understanding of the science of forces is not complete, you may not be able to fully explain your results. But you will use what you have read to come up with your best explanation. Scientists finding out about new things do the same thing. When they only partly understand something, it is impossible for them to form a "perfect" explanation. They do the best they can based on what they understand. As they learn more, they make their explanations better. This is what you will do now and what you will be doing throughout PBIS. You will explain your results the best you can based

SCIENCE

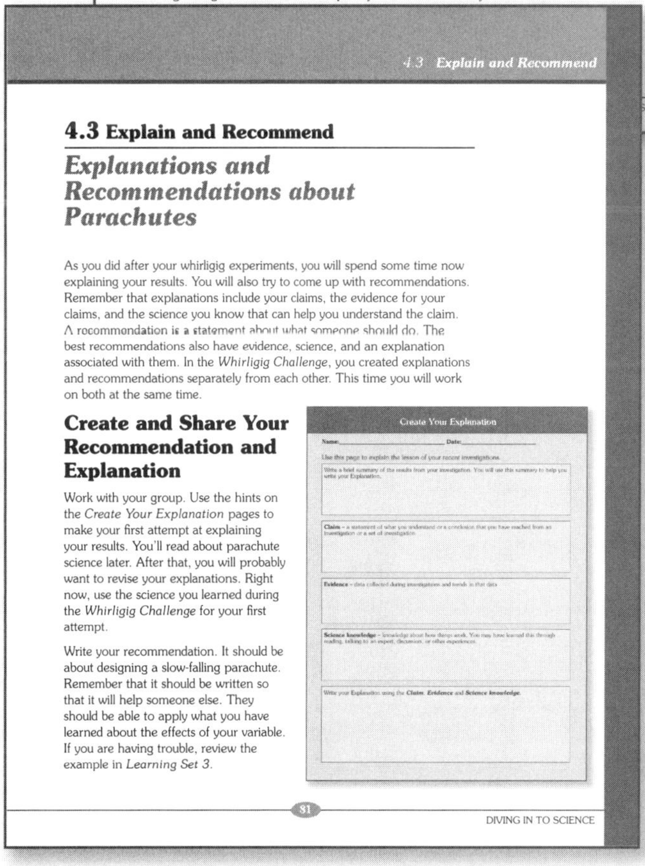
4.3 Explain and Recommend

4.3 Explain and Recommend

Explanations and Recommendations about Parachutes

As you did after your whirligig experiments, you will spend some time now explaining your results. You will also try to come up with recommendations. Remember that explanations include your claims, the evidence for your claims, and the science you know that can help you understand the claim. A recommendation is a statement about what someone should do. The best recommendations also have evidence, science, and an explanation associated with them. In the *Whirligig Challenge*, you created explanations and recommendations separately from each other. This time you will work on both at the same time.

Create and Share Your Recommendation and Explanation

Work with your group. Use the hints on the *Create Your Explanation* pages to make your first attempt at explaining your results. You'll read about parachute science later. After that, you will probably want to revise your explanations. Right now, use the science you learned during the *Whirligig Challenge* for your first attempt.

Write your recommendation. It should be about designing a slow-falling parachute. Remember that it should be written so that it will help someone else. They should be able to apply what you have learned about the effects of your variable. If you are having trouble, review the example in *Learning Set 3*.

Create Your Explanation

Name: Date:

Use this page to explain the lesson of your recent investigations.

Write a brief summary of the results from your investigation. You will use this summary to help you write your Explanation.

Claim – a statement of what you understand or a conclusion that you have reached from an investigation or a set of investigation.

Evidence – data collected during investigations and trends in that data.

Science knowledge – knowledge about how things work. You may have learned this through reading, talking to an expert, discussion, or other experiences.

Write your Explanation using the ***Claim, Evidence*** and ***Science knowledge***.

81

DIVING IN TO SCIENCE

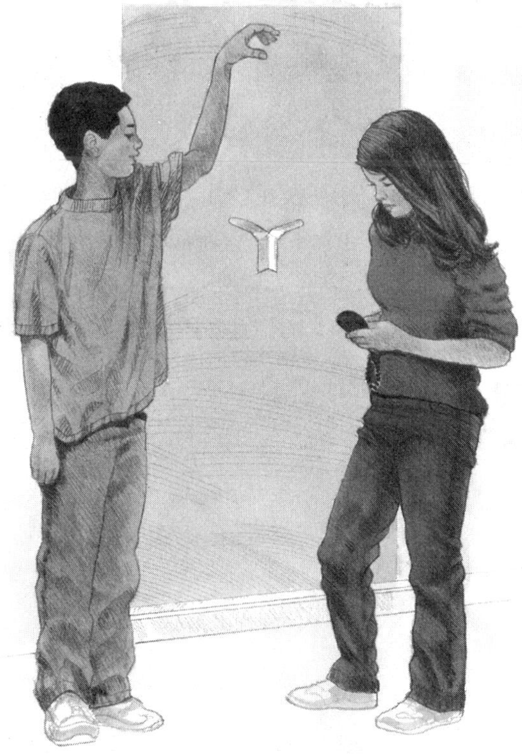

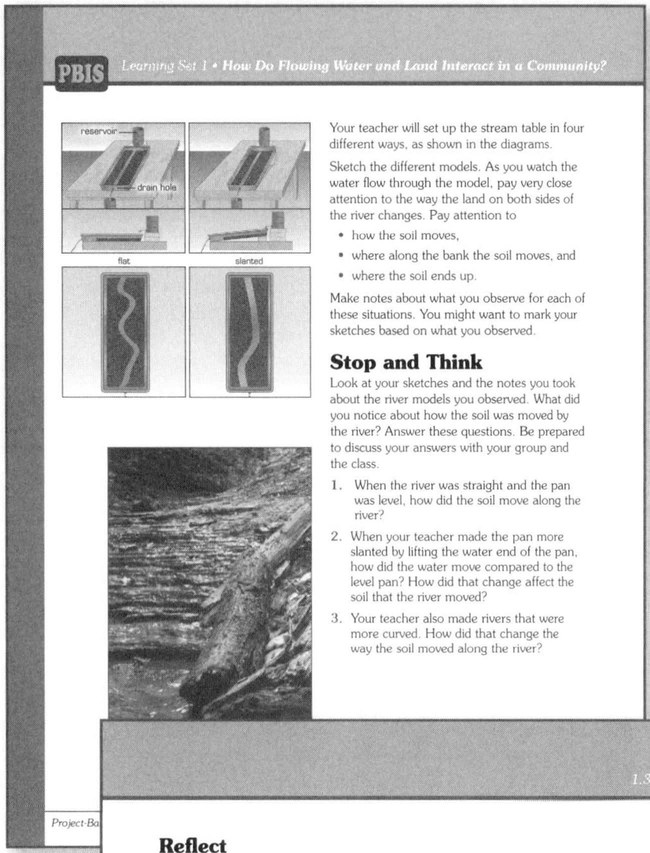

PBIS Learning Set 1 • How Do Flowing Water and Land Interact in a Community?

Your teacher will set up the stream table in four different ways, as shown in the diagrams.

Sketch the different models. As you watch the water flow through the model, pay very close attention to the way the land on both sides of the river changes. Pay attention to

- how the soil moves,
- where along the bank the soil moves, and
- where the soil ends up.

Make notes about what you observe for each of these situations. You might want to mark your sketches based on what you observed.

Stop and Think

Look at your sketches and the notes you took about the river models you observed. What did you notice about how the soil was moved by the river? Answer these questions. Be prepared to discuss your answers with your group and the class.

1. When the river was straight and the pan was level, how did the soil move along the river?
2. When your teacher made the pan more slanted by lifting the water end of the pan, how did the water move compared to the level pan? How did that change affect the soil that the river moved?
3. Your teacher also made rivers that were more curved. How did that change the way the soil moved along the river?

Project-Ba

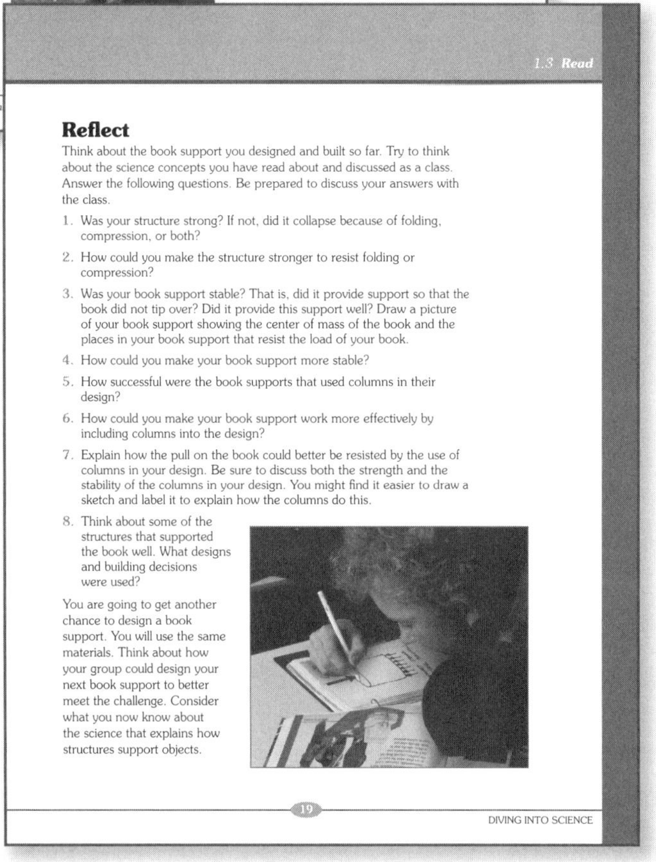

1.3 Read

Reflect

Think about the book support you designed and built so far. Try to think about the science concepts you have read about and discussed as a class. Answer the following questions. Be prepared to discuss your answers with the class.

1. Was your structure strong? If not, did it collapse because of folding, compression, or both?
2. How could you make the structure stronger to resist folding or compression?
3. Was your book support stable? That is, did it provide support so that the book did not tip over? Did it provide this support well? Draw a picture of your book support showing the center of mass of the book and the places in your book support that resist the load of your book.
4. How could you make your book support more stable?
5. How successful were the book supports that used columns in their design?
6. How could you make your book support work more effectively by including columns into the design?
7. Explain how the pull on the book could better be resisted by the use of columns in your design. Be sure to discuss both the strength and the stability of the columns in your design. You might find it easier to draw a sketch and label it to explain how the columns do this.
8. Think about some of the structures that supported the book well. What designs and building decisions were used?

You are going to get another chance to design a book support. You will use the same materials. Think about how your group could design your next book support to better meet the challenge. Consider what you now know about the science that explains how structures support objects.

19

DIVING INTO SCIENCE

3) Scientists...reflect in many different ways.

PBIS provides guidance to help students think about what they are doing and to recognize what they are learning. Doing this often as they are working will help students be successful student scientists.

Tools for Making Sense

Stop and Think

Stop and Think sections help students make sense of what they have been doing in the section they are working on. *Stop and Think* sections include a set of questions to help students understand what they have just read or done. Sometimes the questions will remind them of something they need to pay more attention to. Sometimes they will help students connect what they have just read to things they already know. When there is a *Stop and Think* in the text, students will work individually or with a partner to answer the questions, and then the whole class will discuss the answers.

Reflect

Reflect sections help students connect what they have just done with other things they have read or done earlier in the Unit (or in another Unit). When there is a *Reflect* in the text, students will work individually or with a partner or small group to answer the questions. Then, the whole class will discuss the answers. You may want to ask students to answer *Reflect* questions for homework.

Analyze Your Data

Whenever students have to analyze data, the text will provide hints about how to do that and what to look for.

Mess About

"Messing about" is a term that comes from design. It means exploring the materials to be used for designing or building something or examining something that works like what is to be designed. Messing about helps students discover new ideas—and it can be a lot of fun. The text will usually give them ideas about things to notice as they are messing about.

What's the Point?

At the end of each *Learning Set*, students will find a summary, called *What's the Point?*, of the important information from the *Learning Set*. These summaries can help students remember how what they did and learned is connected to the *Big Question* or *Challenge* they are working on.

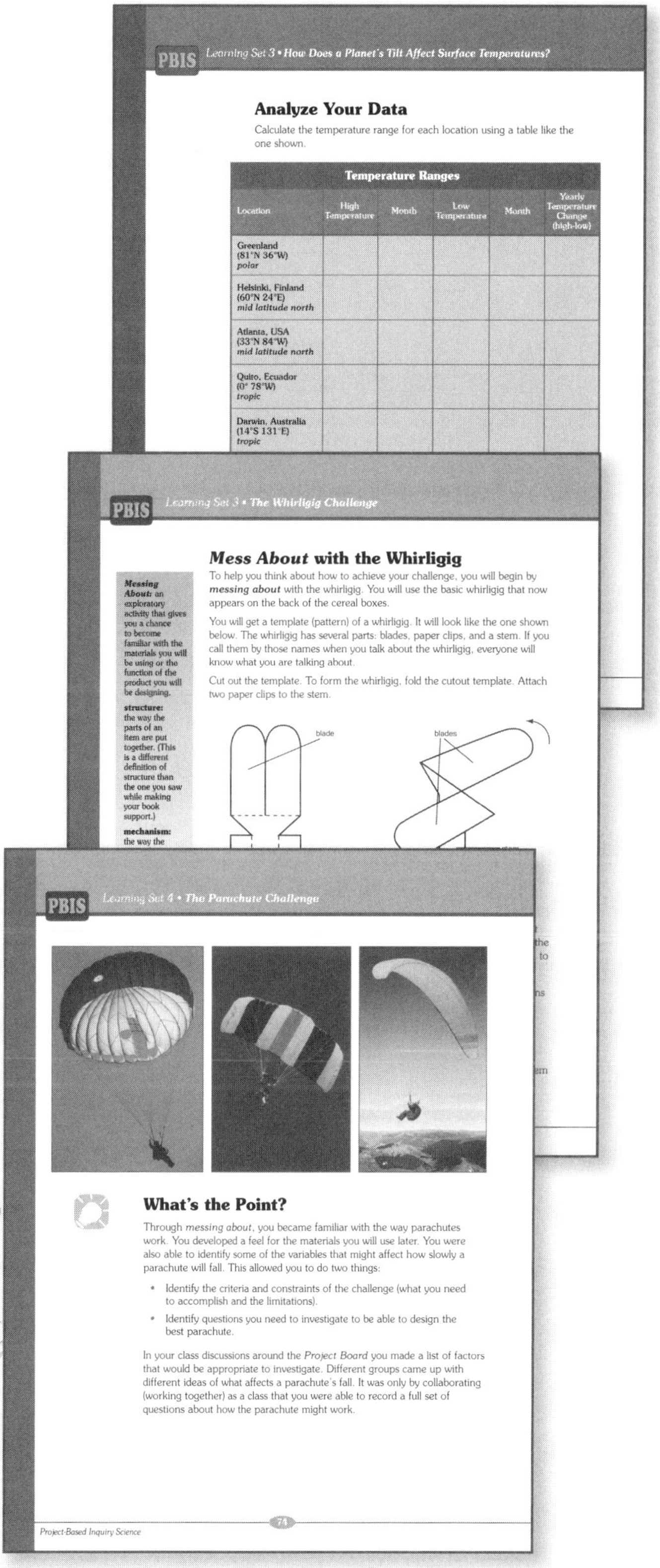
PBIS Learning Set 3 • How Does a Planet's Tilt Affect Surface Temperatures?

Analyze Your Data

Calculate the temperature range for each location using a table like the one shown.

Temperature Ranges

Location	High Temperature	Month	Low Temperature	Month	Yearly Temperature Change (high-low)
Greenland (81°N 36°W) *polar*					
Helsinki, Finland (60°N 24°E) *mid latitude north*					
Atlanta, USA (33°N 84°W) *mid latitude north*					
Quito, Ecuador (0° 78°W) *tropic*					
Darwin, Australia (14°S 131°E) *tropic*					

PBIS Learning Set 3 • The Whirligig Challenge

***Mess About* with the Whirligig**

To help you think about how to achieve your challenge, you will begin by ***messing about*** with the whirligig. You will use the basic whirligig that now appears on the back of the cereal boxes.

You will get a template (pattern) of a whirligig. It will look like the one shown below. The whirligig has several parts: blades, paper clips, and a stem. If you call them by those names when you talk about the whirligig, everyone will know what you are talking about.

Cut out the template. To form the whirligig, fold the cutout template. Attach two paper clips to the stem.

Messing About: an exploratory activity that gives you a chance to become familiar with the materials you will be using or the function of the product you will be designing.

structure: the way the parts of an item are put together. (This is a different definition of structure than the one you saw while making your book support.)

mechanism: the way the

PBIS Learning Set 4 • The Parachute Challenge

What's the Point?

Through *messing about*, you became familiar with the way parachutes work. You developed a feel for the materials you will use later. You were also able to identify some of the variables that might affect how slowly a parachute will fall. This allowed you to do two things:

- Identify the criteria and constraints of the challenge (what you need to accomplish and the limitations).
- Identify questions you need to investigate to be able to design the best parachute.

In your class discussions around the *Project Board* you made a list of factors that would be appropriate to investigate. Different groups came up with different ideas of what affects a parachute's fall. It was only by collaborating (working together) as a class that you were able to record a full set of questions about how the parachute might work.

Project-Based Inquiry Science 74

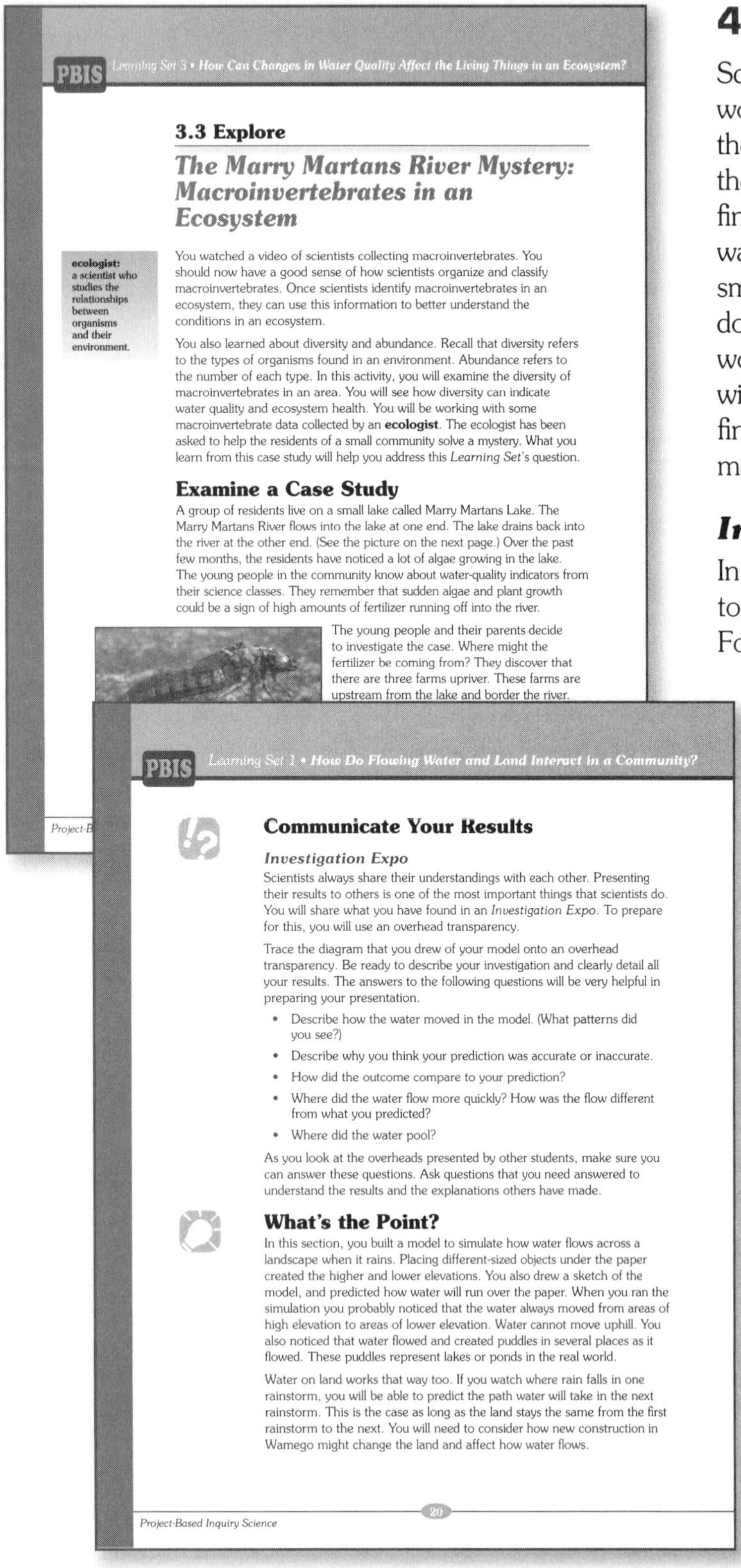

PBIS Learning Set 3 • How Can Changes in Water Quality Affect the Living Things in an Ecosystem?

3.3 Explore

The Marry Martans River Mystery: Macroinvertebrates in an Ecosystem

ecologist: a scientist who studies the relationships between organisms and their environment.

You watched a video of scientists collecting macroinvertebrates. You should now have a good sense of how scientists organize and classify macroinvertebrates. Once scientists identify macroinvertebrates in an ecosystem, they can use this information to better understand the conditions in an ecosystem.

You also learned about diversity and abundance. Recall that diversity refers to the types of organisms found in an environment. Abundance refers to the number of each type. In this activity, you will examine the diversity of macroinvertebrates in an area. You will see how diversity can indicate water quality and ecosystem health. You will be working with some macroinvertebrate data collected by an **ecologist**. The ecologist has been asked to help the residents of a small community solve a mystery. What you learn from this case study will help you address this *Learning Set's* question.

Examine a Case Study

A group of residents live on a small lake called Marry Martans Lake. The Marry Martans River flows into the lake at one end. The lake drains back into the river at the other end. (See the picture on the next page.) Over the past few months, the residents have noticed a lot of algae growing in the lake. The young people in the community know about water-quality indicators from their science classes. They remember that sudden algae and plant growth could be a sign of high amounts of fertilizer running off into the river.

The young people and their parents decide to investigate the case. Where might the fertilizer be coming from? They discover that there are three farms upriver. These farms are upstream from the lake and border the river.

Project-B

PBIS Learning Set 1 • How Do Flowing Water and Land Interact in a Community?

Communicate Your Results

Investigation Expo

Scientists always share their understandings with each other. Presenting their results to others is one of the most important things that scientists do. You will share what you have found in an *Investigation Expo*. To prepare for this, you will use an overhead transparency.

Trace the diagram that you drew of your model onto an overhead transparency. Be ready to describe your investigation and clearly detail all your results. The answers to the following questions will be very helpful in preparing your presentation.

- Describe how the water moved in the model. (What patterns did you see?)
- Describe why you think your prediction was accurate or inaccurate.
- How did the outcome compare to your prediction?
- Where did the water flow more quickly? How was the flow different from what you predicted?
- Where did the water pool?

As you look at the overheads presented by other students, make sure you can answer these questions. Ask questions that you need answered to understand the results and the explanations others have made.

What's the Point?

In this section, you built a model to simulate how water flows across a landscape when it rains. Placing different-sized objects under the paper created the higher and lower elevations. You also drew a sketch of the model, and predicted how water will run over the paper. When you ran the simulation you probably noticed that the water always moved from areas of high elevation to areas of lower elevation. Water cannot move uphill. You also noticed that water flowed and created puddles in several places as it flowed. These puddles represent lakes or ponds in the real world.

Water on land works that way too. If you watch where rain falls in one rainstorm, you will be able to predict the path water will take in the next rainstorm. This is the case as long as the land stays the same from the first rainstorm to the next. You will need to consider how new construction in Wamego might change the land and affect how water flows.

20

Project-Based Inquiry Science

4) Scientists...collaborate.

Scientists never do all their work alone. They work with other scientists (collaborate) and share their knowledge. PBIS helps students by giving them lots of opportunities for sharing their findings, ideas, and discoveries with others (the way scientists do). Students will work together in small groups to investigate, design, explain, and do other science activities. Sometimes they will work in pairs to figure out things together. They will also have lots of opportunities to share their findings with the rest of their classmates and make sense together of what they are learning.

Investigation Expo

In an *Investigation Expo*, small groups report to the class about an investigation they've done. For each *Investigation Expo*, students will make a poster detailing what they were trying to learn from their investigation, what they did, their data, and their interpretation of the data. The text gives them hints about what to present and what to look for in other groups' presentations. *Investigation Expos* are always followed by discussions about the investigations and about how to do science well. You may want to ask students to write a lab report following an investigation.

Plan Briefing/Solution Briefing/ Idea Briefing

Briefings are presentations of work in progress. They give students a chance to get advice from their classmates that can help them move forward. During a *Plan Briefing*, students present their plans to the class. They might be plans for an experiment for solving a problem or achieving a challenge. During a *Solution Briefing*, students present their solutions in progress and ask the class to help them make their solutions better. During an *Idea Briefing*, students present their ideas, including their evidence in support of their plans, solutions, or ideas. Often, they will prepare posters to help them make their presentation. Briefings are almost always followed by discussions of their investigations and how they will move forward.

Solution Showcase

Solution Showcases usually happen near the end of a Unit. During a *Solution Showcase*, students show their classmates their finished products—either their answer to a question or solution to a challenge. Students will also tell the class why they think it is a good answer or solution, what evidence and science they used to get to their solution, and what they tried along the way before getting to their answers or solutions. Sometimes a *Solution Showcase* is followed by a competition. It is almost always followed by a discussion comparing and contrasting the different answers and solutions groups have come up with. You may want to ask students to write a report or paper following a *Solution Showcase*.

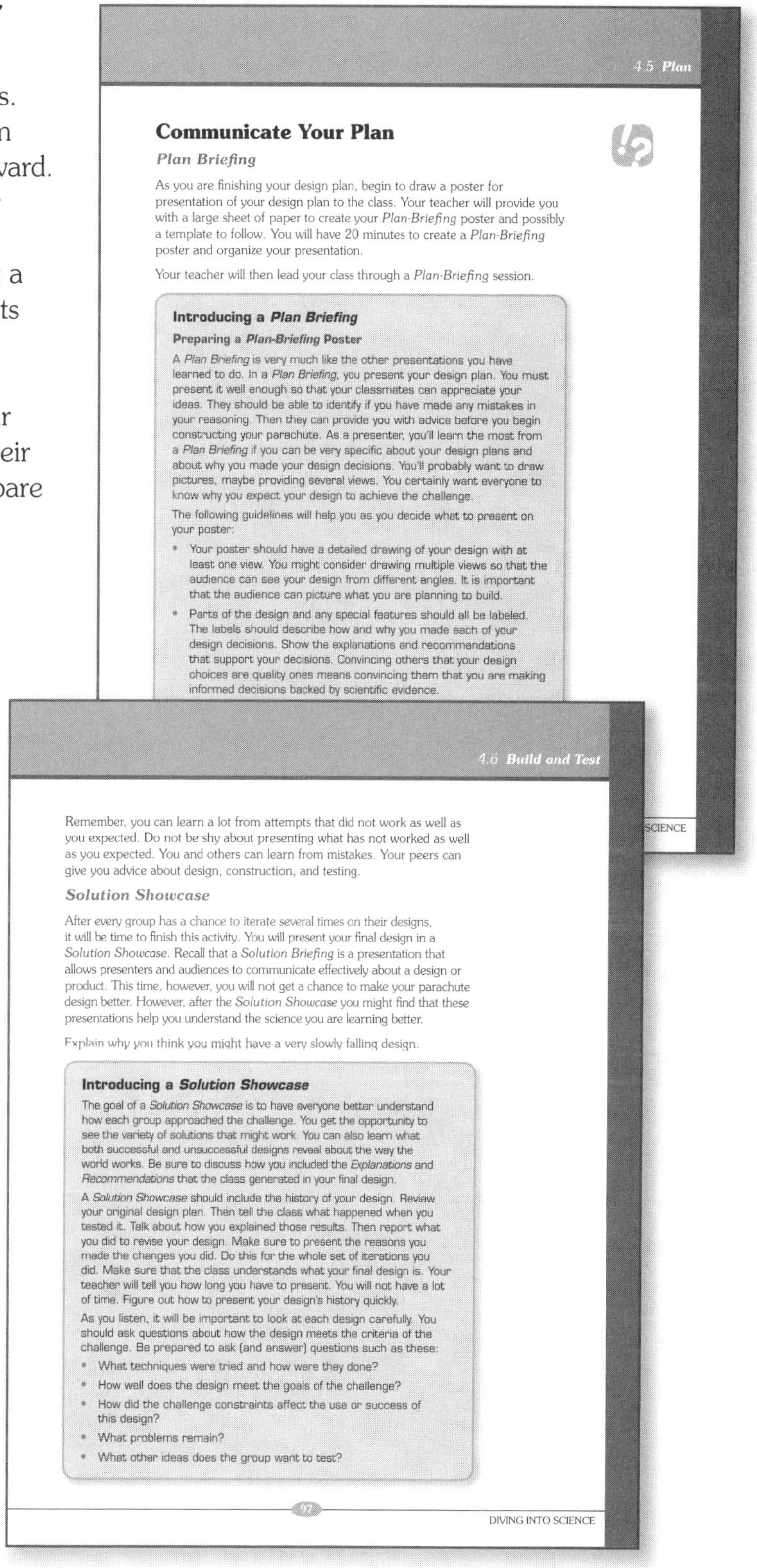

4.5 Plan

Communicate Your Plan

Plan Briefing

As you are finishing your design plan, begin to draw a poster for presentation of your design plan to the class. Your teacher will provide you with a large sheet of paper to create your *Plan-Briefing* poster and possibly a template to follow. You will have 20 minutes to create a *Plan-Briefing* poster and organize your presentation.

Your teacher will then lead your class through a *Plan-Briefing* session.

Introducing a *Plan Briefing*

Preparing a *Plan-Briefing* Poster

A *Plan Briefing* is very much like the other presentations you have learned to do. In a *Plan Briefing*, you present your design plan. You must present it well enough so that your classmates can appreciate your ideas. They should be able to identify if you have made any mistakes in your reasoning. Then they can provide you with advice before you begin constructing your parachute. As a presenter, you'll learn the most from a *Plan Briefing* if you can be very specific about your design plans and about why you made your design decisions. You'll probably want to draw pictures, maybe providing several views. You certainly want everyone to know why you expect your design to achieve the challenge.

The following guidelines will help you as you decide what to present on your poster:

- Your poster should have a detailed drawing of your design with at least one view. You might consider drawing multiple views so that the audience can see your design from different angles. It is important that the audience can picture what you are planning to build.
- Parts of the design and any special features should all be labeled. The labels should describe how and why you made each of your design decisions. Show the explanations and recommendations that support your decisions. Convincing others that your design choices are quality ones means convincing them that you are making informed decisions backed by scientific evidence.

SCIENCE

4.6 Build and Test

Remember, you can learn a lot from attempts that did not work as well as you expected. Do not be shy about presenting what has not worked as well as you expected. You and others can learn from mistakes. Your peers can give you advice about design, construction, and testing.

Solution Showcase

After every group has a chance to iterate several times on their designs, it will be time to finish this activity. You will present your final design in a *Solution Showcase*. Recall that a *Solution Briefing* is a presentation that allows presenters and audiences to communicate effectively about a design or product. This time, however, you will not get a chance to make your parachute design better. However, after the *Solution Showcase* you might find that these presentations help you understand the science you are learning better.

Explain why you think you might have a very slowly falling design.

Introducing a *Solution Showcase*

The goal of a *Solution Showcase* is to have everyone better understand how each group approached the challenge. You get the opportunity to see the variety of solutions that might work. You can also learn what both successful and unsuccessful designs reveal about the way the world works. Be sure to discuss how you included the *Explanations* and *Recommendations* that the class generated in your final design.

A *Solution Showcase* should include the history of your design. Review your original design plan. Then tell the class what happened when you tested it. Talk about how you explained those results. Then report what you did to revise your design. Make sure to present the reasons you made the changes you did. Do this for the whole set of iterations you did. Make sure that the class understands what your final design is. Your teacher will tell you how long you have to present. You will not have a lot of time. Figure out how to present your design's history quickly.

As you listen, it will be important to look at each design carefully. You should ask questions about how the design meets the criteria of the challenge. Be prepared to ask (and answer) questions such as these:

- What techniques were tried and how were they done?
- How well does the design meet the goals of the challenge?
- How did the challenge constraints affect the use or success of this design?
- What problems remain?
- What other ideas does the group want to test?

97

DIVING INTO SCIENCE

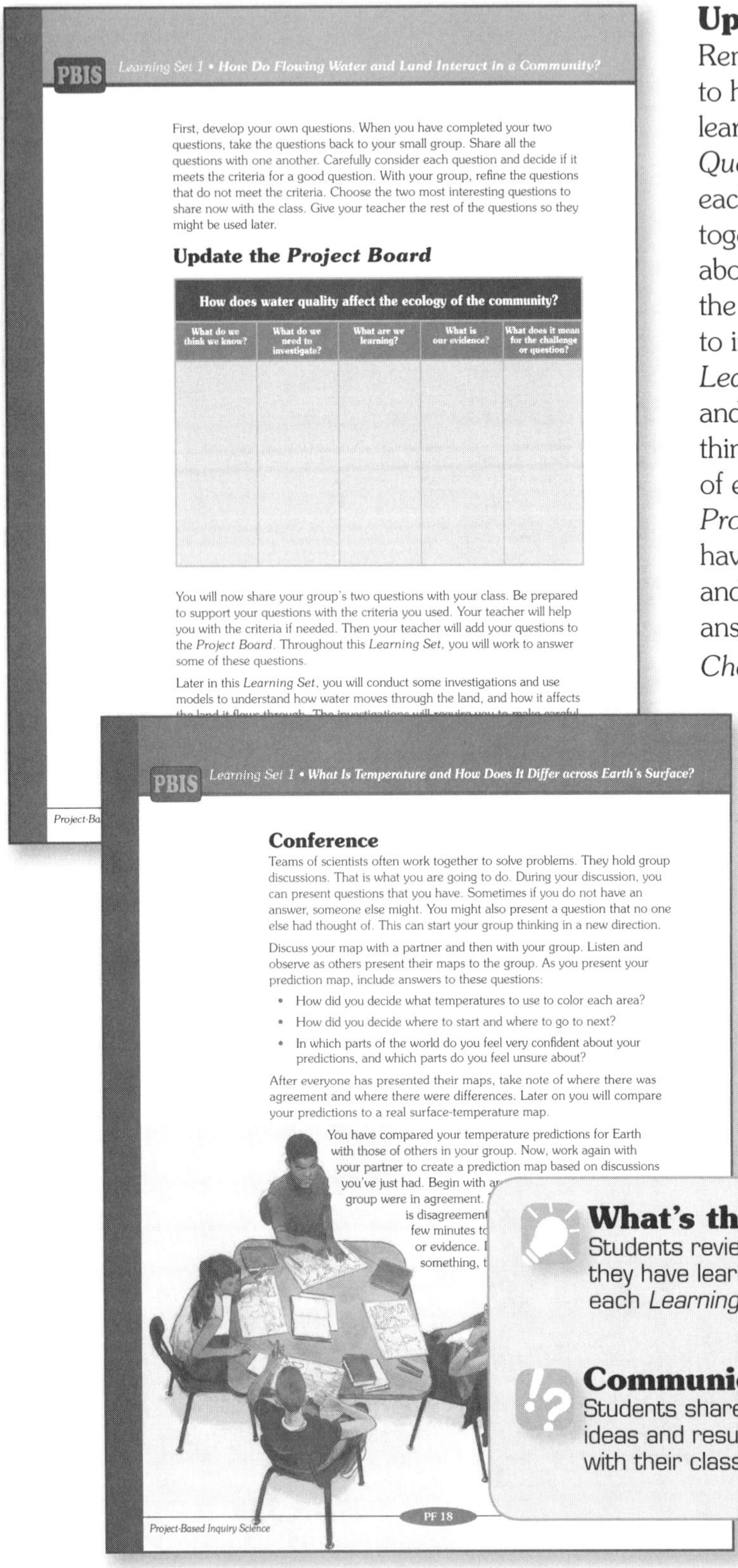

PBIS Learning Set 1 • How Do Flowing Water and Land Interact in a Community?

First, develop your own questions. When you have completed your two questions, take the questions back to your small group. Share all the questions with one another. Carefully consider each question and decide if it meets the criteria for a good question. With your group, refine the questions that do not meet the criteria. Choose the two most interesting questions to share now with the class. Give your teacher the rest of the questions so they might be used later.

Update the *Project Board*

How does water quality affect the ecology of the community?				
What do we think we know?	What do we need to investigate?	What are we learning?	What is our evidence?	What does it mean for the challenge or question?

You will now share your group's two questions with your class. Be prepared to support your questions with the criteria you used. Your teacher will help you with the criteria if needed. Then your teacher will add your questions to the *Project Board*. Throughout this *Learning Set*, you will work to answer some of these questions.

Later in this *Learning Set*, you will conduct some investigations and use models to understand how water moves through the land, and how it affects

PBIS Learning Set 1 • What Is Temperature and How Does It Differ across Earth's Surface?

Conference

Teams of scientists often work together to solve problems. They hold group discussions. That is what you are going to do. During your discussion, you can present questions that you have. Sometimes if you do not have an answer, someone else might. You might also present a question that no one else had thought of. This can start your group thinking in a new direction.

Discuss your map with a partner and then with your group. Listen and observe as others present their maps to the group. As you present your prediction map, include answers to these questions:

- How did you decide what temperatures to use to color each area?
- How did you decide where to start and where to go to next?
- In which parts of the world do you feel very confident about your predictions, and which parts do you feel unsure about?

After everyone has presented their maps, take note of where there was agreement and where there were differences. Later on you will compare your predictions to a real surface-temperature map.

You have compared your temperature predictions for Earth with those of others in your group. Now, work again with your partner to create a prediction map based on discussions you've just had. Begin with a[...] group were in agreement. [...] is disagreement [...] few minutes t[...] or evidence. [...] something, t[...]

Project-Based Inquiry Science PF 18

Update the *Project Board*

Remember that the *Project Board* is designed to help the class keep track of what they are learning and their progress toward a Unit's *Big Question* or *Challenge*. At the beginning of each Unit, the class creates a *Project Board*, and together records what students think they know about answering the *Big Question* or addressing the *Big Challenge* and what they think they need to investigate further. Near the beginning of each *Learning Set*, the class revisits the *Project Board* and adds new questions and information they think they know to the *Project Board*. At the end of each *Learning Set*, the class again revisits the *Project Board*. This time, they record what they have learned, the evidence they have collected, and recommendations they can make about answering the *Big Question* or achieving the *Big Challenge*.

Conference

A *Conference* is a short discussion among a small group of students before a more formal whole-class discussion. Students might discuss predictions and observations, they might try to explain together, they might consult on what they think they know, and so on. Usually, a *Conference* is followed by a discussion around the *Project Board*. In these small group discussions, everybody gets a chance to participate.

What's the Point?
Students review what they have learned in each *Learning Set*.

Stop and Think
Student answer questions that help them understand what they have done in a section.

Communicate
Students share their ideas and results with their classmates.

Record
Students record their data as they gather it.

NOTES

NOTES

NOTES

NOTES

Teacher's Planning Guide

PBIS

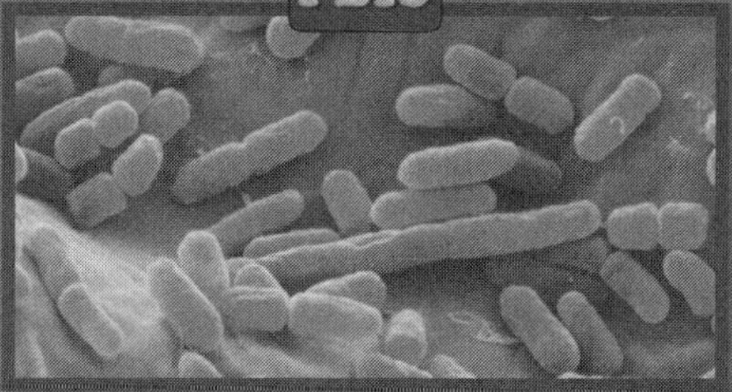

GOOD FRIENDS AND GERMS

As a student scientist, you will...

Ask QUESTIONS

APPLY MEANING

Pursue ANSWERS

Make MEANING

Share ANSWERS

UNIT OVERVIEW

Content

In this unit, the *Big Question* that students focus on is *How can you prevent your good friends from getting sick?* During *Learning Sets,* students investigate smaller questions and update the *Project Board,* create a summary of what they are learning about diseases, disease spread, and disease prevention. Students consider how what they are learning can help them answer the *Big Question.*

Learning Set 1 centers on the question *How do I get sick?* Students learn how disease can be communicated from person to person and the differences between communicable and noncommunicable disease. Assuming the role of student epidemiologists, students identify factors that affect how a disease spreads and are challenged to trace an infection to its initial carrier.

In *Learning Set 2: What kinds of things make you sick?,* students are introduced to the idea that many diseases are caused by bacteria and viruses. They read about bacteria, their structure, life processes, and how they can be helpful or harmful. Through an iterative process of reading and observing, students then develop the understanding that all living things are composed of cells and the structure of these cells. Students are shown microscope techniques such as: preparing wet-mount slides, staining specimens, and estimating their size. Students then use the microscope to observe cells, identify cell structures, and differentiate between plant and animal cells. By examining a case study of food poisoning, students learn about the link between hygiene, sanitation, and disease transmission. Students are shown how to safely culture bacteria and use this technique to design and carry out experiments to identify places in their classroom that harbor bacteria. Students then learn how viruses differ from bacteria, how viruses reproduce, and the various diseases caused by viruses. Students examine the case studies of a variety of diseases caused by bacteria and viruses. As students present what they have learned about different diseases to their peers, they develop a *Communicable Disease Information Table* on which they summarize what they have learned about the causes of diseases, how they spread, how they affect body systems, and their symptoms and treatments. Students will continue to record what they learn about diseases on this chart throughout the Unit, and it will serve as an important resource that students draw upon as they prepare their presentations when addressing the first challenge of answering the *Big Question.*

47 class periods

A class period is considered to be one 40 to 50 minute class.

In *Learning Set 3: What happens to you when you get sick?,* students investigate and read about the structural and functional levels of organization in living things from cells to organ systems. Throughout this *Learning Set,* students build and run models of organ systems and connect the model to what they are learning by reading about the organ systems. In this way, students steadily build their understanding of how human body systems work and how they are interrelated. As students read about diseases that affect various organ systems, they develop the understanding that disease is a breakdown in structures or functions of a living thing and that some diseases are the result of damage by infection and others result from the inherent failure of a body system. Within this *Learning Set*, they repeatedly return to the *Communicable Disease Information Table* created in *Learning Set 2* and add their new understandings about diseases and how they affect body systems.

Learning Set 4 centers on the question *How do scientists identify and stop disease outbreaks?* Students again assume the role of student epidemiologists, focusing this time on how diseases are spread in a community and how to prevent their spread. Students are introduced to the idea that disease can spread through both direct and indirect interactions. As students read about the challenges facing epidemiologists in tracking disease through complex interactions, they develop an appreciation for how difficult it is to identify the source of a disease outbreak and prevent the spread of a disease.

The Unit concludes with students returning to the *Big Question: How can you prevent your good friends from getting sick?* Each group focuses on one communicable disease, describes how to prevent the disease from being spread to other people, and makes recommendations about what can be done to avoid transmitting the disease to other people. Each group then presents their findings to the class in a *Solution Showcase*. After evaluating the recommendations made by each group, the class develops a set of recommendations for staying healthy, helping others to stay healthy, and how to make their school a healthier place.

Investigations

In *Learning Set 1*, students are given the *Big Question*, they set up the *Project Board* and simulate disease transmission, first by passing around a beanbag covered in an undetectable powder that is sensitive to ultraviolet light (UV) and then through a water-sharing activity in which some "infected" water is spread during interactions with other students. Students collect and analyze data to determine the initial carrier of the "disease."

In *Learning Set 2*, students use the microscope to examine plant and animal cells. They prepare wet-mount slides, stain them, identify cells and cell structures, and estimate their size. They examine a case study of food poisoning, analyze the data and present their conclusions about the source of the bacteria to the class in an *Investigation Expo*. Groups design and carry out an experiment in which they culture bacteria to determine places in the classroom that harbor bacteria. Based on a surprising result of this investigation, they then develop a research question that can be answered by gathering, growing, and judging the amount of bacteria that has grown. Groups then design an experiment to answer the research question, present their experimental design and hear about other groups' designs during class presentations. They revise their plan, carry out their experiment and report their findings to the class in an *Investigation Expo.* Students then examine a series of case studies and construct a *Communicable Disease Information Chart* that summarizes what they are learning about different diseases.

The investigations in *Learning Set 3* help students develop their understanding of how human body systems work and interact. Students measure their breathing rate while resting and after exercise. They then build a working model of the respiratory system. After reading about the respiratory system, they connect their model to the respiratory system. Next, students locate pulse points and measure their heart rate while resting and after exercise. After building and running a model for getting oxygen from their heart to their big toe, they connect the model to their circulatory system. After sketching an outline of their body, students construct a model of the digestive system as each organ is described in a story read to them. Students evaluate the model and discuss its strengths and weaknesses. Students read about the nervous system and the immune system and then each group constructs a poster about one part of the immune system. After all groups have presented their poster to the class, the class works collaboratively to describe how the immune system organs work together to keep you healthy. At the end of *Learning Set 3*, students make transparencies of organ systems that overlay and show connections between the organ systems.

In *Learning Set 4*, students investigate tracking a disease through simple and complex interactions. They draw interaction diagrams which help them identify the sentinel case.

The Unit concludes with students collaborating to develop a set of recommendations for staying healthy, helping others to stay healthy, and how to make their school a healthier place.

Nature of Science

Through repeated practice, students experience how scientists work together to refine their understanding and use what they know to answer important questions. To help students reflect on how their understanding grows through investigation just as scientists' understanding does, they set up a *Project Board* to keep a running record of what they know, what they need to investigate, their claims with the evidence they have to back these up, as well as how the information fits together to answer the *Big Question.*

Through iterative investigation of disease transmission in *Learning Set* 1, students develop the ability to support their ideas with evidence and logical arguments. The challenge of interpreting the data in the first water-sharing investigation helps students understand the importance of systematic observation, accurate record-keeping, deciding which evidence should be used, and accounting for anomalous data. It also provides an opportunity to critique procedures and redesign the investigation to generate reliable data. The second, redesigned water-sharing investigation provides the opportunity to interpret data, identify cause-and-effect relationships, use evidence to generate explanations, and propose alternative explanations.

In *Learning Set 2*, students use the microscope and culturing investigations to gain experience using appropriate tools and technologies to gather, and to analyze and interpret data. While students design their own bacteria experiment, they develop the ability to formulate questions, identify and control variables, design and execute investigations, interpret data and use evidence to generate explanations. Working collaboratively to communicate experimental methods and critique procedures helps students understand the value of sharing and building on one another's ideas. Through their examination of case studies, students develop the ability to review and summarize data, and then apply logic as they construct an argument for their proposed explanations. These experiences all help students understand that an important aspect of scientific inquiry is to evaluate the results of scientific investigations and the explanations proposed by other scientists.

In *Learning Set 3*, as students construct and investigate a series of models, they develop an understanding of the usefulness as well as the limitations of models in scientific inquiry. When students interpret the data collected in the breathing rate and heart-rate investigations, they again identify cause-and-effect relationships, and use evidence to generate explanations.

In *Learning Set 4*, students develop an appreciation of the challenges that face scientists in developing explanations of events involving complex interactions. As groups collaborate to develop recommendations for staying healthy, helping others to stay healthy, and how to make their school a healthier place, they learn how scientific understandings can help them develop good solutions to everyday problems.

Artifacts

In *Learning Set 1*, students set up the *Project Board* to help them keep track of what they know, what they need to investigate, their claims with the evidence they have to back these up, as well as how the information fits together to answer the *Big Question*. They keep records of their individual interactions in the water-sharing activities as well as data from the entire class. Students should also keep any interaction diagrams or other record of their work in determining who the initial carrier was.

In *Learning Set 2*, students set up a *Communicable Disease Information Table* on which they keep a running record of what they are learning about the causes of diseases, how they spread, how they affect body systems, and their symptoms and treatments. When students use the microscope, they keep a record of their observations in the form of diagrams, written descriptions, and size calculations. They also create posters of cells consisting of a labeled drawing, and written descriptions of the functions of organelles. When students examine the case study, they keep records of their organization of the data, and conclusions they drew from the data. Students perform two investigations in which they grow bacterial cultures and keep diagrams and written descriptions of the cultures that grow. The *My Experiment* pages are a record of their thinking and experimental design. They also make a poster to communicate their experimental design and results in an *Investigation Expo*. Groups examine case studies of communicable diseases. Students should keep written records of their disease case study presentation. These may consist of notes, note cards, presentation software, or posters.

In *Learning Set 3*, students keep records of data collected during the breathing-rate and heart-rate investigations. They also create and keep a class table of results. Models of the respiratory and digestive systems constructed by groups may also be kept or photographed. Students also keep a record of their recommendations for improving the circulatory system model. Students produce posters showing part of the immune system which they present to the class. They also keep notes of other groups' presentations about how the different parts of the immune system work together. For the *Solution Showcase* at the end of *Learning Set 3*, students make transparencies of organ systems that overlay and show connections between the organ systems.

In *Learning Set 4*, students keep records of the simple and complex interaction diagrams they created in the *Track a Disease* investigation. They also write a story describing the possible context in which the interactions took place.

In *Answer the Big Question*, students engage in a *Solution Showcase*, creating presentations that focus on answering the *Big Question* for one communicable disease. (presentation artifacts may include written notes, presentations software, posters, charts, diagrams, etc.). As they prepare their presentations, students complete a *Create Your Explanation* page for three types of recommendations made concerning their assigned disease. Students' final disease prevention posters made for display in various places around the school are also kept.

Targeted Concepts, Skills, and Nature of Science	Section
Scientists often collaborate and share their findings. Sharing findings makes new information available and helps scientists refine their ideas and build on others' ideas. When another person's or group's idea is used, credit needs to be given.	*All Learning Sets*
Scientists must keep clear, accurate and descriptive records of what they do so they can share their work with others, consider what they did, why they did it, and what they want to do next.	1.1, 1.2, 2.3, 2.5, 2.6, 2.7, 3.3, 3.5, 4.3, 4.4

Targeted Concepts, Skills, and Nature of Science	Section
Scientific questions are directed toward objects and events that can be described, explained, or predicted by scientific investigations.	1.1, 2.1, 2.6, 2.7, 3.1, 4.1, 4.3, 4.4
Identifying factors that lead to variation is an important part of a scientific investigation.	2.5, 2.6, 2.7, 3.3, 3.5, 4.3, 4.4
Scientific investigations and measurements are considered reliable if the results are repeatable by other scientists using the same procedures.	1.2, 2.6, 2.7, 3.3, 3.5, 4.3, 4.4
In a fair test, only the manipulated (independent) variable and the responding (dependent) variable change. All other variables are held constant.	2.5, 2.6, 2.7
Scientists measure surface temperature a short distance above the ground in order to measure the temperature of the environment.	1.2
Scientists make claims (conclusions) based on evidence obtained (trends in data) from reliable investigations.	1.2, 2.5, 2.6, 2.7, 3.3, 3.5, 4.3, 4.4
Explanations are claims supported by evidence, accepted ideas, and facts.	1.2, 2.5, 2.6, 2.7, 3.3,
Models are a representation of something in the world. Simulations use a model to imitate, or act-out, real-life situations.	1.1, 1.2, 3.4, 3.6, 3.7,
Students should be able to manipulate a compound microscope to view microscopic objects.	2.3
Students should be able to prepare a wet-mount slide, use appropriate staining techniques, and determine the size of a microscopic object, using a compound microscope.	2.3
Students should be able to recognize and analyze patterns and trends.	1.1, 1.2, 2.5, 2.6, 2.7, 3.3, 3.6, 4.3
Students should be able to sequence events.	1.1, 1.2, 2.5, 3.4, 3.6
Students should be able to identify pulse points and pulse rates.	3.5

Targeted Concepts, Skills, and Nature of Science	Section
Students should be able to identify structure/function relationships in organisms.	3.2, 3.4, 3.6, 3.7, 3.8
The cell is the basic unit of structure and function in all living things. Cells are usually microscopic in size.	2.3, 3.2, 3.8
Cells have specialized structures that perform specific functions. Most cells have a cell membrane, cytoplasm and genetic material. Many cells have a nucleus. Some cells have cell walls and/or chloroplasts.	2.3, 3.2
Some living things are single cells; others, including humans, are multicellular.	2.4, 3.2
In multicellular living things, cells are organized for more effective functioning into tissues, organs, and organ systems. Tissues, organs, and organ systems help to provide all of the cells in the organism with nutrients, oxygen, and to remove wastes.	3.2, 3.4, 3.6, 3.7, 3.8
Organ systems are composed of organs and tissues which perform specific functions and interact with each other.	3.2, 3.4, 3.6, 3.7, 3.8
Disease is a breakdown in the structures or functions of an organism. Some diseases are the result of failures of a body system. Other diseases are the result of damage by infection from other organisms (germs).	2.2, 2.4, 2.8, 2.10, 3.1, 3.4, 3.6, 3.7, 3.8
Communicable diseases are diseases that spread from one person to another. Noncommunicable diseases cannot be passed from the person who is sick to other people.	1.1, 1.3, 2.2, 2.5, 2.8, 2.9, 2.10, 4.2, 4.3, 4.4
Communicable diseases may be transmitted by direct contact with an infected person or by indirect contact with an infected object or substance.	1.1, 1.2, 1.3, 2.2, 2.5, 2.8, 2.9, 2.10, 4.2, 4.3, 4.4
Communicable diseases are commonly spread through hand-to-hand contact. Hand-washing is a simple habit that can help keep you healthy.	1.1, 2.2, 2.9, 2.10, 3.7,
Many diseases are caused by bacteria or viruses.	2.4, 2.5, 2.8, 2.9, 2.10, 3.4, 3.6, 3.7, 3.8
An antibiotic is a drug that kills or prevents the growth of bacteria.	2.4, 2.10, 4.4

Targeted Concepts, Skills, and Nature of Science	Section
Immunization is a medical treatment that helps protect you from disease.	2.8, 2.9, 3.8, 4.4
Vaccination is the process by which a person is given a vaccine – a substance that protects a person from a disease.	2.9, 2.10, 3.7, 3.8
Certain specialized cells protect the body from germs that cause communicable diseases. These cells produce chemicals that can identify and destroy germs that enter the body.	2.10, 3.8

Unit Material List

Quantities based on 4-6 students		
Unit Durable Group Items	**Section**	**Quantity**
Prepared animal cell slides	2.3	1
Plastic microscope slides, pkg. of 144	2.3	1
Microscope slipcovers	2.3	1
Poster of respiratory system	3.4	1
Poster of circulatory system	3.6	1
Poster of digestive system	3.7	1
Poster of lymphatic system	3.8	1
UV LED flashlight	ABQ	1
Microbe doll - white blood cell	ABQ	1
Microbe doll - flu germ	ABQ	1
Stopwatch	3.3,3.5	2
Transparency of body outline	3.1, 3.4, 3.6, 3.7, 3.8	1

Unit Material List

Quantities based on 4-6 students		
Unit Durable Group Items	**Section**	**Quantity**
Tweezers	2.3	2
Compound microscope, 400X	2.3	1
Safety Scissors	3.7	1

Quantities for 5 classes of 8 groups		
Unit Durable Group Items	**Section**	**Quantity**
Project Board, laminated	ABQ, 1.1, 2.1, 2.3, 2.5, 3.1, 3.4, 3.6, 3.7, 3.8, 4.1	5
Project Board transparency	ABQ, 1.1, 2.1, 2.3, 2.5, 3.1, 3.4, 3.6, 3.7, 3.8, 4.1	1
Long balloons, 12" long (to represent small intestines)	3.7	200
Long, squiggly balloons (to represent large intestines)	3.7	40
Box of modeling clay, 1 lb	3.4	1

Quantities for groups of 4-6 students		
Unit Consumable Group Items	**Section**	**Quantity**
Set of colored markers		6
Bottle of .1M NaOH (sodium hydroxide), 8 oz	1.2	1
Phenolphtalein dropper bottle, 50 mL	1.2	1
Iodine dropper bottle, 50 mL	2.3	1

Quantities for 5 classes of 8 groups		
Unit Consumable Group Items	**Section**	**Quantity**
Beral pipettes, 3 mL, pkg. of 10	1.2	100
Glowpowder	ABQ	1
Battery, AA type	ABQ	1
Pack of rubber band	3.4	1
Straws, large diameter	3.4	100
Masking tape, 1 roll	3.7	1
Round balloons (to represent the stomach)	3.7	96
9" x 12" Blue construction paper, pkg. of 50	3.6	2
9" x 12" Red construction paper, pkg. of 50	3.6	2
Sterile 3 pack of Petri dishes with powdered nutrient agar	2.6, 2.7	28
Invisible tape	2.6	1
Cotton swabs, pkg. of 100	2.6, 2.7	4
Plastic cup, 10 oz	1.2	200
Restickable Easel Pad	ABQ, 2.3, 2.5, 2.7, 3.7, 4.3	4
Colored pencils	ABQ, 2.3, 2.5, 2.7, 3.7, 4.3	6
Rubber gloves, pkg. of 100	2.3	2
China marker, black	2.6	4
Plastic wrap, roll	3.4	1

Unit Material List

Additional Items Needed Not Supplied	Section	Quantity
Antibacterial wipes	ABQ, 1.2	1 per classroom
Distilled water, 1 gal (*option 2)*	1.2	1 per classroom
Onion (sliced)	2.3	1 per classroom
Antibacterial cleaner (for table)	2.3	1 per classroom
Bleach, 1 gal	2.6, 2.7	1 per classroom
Incubator (suggested)	2.6, 2.7	1 per classroom
Access to water	2.3	1 per classroom
Safety goggles	2.3	1 per student
Roll of paper towels	2.3	1 per classroom
Plastic bottle, with bottom cut off, 2 L	3.4	1 per group

NOTES

UNIT INTRODUCTION

What's the Big Question?

How can you prevent your good friends from getting sick?

Overview

Students are introduced to the *Big Question* of *Good Friends and Germs: How can you prevent your good friends from getting sick?* Students are introduced to the idea that diseases are communicated from person to person, first by passing around a beanbag covered with an undetectable powder that is sensitive to ultraviolet light and then through a water-sharing activity in which some "infected" water is spread during interactions with other students. Both activities highlight the spread of germs and help students define communicable diseases.

Targeted Concepts, Skills, and Nature of Science	Performance Expectations
Scientists often collaborate and share their findings. Sharing findings makes new information available and helps scientists refine their ideas and build on others' ideas. When another person's or group's idea is used, credit needs to be given.	Students should share their individual ideas with their group and their group's ideas with the class as they consider what they know about preventing their friends from getting sick.
Communicable diseases may be transmitted by direct contact with an infected person or by indirect contact with an infected object or substance.	Students should be able to identify examples of the spread of germs (represented by the glow powder) through direct and indirect contact.
Communicable diseases are commonly spread through hand-to-hand contact. Handwashing is a simple habit that can help keep you healthy.	Students should be able to describe how washing could help prevent the spread of germs.

Materials	
2 per classroom	beanbags

Materials	
Enough to coat beanbags	Glow powder
1 per classroom	UV LED flashlight
1 per student	Hand soap or antibacterial wipes

Activity Setup and Preparation

This activity begins with a brainstorming session and ends with beginning the Project Board. If this is the first introduction of the *Project Board* you should spend time thinking about how you are going to introduce it to students.

Students should not turn to the page in their books before completing this first activity. During the brainstorming session, students will be passing around a beanbag to help them share ideas in an orderly fashion. Arrange your room to facilitate a whole-class discussion. It is suggested that students write down ideas shared by their classmates. You may want to have them write each item in their notebook during the beanbag passing, or you may want to keep track of it on a public list you create that they can copy.

Before class starts, dust the beanbag with glow powder (to simulate germs), but not so much that it is noticeable. You may want to put glow powder on your hands and shake hands with some students as they walk into the room. (Don't worry if you don't shake hands with everyone. You don't want the students to notice that there is anything unusual going on.)

If soap and water are not available in your classroom for the hand-washing portion of the activity, pre-packaged moist towelettes or wipes may be used.

Homework Options

Reflection

- **Science Content:** Imagine you are asked to teach a kindergarten class about the importance of hand-washing. What important points would you try to make to the class? What would you want to demonstrate to the class?

Preparation for 1.1

- **Science Content:** Your school is suffering an outbreak of the flu and many of your classmates are home sick. How do you think this outbreak might have started? What information would help you answer this question?

UNIT INTRODUCTION IMPLEMENTATION

What's the Big Question?

How can you prevent your good friends from getting sick?

This year you may get five or six colds. Colds make you feel sick. You might get a stuffy nose, watery eyes, and a cough. How did you get that cold? What causes you to be sick when you have a cold or the flu? What changes take place in your body? These are some of the questions you might ask as part of this science Unit.

Why is it important to be able to answer these questions? You probably have heard people say that the flu or another illness is "going around." If you know how the flu spreads, you can try to protect yourself from getting sick. The more you know, the healthier you can keep your own body.

Look at the *Big Question* of this science Unit: *How can you prevent your good friends from getting sick?* All of the investigations and readings you will be doing during the next few weeks will help you answer this question.

It is a very big question. You will be able to answer the question more easily if you break it up into smaller questions. You can use the smaller questions to begin investigations and read about science concepts. As you answer each of these smaller questions, you will be solving pieces of a puzzle that will help you answer the *Big Question*. You will combine everything you learn in answering smaller questions to answer the *Big Question* at the end of the Unit. In doing so, you will be doing what scientists do. During this Unit, you will also use many of the science practices you learned about and used in other Units.

Welcome to *Good Friends and Germs.*
Enjoy your journey as a student scientist.

GF 3

GOOD FRIENDS and GERMS

◀ *2 class periods*

Unit Introduction

What's the Big Question?

5 min.

Introduce the Big Question: How can you prevent your good friends from getting sick?

○ Engage

Begin by eliciting students' ideas about how sickness spreads and how they could prevent their friends from becoming sick.

TEACHER TALK

"Who has been sick during the last month?

What sickness did you have?

What do you think caused the sickness?"

*A class period is considered to be one 40 to 50 minute class.

Think about the Big Question

5 min.

Introduce the Big Question and discuss students' ideas about how people get sick.

Think about the Big Question

Before you start, it's a good idea to make sure you understand what you know about the *Big Question*. What do you think you know about preventing your friends from getting sick? What are you not sure about?

Get Started

You are going to share what you know about how you get sick. You will pass a beanbag around the classroom. When you receive the beanbag, hold onto it and say aloud one thing you know about how you get sick. Do not worry if others have already mentioned your idea. Listen carefully to what your classmates have to say when they receive the beanbag. You may wish to record some of the ideas you hear.

△ Guide

Introduce the *Big Question* by reading the introductory page with students. Make sure students understand how this question will serve as the framework for all of the investigations and readings in the Unit.

After you complete the beanbag-sharing activity, use an ultraviolet flashlight to look around the classroom. What do you see? Do you notice a glow around the classroom? A powder that glows under ultraviolet light was put on the beanbag. As you passed the bag around, all of you touched the powder. When you touched different places with powder on your hands, the powder rubbed off onto the surfaces.

Stop and Think

1. Which surfaces were touched the most in your classroom?
2. Which parts of your hands and head have powder on them?
3. What is the farthest point in your classroom that the powder was moved?
4. What surprised you about where you found the powder?

Germs Can Be Anywhere

The powder on the beanbag is like germs. When you get germs on your hands, you move them from one place to another, just like you moved the powder. You touch your face, the doorknob, your pencil, and your friends. Like the powder, the germs move from your hands to your face, your friends' hands, and all over anything you touch. When you put your hands on your face, you move the germs there. The germs on your face, especially near your eyes, nose, or mouth can then get inside your body.

Could you tell that the beanbag, your hands, and other people's hands had powder on them before you started passing the beanbag? You probably did not know the beanbag was covered with powder because the powder is very hard to see. It is not possible to see germs with your eyes either.

Germs can be on the things you touch. Doorknobs, handles, chairs, beanbags, and many other things can be covered with germs. When you touch these things, germs can be moved from the object to your hands, just like the glowing powder moved from the beanbag to your hands, and then from your hands to other surfaces.

GF 5

GOOD FRIENDS and GERMS

Get Started

15 min.

Students pass a beanbag "infected" with glow powder as they share ideas they have about how you get sick.

△ Guide

Briefly describe how, as the beanbag is being passed around, each person will share one thing they know about how you get sick. Have the students write down some of the ideas shared by their classmates.

⬡ Get Going

Have the students pass the beanbag and share what they think they know about preventing their good friends from getting sick and what they are not sure of. After they have shared their ideas, shine the UV flashlight around the room and have students observe the glow powder. Let students suggest places to check with the UV flashlight. Describe how you introduced the harmless, fluorescent powder into the classroom on your hands and on the beanbags.

Stop and Think

5 min.

Discuss how the glow powder has spread from person to person and throughout the classroom.

After you complete the beanbag-sharing activity, use an ultraviolet flashlight to look around the classroom. What do you see? Do you notice a glow around the classroom? A powder that glows under ultraviolet light was put on the beanbag. As you passed the bag around, all of you touched the powder. When you touched different places with powder on your hands, the powder rubbed off onto the surfaces.

Stop and Think

1. Which surfaces were touched the most in your classroom?
2. Which parts of your hands and head have powder on them?
3. What is the farthest point in your classroom that the powder was moved?
4. What surprised you about where you found the powder?

Germs Can Be Anywhere

The powder on the beanbag is like germs. When you get germs on your hands, you move them from one place to another, just like you moved the powder. You touch your face, the doorknob, your pencil, and your friends. Like the powder, the germs move from your hands to your face, your friends' hands, and all over anything you touch. When you put your hands on your face, you move the germs there. The germs on your face, especially near your eyes, nose, [illegible] then get [illegible] your body.

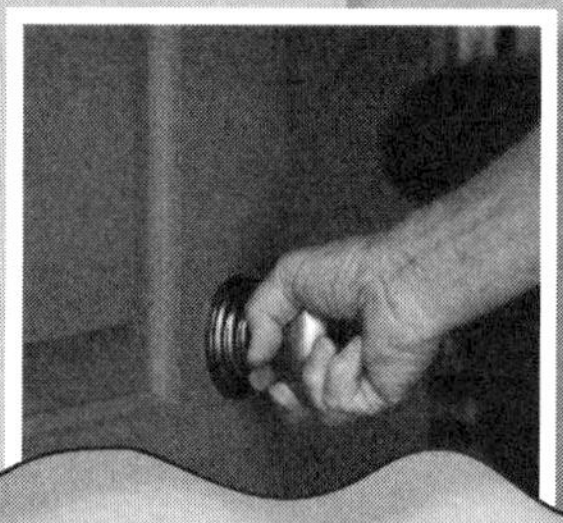

META NOTES

Having students share what they think they know as they pass the beanbag elicits prior knowledge.

It is critical to access students' understanding before teaching for two reasons.

First, it enables the teacher to assess where students are in their understanding of disease, disease spread, the effects of disease, and disease prevention; and second, it provides a basis of comparison for students to use when identifying what they have learned.

Record students' ideas so they can later be used in the *Project Board*.

○ Get Going

Challenge students to identify any patterns that can be observed in the distribution of the glow powder in the classroom and/or on their bodies. Help students focus their search for patterns through questioning. Encourage students to record their observations.

TEACHER TALK

"Where is the most glow powder? What does that mean?
What were the most touched surfaces in the classroom?
How do you know?
Who has the most powder on their hands?"

TEACHER TALK

"Where else on people's bodies is there glow powder?

Where are the strangest places the glow powder showed up? How did it get there?

Where is the furthest location from where you were passing the beanbag that the glow powder showed up? How did it get there?

Was there anything about where you found the glow powder that surprised you? Why?"

△ Guide

Once students have recorded their observations, either have them record their answers to the *Stop and Think* questions and then discuss them with the class, or simply initiate a discussion of the questions.

△ Guide and Assess

Facilitate a brief class discussion checking for student comprehension of what they have read. Emphasize the relationship between the powder and germs — that the powder, like germs, is hard to see. Check to see if students are making the connection between germs and glow powder and between the spread of germs and the spread of the glow powder.

Reflect

5 min.

Students consider how the spread of the glow powder relates to the spread of disease germs.

NOTES

Procedure

10 min.

Students try to remove the "germs" from their hands by washing. Some students find that all of the glow powder "germs" were not removed and need to wash again.

Wash your hands thoroughly to remove all the powder.

Procedure

Remove the powder from your hands. Wash your hands, and then use the light to see if you have removed all the powder. Look especially around your fingernails and in the creases of your skin on the palms of your hands. How well did you wash your hands? If you can still see powder on your hands, wash them again.

Your Challenge

In this Unit, you will need to respond to a challenge to answer the *Big Question, How can you prevent your good friends from getting sick?* The challenge has two parts. In the first part, the class will study several diseases. Your group will study one disease. You will describe the disease and how you know somebody has it. You will report about how the disease affects the body. You will explain how it is spread and how it is treated. You will then make recommendations about how to prevent spreading the disease to others. Other groups will study different diseases. Each group will report to the class about their disease.

In the second part of the challenge, you will make recommendations about how you can stay healthy and prevent others from getting sick. Then, as a class, you will choose recommendations to present to other students.

Conference

Remember that the *Big Question* for this Unit is *How can you prevent your good friends from getting sick?* You need to break down this question into smaller questions that you can investigate. As a group, review what you and your classmates said you thought you knew. You probably have a lot of questions that need to be answered before you can fully answer the *Big Question*. The powder activity may have reminded you of some things that you think you know, or don't know, about germs. Record these things so that you will remember what you discussed when you share again with the class. Which questions that you would like answered are important for answering the *Big Question?*

○ Engage

Ask students how they might remove these "germs" from their hands. Elicit that washing their hands will remove the glow powder "germs."

⬡ Get Going

Have students wash their hands or wipe them with wipes. After washing, have students check their hands using the UV flashlight. After students observe there is still glow powder in certain places on their hands, have them wash their hands again.

△ Guide

Help students develop the understanding that it is not easy to remove glow powder from their hands because the small particles lodge in crevices in their skin or under fingernails. Germs are even smaller than the particles of glow powder, and can be even more difficult to remove from hands than the glow powder.

TEACHER TALK

"Where did the glow powder remain?

After washing the second time, did the glow powder still remain?

Why do you think it is important to wash our hands?

Why do you think it is important for doctors to wash their hands?"

NOTE: This is a good place to end the first class period of the session.

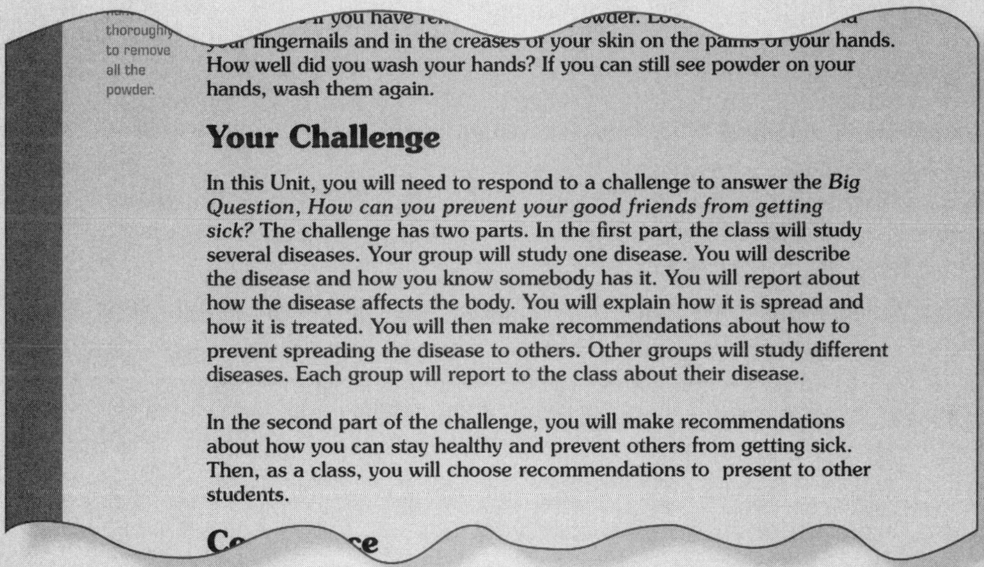
thorough to remove all the powder.

your fingernails and in the creases of your skin on the palms of your hands. How well did you wash your hands? If you can still see powder on your hands, wash them again.

Your Challenge

In this Unit, you will need to respond to a challenge to answer the *Big Question, How can you prevent your good friends from getting sick?* The challenge has two parts. In the first part, the class will study several diseases. Your group will study one disease. You will describe the disease and how you know somebody has it. You will report about how the disease affects the body. You will explain how it is spread and how it is treated. You will then make recommendations about how to prevent spreading the disease to others. Other groups will study different diseases. Each group will report to the class about their disease.

In the second part of the challenge, you will make recommendations about how you can stay healthy and prevent others from getting sick. Then, as a class, you will choose recommendations to present to other students.

Your Challenge

5 min.

Students are introduced to the two parts of their challenge—first, to study one disease in depth and report on it to the class, and second, to make recommendations about how to prevent spreading the disease to others.

△ Guide

Read this section together with the class and lead a discussion of the two parts of the challenge. Check that students understand what they will be responsible for producing for each part of the challenge.

Conference

10 min.

Students rethink what they know and what they think they know about germs with the glow powder in mind. By sharing their ideas, student groups can decide on the most important ideas based on what they observed with the glow germ simulation and their experiences.

ach group wi… …lass abou…

In the second part of the challenge, you will make recommendations about how you can stay healthy and prevent others from getting sick. Then, as a class, you will choose recommendations to present to other students.

Conference

Remember that the *Big Question* for this Unit is *How can you prevent your good friends from getting sick?* You need to break down this question into smaller questions that you can investigate. As a group, review what you and your classmates said you thought you knew. You probably have a lot of questions that need to be answered before you can fully answer the *Big Question*. The powder activity may have reminded you of some things that you think you know, or don't know, about germs. Record these things so that you will remember what you discussed when you share again with the class. Which questions that you would like answered are important for answering the *Big Question?*

△ Guide

Ask students to form small groups and discuss the following questions:

- Which ideas about how you get sick is the class most sure of? Most unsure?
- Which ideas about how you can keep others from getting sick is the class most sure of? Most unsure?
- What do you think the class needs to learn more about in order to understand better how we get sick and how to prevent others from getting sick?

Lead a class discussion to develop smaller questions that could guide their inquiries into getting sick and preventing others from getting sick. Guide the discussion in such a way that students identify the questions that lie at the core of each of the four *Learning Sets.*

- How do I get sick?
- What kinds of things make you sick?
- What happens to you when you get sick? How do scientists identify and stop disease outbreaks?

Create a *Project Board*

When you work on a project, it is useful to keep track of your progress and what you still need to do. You will use a ***Project Board*** to do that. It is designed to help your class organize its questions, investigations, results, and conclusions. The *Project Board* will also help you to decide what you are going to do next. During classroom discussions, you will record the class's ideas on a class *Project Board*. At the same time, you will keep track of what has been discussed on your own *Project Board* page.

The *Project Board* has space for answering five guiding questions:

- What do we think we know?
- What do we need to investigate?
- What are we learning?
- What is our evidence?
- What does it mean for the challenge or question?

How can you prevent good friends from getting sick?

What do we think we know?	What do we need to investigate?	What are we learning?	What is our evidence?	What does it mean for the challenge or question?

Create a *Project Board*

20 min.

Students create a Project Board *and start to record their ideas.*

⬡ Get Going

Tell the class that now that they have thought about the *Big Question*, and have identified some of the smaller questions that need to be investigated in order to answer it, they are going to need a way to keep track of their progress.

Introduce students to the idea of a *Project Board* by having them read the selection in the text.

Go over the five headings used in the *Project Board* and describe how the *Project Board* will be ongoing record of their inquiry —what they are asking and what they are learning, the supporting evidence, and how it all contributes to answering the *Big Question*. Begin to add students' ideas to the first two columns.

△ Guide

Help students understand why it is important to articulate what they think they know before getting started—specifying what they think they know will help them to recognize what they are learning later on.

Discuss ideas they are confused about and ideas they think they should investigate in order to assist them in answering the *Big Question*. This should transition students to the next activity where they model the spread of disease.

Students may need help formulating questions from what they are confused about. An important part of managing discussion about this part of the *Project Board* is to provide that help. You have the opportunity here to model for your students how to ask good questions and point out what makes a question good. In this case a good question is an answerable question that will help us to answer the *Big Question*.

NOTES

PBIS

To begin, write the *Big Question, How can you prevent your good friends from getting sick?* on the *Project Board*.

The activity you just completed was meant to help you recall what you think you know about how you can prevent your good friends from getting sick. Discuss the things you and your classmates think you know about how you get sick and what happens when you get sick. Record these things in the first column of the *Project Board*.

In the second column, record the things you need to investigate to answer the *Big Question*. During your discussions, you and others in your class may have disagreed about some ideas. This second column is designed to help you keep track of things that are debatable or that you do not know enough about yet.

What's the Point?

Often, people try to solve problems without first understanding what they need to do. You have just spent some time thinking about what you need to learn to answer the *Big Question*.

In the activity, you were probably surprised by how easily the powder spread. The powder is similar to germs that spread disease. Understanding how germs spread will help you understand why you get sick.

You started a *Project Board* to help track what you understand. You also added questions about what things you need to learn to prevent your friends from getting sick. The *Project Board* is a space to help the class work together to understand and solve problems. Using it will help you have good science discussions as you work on a project.

GF 8

Project-Based Inquiry Science

What's the Point?

5 min.

Teacher Reflection Questions

- In this segment, students observed a possible mechanism by which germs might spread. Students need to realize that we can't easily detect germs and that they can be spread over a large area very easily before moving on to the rest of the Unit. What issues did students encounter when they were developing these understandings?
- This was an introductory activity for this Unit. The goal of this activity is to activate students' prior knowledge and prepare them for further investigation. How did the activity ideas accomplish those things?
- As students carried out the glow-powder activity, they were asked to track their own and others' ideas. How did the writing task affect the students' ability to attend to the other students' ideas? What else might you do so that students attend to each other and make a record of the ideas?

NOTES

LEARNING SET 1 INTRODUCTION

Learning Set 1

How Do You Get Sick?

◀ ***5 class periods***

A class period is considered to be one 40 to 50 minute class.

Students simulate the spread of disease, learn to distinguish models from simulations, and communicable diseases from noncommunicable diseases.

Overview

Section 1.1 initiates and sets the objective for this *Learning Set*, answering the question, *How do we get sick?* Students formulate questions that can be answered through scientific inquiry. Working in small groups, students evaluate one another's questions to be sure that they meet the criteria for a good question. Each group comes up with two questions that will help guide their investigation into how diseases are spread.

In *Section 1.2*, students read about the nature of a simulation and engage in a series of water-sharing activities that simulate the spread of a disease. Students are given cups of water, one of which has been "infected" with a harmless chemical. Through a series of water-sharing interactions, the chemical spreads from one student to another. An indicator that changes color in the presence of the chemical is then added to each student's cup of water to reveal who has become "infected." Students are challenged to identify the initial carrier; however, since no specific directions were given to keep a record of each interaction, students will find it difficult to recall sufficient information to enable them to determine the initial carrier.

Next, students develop a strategy for keeping track of their interactions and then they repeat the water-sharing investigation. Using the data collected during this second iteration of the water-sharing activity, students work collaboratively to determine who the initial carrier was. Any students' claims must be supported with data collected during the investigation.

In *Section 1.3*, students read about the distinction between a model and a simulation and apply their understandings to both the glow powder activity done earlier and the water-sharing activity they just completed.

The *Learning Set* concludes with students comparing and contrasting communicable and noncommunicable diseases.

LOOKING AHEAD

Prepare the solutions needed for the water-sharing investigations ahead of time. Instructions for preparing the solutions are *given* in the *Activity Setup and Preparation* section for *Section 1.2*.

Targeted Concepts, Skills, and Nature of Science	Section
Scientists often collaborate and share their findings. Sharing findings makes new information available and helps scientists refine their ideas and build on others' ideas. When another person's or group's idea is used, credit needs to be given.	1.1, 1.2, 1.3
Scientists must keep clear, accurate and descriptive records of what they do so they can share their work with others; consider what they did, why they did it, and what they want to do next.	1.1, 1.2
Scientific questions are directed toward objects and events that can be described, explained, or predicted by scientific investigations.	1.1
Explanations are claims supported by evidence, accepted ideas, and facts.	1.1, 1.2
Models are a representation of something in the world. Simulations use a model to imitate, or act out, real-life situations.	1.1, 1.2
Communicable diseases are diseases that spread from one person to another. Noncommunicable diseases cannot be passed from the person who is sick to other people.	1.3

Understanding for Teachers

The first iteration of the water-sharing investigation intentionally lacks any instructions to the student to keep a record of his or her interactions. The difficulty experienced in trying to recall details of their interactions helps students recognize that clear and reliable observation notes are important during scientific investigations.

Before students engage in the second iteration of the water-sharing investigation, they should carefully consider what data is needed to determine the initial carrier. Check that students understand that in order to identify the initial carrier, they need to keep track of who they interacted with, and also of information that will allow them to determine the order in which all of the interactions in the class occurred. They should also consider how the data they collect should be organized so that it can be easily shared. Realizing that the initial carrier cannot be identified unless everyone shares their data helps students understand the importance of collaboration and sharing ideas in science.

LEARNING SET 1 IMPLEMENTATION

◀ *5 class periods**

Learning Set 1

How Do You Get Sick?

The *Big Question* that you have to answer in this Unit is *How can you prevent your good friends from getting sick?* This is a very big question. To answer this *Big Question*, you need to break it down into smaller questions. There are many smaller questions you might need to ask. You already recorded some of these questions on the *Project Board*.

Think about the last time you were sick. Were any of your friends sick at the same time? Did you get sick after someone in your class or family was sick? The smaller question you will be looking at in this *Learning Set* is *How do you get sick?*

GF 9

GOOD FRIENDS and GERMS

Learning Set 1

How Do You Get Sick?

15 min.

Introduce the driving question for Learning Set 1: How Do You Get Sick?

○ Engage

Ask students to think about the last time they were sick and to share their experiences. Together with the students, read the introductory section and lead a class discussion in which students share their answers to the questions:

- The last time you were sick, were any of your friends sick at the same time?
- Did you get sick after someone in your class or family got sick?

Tell students that they will be investigating a smaller question in this *Learning Set: How do you get sick?*

*A class period is considered to be one 40 to 50 minute class.

NOTES

SECTION 1.1 INTRODUCTION

1.1 Understand the Challenge

Thinking about How You Get Sick

◀ ***1 class period***

A class period is considered to be one 40 to 50 minute class.

Overview

This section is designed to engage students in active reflection on how diseases are spread in preparation for answering the *Big Question*. Working collaboratively, they identify what they think they know about how they get sick, what they are unsure of, and what needs to be investigated. Students formulate questions that might help them answer the *Learning Set* question, and share their questions with their groups. Groups evaluate whether their questions meet the criteria, in that the questions must be: interesting, require several resources to answer, relate to the *Big Question*, and require collecting and using data. Each group then selects their two most interesting questions to share with the class and the *Project Board* is updated.

Targeted Concepts, Skills, and Nature of Science	Performance Expectations
Scientists often collaborate and share their findings. Sharing findings makes new information available and helps scientists refine their ideas and build on others' ideas. When another person's or group's idea is used, credit needs to be given.	Students should share their individual questions with their group and their group's questions with the class as they consider what they think they need to learn and what they need to investigate as they fill out the class *Project Board*.
Scientists must keep clear, accurate and descriptive records of what they do so they can share their work with others; consider what they did, why they did it, and what they want to do next.	Students participate in a class discussion while updating their *Project Board*, which is an organized record of their ideas and what they are learning.

Materials	
1 per class	Class *Project Board*

Activity Setup and Preparation

- This activity begins with students working individually to formulate questions, and ends with a class discussion leading to updating the *Project Board*. Decide how you would like students to keep a record of individual work — will it be in the form of a science notebook, a learning journal, or some other format? Make sure the class *Project Board* is posted where everyone can see it.

Homework Options

Reflection

- **Science Process:** What is your long-term goal in this Unit? *(Your long-term goal is to answer the* Big Question *and complete the two-part challenge.)*

What were some of the challenges you faced in writing your questions in this activity? *(Answers may vary, but should reflect issues related to meeting the five criteria for questions described on page 10 in the text. For example: My questions were not directed toward something that could be described, explained, or predicted by a scientific investigation. My questions did not require collecting and using data. My questions did not relate to the* Big Question. *My questions had yes/no answers.)*

- **Science Process:** What is your short term goal in this Learning Set? *(Your short term goal is answering the smaller questions.)*

How will the smaller question for this *Learning Set- How do you get sick?* – help you to answer the two challenges of the Unit *(Answering the question* "How do you get sick?" *will help me to explain how the disease my group is assigned to study spreads. It will also help me make recommendations about how I could prevent a friend from getting sick.)*

- **Science Process:** How is the *Project Board* helpful in reaching the long-term goal of answering the *Big Question? (The* Project Board *helps us to organize the smaller questions, investigations, results, and conclusions that are gathered. Putting all this information together on the* Project Board *organizes the information we need to reach our long term goal of answering the* Big Question.*)*

Preparation for 1.2

- **Science Process:** How was the glow-powder activity similar to how you might catch and share a cold? *(Colds are caused by germs which may be spread by contact with an infected person.)*

SECTION 1.1 IMPLEMENTATION

1.1 Understand the Question

Thinking about How You Get Sick

The question for this *Learning Set* is *How do you get sick?* It is a good idea to think about what you already know about how you get sick. It is also important to think about what you are unsure about and what you would like to investigate.

Get Started

You thought about how you get sick during the glow-powder activity. Share any additional ideas you have with your group. Listen carefully to all the ideas presented. You may want to record some of the ideas you hear. Then on your own, think of several questions that might help you answer the question for this *Learning Set*. Develop two questions that might help you understand how you get sick. When you write your questions, keep in mind that your questions should

- be interesting to you;
- require several resources to answer;
- relate to the *Big Question*; and
- require collecting and using data.

Your questions should not have yes/no or one-word answers.

When you have completed your questions, take the questions back to your small group. Share all the questions with each other. Carefully consider each question and decide if it meets the criteria for a good question. With your group, refine the questions that do not meet the criteria. Choose the two most interesting questions to share with the class. Give your teacher the rest of the questions so they might be used later.

Update the *Project Board*

You will now share your questions with your class. Be prepared to support your questions with the above criteria. Your teacher will help you with the criteria if needed. Then, your teacher will add your questions to the *Project Board*. In this *Learning Set*, you will work to answer some of these questions.

1.1 Understand the Question

Thinking about How You Get Sick

20 min.

Students share their ideas about what they learned from the glow-powder activity and identify what they are unsure of and what needs to be investigated.

△ Guide

Lead a class discussion of ideas that students are confused about and ideas that they think they should investigate in order to assist them in answering the *Big Question*.

Get Started

15 min.

Students formulate questions about How You Get Sick that can be answered through scientific inquiry. Working in small groups, students evaluate one another's questions to be sure that they meet the criteria for a good question. Each group comes up with two questions that will help guide their investigation into how diseases are spread.

idea to think about what you ... know about how ... also important to think about what you are unsure about and what you would like to investigate.

Get Started

You thought about how you get sick during the glow-powder activity. Share any additional ideas you have with your group. Listen carefully to all the ideas presented. You may want to record some of the ideas you hear. Then on your own, think of several questions that might help you answer the question for this *Learning Set*. Develop two questions that might help you understand how you get sick. When you write your questions, keep in mind that your questions should

- be interesting to you;
- require several resources to answer;
- relate to the *Big Question*; and
- require collecting and using data.

Your questions should not have yes/no or one-word answers.

When you have completed your questions, take the questions back to your small group. Share all the questions with each other. Carefully consider each question and decide if it meets the criteria for a good question. With your group, refine the questions that do not meet the criteria. Choose the two most interesting questions to share with the class. Give your teacher the rest of the questions so they might be used later.

Update the *Project Board*

You will now share your questions with your class. Be prepared to support your questions with the above criteria. Your teacher will help you with the criteria if needed. Then, your teacher will add your questions to the *Project Board*. In this *Learning Set,* you will work to answer some of these questions.

GF 10

Project-Based Inquiry Science

Get Going

Read the *Get Started* section together with the class and discuss the criteria for the questions to be written. Check that students understand each of the criteria.

Then, have students work individually to formulate their questions.

Next, have students share their questions with the group. As questions are shared, the group should evaluate whether each question meets the stated criteria.

Your questions should not have yes/no or one-word answers.

When you have completed your questions, take the questions back to your small group. Share all the questions with each other. Carefully consider each question and decide if it meets the criteria for a good question. With your group, refine the questions that do not meet the criteria. Choose the two most interesting questions to share with the class. Give your teacher the rest of the questions so they might be used later.

Update the *Project Board*

You will now share your questions with your class. Be prepared to support your questions with the above criteria. Your teacher will help you with the criteria if needed. Then, your teacher will add your questions to the *Project Board*. In this *Learning Set*, you will work to answer some of these questions.

Update the Project Board

5 min.

Take a few moments to review the Project Board *with the class. Add the questions generated by this exercise to the* Project Board.

△ Guide and Assess

Lead a discussion in which groups share their two most interesting questions with the class. Encourage the class to evaluate whether the questions meet the stated criteria. Challenge students to support their questions by describing how they met the criteria.

TEACHER TALK

"What were the five criteria for questions that were described in the *Get Started* section for this activity?

Consider the following question as an example before you begin to look at each other's questions. Can you get a cold by sharing a cup with someone who has a cold?

Does this question meet all of the criteria? *(No, it has a yes/no answer and answering the question doesn't require collecting and using data.)*

As each question is shared by a group, let's check to see if it meets each of the five criteria listed on the board. If you think a question does not meet the criteria, share your reasoning, and then allow the group to respond to your analysis."

> **META NOTES**
>
> If students are having difficulty generating questions, it may be helpful to have them consider framing questions along these lines:
>
> - "What do I want to find out about...?"
> - "If I do this, then what do I expect will happen?"
> - "What are the best observations or measurements to make?"
> - "Is this the best way to answer my question?"
> - "How can I make more accurate observations?"
> - "What data will best answer my question?"

> **META NOTES**
>
> The *Project Board* helps the students and the class focus on what is important (*the Big Question*) and keeps them organized as they grapple with how to reach the goal. By the end of this *Learning Set*, the *Project Board* should have on it a wide variety of knowledge the students have gained about diseases, symptoms of diseases, spreading diseases, and preventing their spread. The guiding questions you ask should lead them to think of smaller questions they need to ask and investigate to answer the *Big Question.*

Assess students' ability to generate questions that meet the criteria and are suitable for scientific investigation. Students often need help generating questions from what they are confused about. An important part of managing this discussion is to provide that help. There is an opportunity here to model questions that are suitable for scientific investigation and to point out what makes a question suitable.

Guide students to understand that throughout the *Learning Set*, they will work on answering some of these questions.

Teacher Reflection Questions

- What misconceptions about disease have come up during class discussions? Where during the Unit might you challenge some of these misconceptions?
- What did you do to facilitate students' thinking about questions that are suitable for scientific investigation? What ideas do you have for the future?
- How were you able to keep students focused on the *Big Question* as you updated the *Project Board?*
- How did you implement the *Project Board* discussion to include students who feel inhibited about sharing their ideas?

NOTES

SECTION 1.2 INTRODUCTION

1.2 Investigate

Simulating the Spread of Disease

◀ ***2 class periods***

A class period is considered to be one 40 to 50 minute class.

Overview

Students assume the role of student-epidemiologists as they try to determine the initial carrier of a disease after a series of water-sharing interactions. In the first iteration of this investigation, students are given no specific instructions to keep detailed, written records of their interactions. The challenges they face in trying to recall sufficient details afterwards to determine the initial carrier underscores the need for keeping detailed, accurate records during an investigation.

In order to determine the initial carrier and design a method for recording and organizing their data, students are responsible for identifying the nature of the data needed to reconstruct the interactions in sufficient detail. They then repeat the investigation, putting their ideas into action. Upon completion of the second iteration, students share their data with the class and collaborate to analyze the data and try to determine the initial carrier.

Targeted Concepts, Skills, and Nature of Science	Performance Expectations
Scientists often collaborate and share their findings. Sharing findings makes new information available and helps scientists refine their ideas and build on others' ideas. When another person's or group's idea is used, credit needs to be given.	Students should share their individual questions with their group and their group's questions with the class as they consider what they think they need to learn and what they need to investigate as they fill out the class *Project Board.*
Scientists must keep clear, accurate and descriptive records of what they do so they can share their work with others, consider what they did, why they did it, and what they want to do next.	Students experience challenges when trying to reconstruct data from recall after engaging in the first water-sharing investigation. Students design a method of recording and organizing data so that it can be shared with the class after the second water-sharing investigation. Students experience that success in identifying the initial carrier is only possible when accurate and complete data from the entire class is shared.

Targeted Concepts, Skills, and Nature of Science	Performance Expectations
Explanations are claims supported by evidence, accepted ideas, and facts.	Students support claims concerning the identity of the initial carrier with evidence drawn from records of interactions shared by the class.
Models are a representation of something in the world. Simulations use a model to imitate, or act out, real-life situations.	Students are able to differentiate between the parts of the glow-powder and water-sharing investigations that are models and those that are simulations.

Materials	
2 per student, per class period	Clear disposable plastic cups
1 per class	Distilled water (approximately one gallon)
1 per class	A dilute solution of sodium hydroxide (0.1M NaOH) or a solution of baking soda in bottled distilled water
1 per class	Phenolphthalein pH indicator
2 per student per class	Beral pipettes; if pipettes are rinsed between uses, they can be reused.
1 per student	Goggles and apron
1 per student	Antibacterial wipes

NOTE: This experiment is to be repeated. Students will need to be prepared with another set of cups for the second simulation.Homework Options

Activity Setup and Preparation

Prepare the solutions needed for the water-sharing investigations. You will need a basic solution (for the disease source) and a pH indicator solution (to indicate who is sick). The basic solution can be made using either sodium bicarbonate (baking soda) or sodium hydroxide (NaOH). You will use phenolphthalein as your indicator. Directions for making these solutions are listed on the next page.

Baking soda basic solution

Dissolve 15-20 teaspoons of baking soda in 1L of distilled water. Stir or shake the solution to dissolve as much of the baking soda as possible. Let the un-dissolved baking soda settle to the bottom, and then decant the clear solution from the top. This is done so that the "disease source" looks the same as the other cups of water. Students should not know that there are two solutions present.

One liter should provide more than enough basic solution for all of your classes since you will only need a single cup filled $\frac{1}{4}$ to $\frac{1}{3}$ with solution each time you run the procedure. (5 classes × 2 runs × $\frac{1}{3}$ cup = 3.3 cups)

NaOH basic solution

The molecular weight of NaOH is 40 g/mol. To make 1 M stock solution, dissolve 40 g NaOH into a total of 1 L of distilled water. This 1 M stock solution can then be diluted to a 0.1 M solution that you can use as a disease source in the activity by taking 10 mL of the 1 M solution and adding it to 90 ml of distilled water.

Alternatively, you can make a 0.1 M solution by dissolving 4 g of NaOH into 1 L of distilled water. Here again, one liter should provide more than enough basic solution for all of your classes.

Instruct students not to drink the NaOH or phenolphthalein solutions as they are harmful. They should immediately wash with soap and water if any solution is spilled on their hands or clothing. Students should use eye goggles when using the NaOH solution or phenolphthalein.

Phenolphthalein pH indicator solution

Use the phenolphthalein indicator that is packaged in the kit. Only one drop per cup is needed during the test phase. Care should be taken not to ingest this solution as phenolphthalein is potentially hazardous. Refer to the packaging of the solution for proper handing and disposal procedures.

- For each student, fill two sets of cups $\frac{1}{4}$ of the way with a transparent liquid. In each class set, all of the cups should contain distilled water except one. A single cup in each set should contain a basic solution instead of water. Do not let students see which cup you fill with basic solution and check to be sure there is no visible difference between the distilled water and the basic solution. Important! Keep track of which cup contains the basic solution — this is the initial carrier.
- You may want to fill sets of cups for several classes ahead of time and store on trays in a prep room or other storage area.
- **DO NOT** reuse cups or pipettes without thoroughly washing them as even a tiny bit of cross-contamination will invalidate results and cause confusion in the activity.

Homework Options

Reflection

- **Science Content:** Explain the difference between a model and a simulation. *(Models are a representation of something in the world. Simulations use a model to imitate, or act out, real-life situations.)*

  Both the glow-powder and water-sharing activities made use of models and simulations. Identify the model and the simulation in each activity. *(In the glow-powder activity, the glow powder was a model of germs; the beanbag was a model of something that had germs on it. As you passed the beanbag and touched objects in the classroom, you simulated how germs can spread from place to place. In the water-sharing activity, the cup of water was a model of a person and the chemical in the water was a model of disease germs. The water-sharing interactions simulated how diseases are transmitted using models of people with and without a disease.)*

- **Science Process:** Why was it difficult to identify the initial carrier after doing the first water-sharing activity? *(It was difficult to recall the details of every exchange that was made.)* How did you change the way in which you did the investigation the second time to solve this problem? *(By keeping careful, detailed records of every interaction.)*

- **Science Process:** Describe the steps you used to identify the person who started the spread of the disease (initial carrier). *(The class constructed a chart showing all of the interactions. Those who were not infected at the end were eliminated. Then, those who interacted with the uninfected in the previous interaction were eliminated. By a continued process of elimination, the possibilities were narrowed down to two individuals.)* Why could you not identify which of these two individuals was the initial carrier? *(Once the initial carrier exchanged with another person, both were infected. Without a previous exchange, there was no basis for further elimination.)*

- **Science Process:** What could the class have done at the beginning of the activity so that at the end of the activity, they would be able to confirm which person was the initial carrier? *(Each person could have kept a sample of the original liquid in their cup to be tested.)*

- **Science Process:** How would the results of this investigation have been different if some students chose not to exchange liquids at all? *(Fewer students would have become infected.)*

Preparation for 1.3

- **Science Content:** Suppose you visited a hospital in which the doors were labeled with the name of the disease that the person inside was suffering from. Which rooms would you feel confident entering because you wouldn't have to worry about getting the disease?

NOTES

NOTES

SECTION 1.2 IMPLEMENTATION

1.2 Investigate

Simulate the Spread of Disease

To begin to answer the question, *How do you get sick?,* you will run a simulation. You will **simulate** what could happen when you make contact with others. This investigation is going to require a lot of observing.

Be a Scientist

Simulations

At the beginning of this Unit, you passed a beanbag around the classroom. There was glow powder on the beanbag. The glow powder passed from one person to another and around the room. This activity was used to show you how germs might spread. The powder was used because it would not make sense to use real germs that could make you sick. This type of activity is called a **simulation**. Simulations imitate, or act out, what happens in real life. Scientists often use simulations. They use them when what they want to study is too big, too small, too fast, too slow, or too dangerous to investigate directly. The glow-powder activity simulated how disease might spread from person to person, but did so without causing any harm or danger to anyone.

simulation: the process or act of imitation or acting out.

simulate: to imitate how something happens in a real-world situation by acting it out with a model.

Materials

- clear plastic cups containing a liquid
- plastic pipettes

Procedure

1. Get a cup with some liquid in it and a pipette. Fill your pipette half full of liquid.
2. Walk around the room and introduce yourself to your classmates (friends). As you do this, put the liquid in your pipette into your friend's cup. Your friend will also share the liquid in his/her pipette with you.
3. Continue to meet your classmates until time is up. It is O.K. to meet the same person twice.
4. After you have interacted with several classmates, your teacher will test your liquid. If your liquid changes color, you are sick.

Do not drink the liquid. It may not be water.

Be careful not to spill any liquid. Wipe up any spills immediately.

GF 11

GOOD FRIENDS and GERMS

◀ *2 class periods**

1.2 Investigate

Simulate the Spread of Disease

5 min.

Students simulate the exchange of body fluids and then test whether they got infected with a disease. Students are all given identical-looking cups of liquid. All cups contain water except one, which contains a basic solution of either sodium bicarbonate (baking soda) or sodium hydroxide (NaOH). After exchanging liquids with several other students, students test their liquid using a pH indicator and observe the spread of the basic solution. Students have difficulty tracing the source of the "infection" because they have not kept records of their interactions.

○ Engage

Review the glow powder activity and the *Project Board,* and discuss the questions relating to how you get sick that students listed in the *Project Board.*

*A class period is considered to be one 40 to 50 minute class.

TEACHER TALK

"You've already simulated one way in which germs may spread in the glow-powder activity. In that simulation, how were germs spread? Look at the questions you have listed as things to investigate on the *Project Board*. Today, you are going to consider your questions about disease transmission and simulate another way in which this might occur.

Have you ever shared a drink with another person? If so, you probably shared body fluids with that person. In this investigation, you will share liquids with one another to simulate the spread of a disease through a population."

△ Guide

Ask students to read the *Procedure.* As they read, have them focus on the steps they will complete. Ask if they have any questions about how they are to perform the activity.

□ Assess

Assess whether students are able to carry out the procedure by having students restate the procedure to determine if they understand the steps they are to follow.

Procedure

20 min.

Guide students through running the procedure.

clas... glow powder on... glow powd... passed ... person to another and aro... oom. This activity was used to show you how germs might spread. The powder was used because it would not make sense to use real germs that could make you sick. This type of activity is called a **simulation**. Simulations imitate, or act out, what happens in real life. Scientists often use simulations. They use them when what they want to study is too big, too small, too fast, too slow, or too dangerous to investigate directly. The glow-powder activity simulated how disease might spread from person to person, but did so without causing any harm or danger to anyone.

...ate: to imitate how something happens in a real-world situation by acting it out with a model.

Materials

- clear plastic cups containing a liquid
- plastic pipettes

Procedure

1. Get a cup with some liquid in it and a pipette. Fill your pipette half full of liquid.
2. Walk around the room and introduce yourself to your classmates (friends). As you do this, put the liquid in your pipette into your friend's cup. Your friend will also share the liquid in his/her pipette with you.
3. Continue to meet your classmates until time is up. It is O.K. to meet the same person twice.
4. After you have interacted with several classmates, your teacher will test your liquid. If your liquid changes color, you are sick.

⚠ Do not drink the liquid. It may not be water.

Be careful not to spill any liquid. Wipe up any spills immediately.

⬡ Get Going

- Distribute the materials. Make sure you know which cup contains the basic solution. You might want to keep the basic solution yourself and be part of the interactions.
- CAUTION: Students should not to drink the liquid.
- Describe how students will exchange liquid with other students to simulate the spread of disease.
- Inform students how long you will run the investigation. Five to ten minutes should be sufficient.
- Run the simulation.

TEACHER TALK

"Now that everyone has their materials, you are ready to begin the simulation. Remember that when you exchange fluids, each person must put a half pipette full of their liquid in the other person's cup. After you exchange liquids, stir your cup to evenly mix the liquids. You may visit the same person twice, and you don't have to interact if you do not wish to. You may begin."

META NOTES

Encourage students to interact by telling them to ask each other questions before or during the sharing of their liquid such as "What is your favorite after-school activity?" or "What did you do yesterday after school?" This makes the interaction similar to an everyday situation and serves as a large group-building exercise.

☐ Assess

During the simulation, make sure both students in an interaction are sharing their liquid. Also check that students are exchanging the correct amount of liquid and that they are stirring after each exchange to evenly mix the liquids.

⬡ Get Going

After the allotted time has elapsed, put a drop of phenolphthalein pH indicator into each student's cup. An infected solution will turn red-pink and that of an uninfected individual will stay clear.

Inform students that those whose liquid turned pink have been infected.

△ Guide

Ask students to answer the following questions at the end of the procedure.

- What happened to your liquid?
- Were you infected or did the disease miss you?
- Who do you think was the initial carrier?

Briefly survey the class for answers. Consider making a chart of student responses, noting how many students in the class became infected, listing who they think may have been the initial carrier of the infection and any evidence they have to support that claim. This is a good opportunity to discuss the importance of supporting claims with evidence.

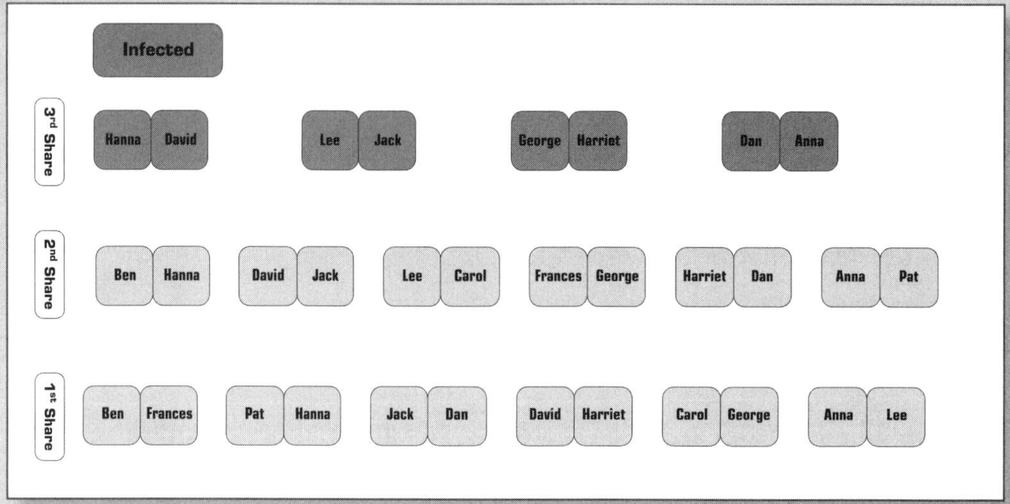

TEACHER TALK

"Let's begin by considering one pair of students who were both infected after they exchanged liquids in the third interaction: Hanna and David.

How might you determine whether Hanna infected David, David infected Hanna, or whether both were already infected when they interacted?

What evidence do you have that Hanna was not infected before she exchanged liquid with David?"

△ Guide

Guide students in using logical arguments based on previous interactions to determine whether a person was already infected when they engaged in an exchange of liquids.

For example, in the previous interaction, Hanna exchanged with Ben. Since Ben did not end up infected, neither Ben nor Hanna was infected when they interacted. Therefore, David infected Hanna. Since David was the one who infected Hanna, it is logical to conclude that in the previous exchange, both David and Jack ended up infected. However, Jack had previously exchanged with Irene, and Irene did not end up infected. Therefore, Jack was not infected before he exchanged with David, and it was David who infected Jack. Using similar reasoning, the infection can be traced back to David and Harriet. However, note that it is not possible to determine whether David infected Harriet or Harriet infected David without more evidence.

META NOTES

This is an opportune time to conclude the first of the two class periods of this section. It will make it even more difficult for students to recall the details of their interactions when they next meet.

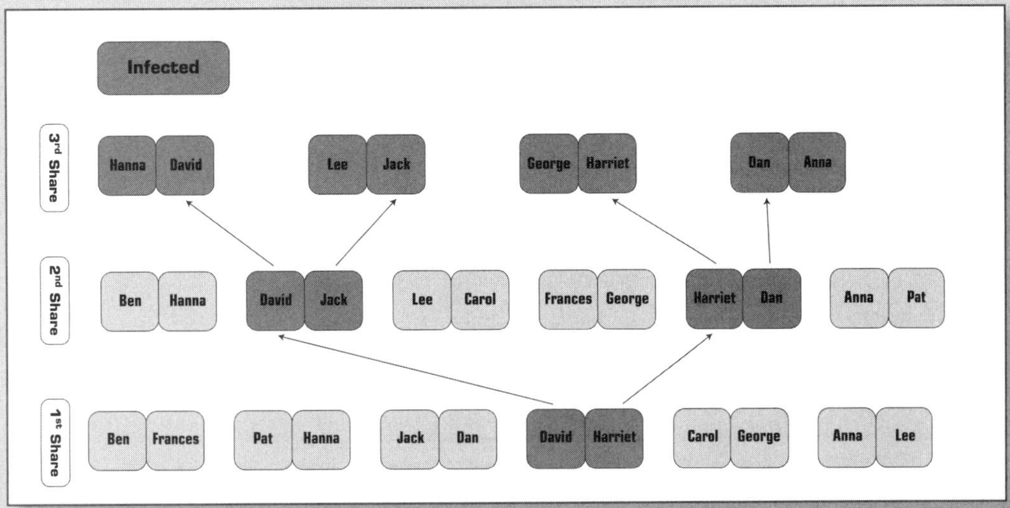

How can you prevent good friends from getting sick?

What do we think we know?	What do we need to investigate?	What are we learning?	What is our evidence?	What does it mean for the challenge or question?
Illnesses are caused by germs.	What makes me sick? What are germs? What kinds of germs are there?			
Germs can spread easily.	How do germs spread? How can we prevent spreading germs? How can we get rid of germs?			
A list of many different illnesses.	What kinds of illnesses are there? Are all illnesses caused by germs?			
Lots of statements about the symptoms of specific illnesses.	How do germs affect my body?			
Statements about bacteria and viruses may come up.	What are bacteria and viruses?			
There are "shots" that can prevent you from getting certain diseases.	What is in a "shot"? How does a "shot" "prevent" you from getting a disease?			

Recording Your Data

10 min.

Students may have difficulty tracing the source of the "infection" because they haven't kept records of their interactions. By conducting this experiment twice, students will be able to compare methods of observation. They will be able to decide at which time it is better to fill in the Record Your Data *chart, once the simulation has occurred or during the simulation.*

5. Based on your observations, answer these questions.
   a) What happened to your liquid?
   b) Were you infected, or did the disease miss you?
   c) Who do you think was the **initial carrier**?

initial carrier: the first person in a group to get a contagious disease.

Recording Your Data

In this activity, you and your classmates interacted with each other and shared your liquids. The interaction between you and your classmates was like meeting each other in a real-life situation. One of the liquids was "sick" (infected).

It might be easier to identify the initial carrier if you record with whom you came in contact. Use a table similar to the one shown below. Write the names of the people you interacted with first, then second, then third, all the way to the last person.

Liquid Interactions Table

Name: ______________________ Date: __________

Interaction	Name	Time of Interaction
1		
2		
3		
4		
5		
6		
7		
8		
9		
10		

Engage

Tell students that the cup represented their body and that one student was infected with a communicable disease. Point out that this activity showed how one person who is infected with a disease can infect other people, who in turn, infect others. Ask students to suggest ways in which the disease could be traced back to its source.

How do you think we could go about tracing the disease back to its source?

What information do we need in order to do this?

△ Guide

Ask students to identify the specific information they would need to know in order to trace the disease back to its source. Lead a class discussion focusing on the need to know both the persons with whom each student interacted, and the order in which those interactions occurred.

⬡ Get Going

Ask students to construct a chart like the one shown in the book, and to record the names of the people with whom they exchanged liquids from the

NOTES

Reflect

10 min.

Have students answer the Reflect *questions and discuss their answers with the class.*

Reflect

Answer the following questions. Be prepared to discuss your answers with the class.

1. How easy was it for you to fill in the data table?
2. How reliable do you think your data is?
3. Can you identify the initial carrier now?
4. If you were to repeat the activity, what could you do to improve your data collection?

Procedure: Repeat the Simulation

You may have decided that it would have been easier to record your data while you were having your interactions. Repeat the simulation. This time record your data as you do the activity.

Compile your class data. As a class, try to figure out who was the initial carrier.

Stop and Think

Answer the following questions. Be prepared to discuss your answers with the class.

1. It is possible to begin finding the initial carrier by looking at whose cups are not infected. How can you eliminate others?
2. Who do you think was the initial carrier? Why? What made the person easier to find this time?
3. Which cup(s) can you rule out as being the one that first had the disease? How do you know?
4. Now that you have finished the activity, how would you explain how a disease spreads through a community?
5. This investigation was a simulation. How do you think this investigation is like spreading a real disease? How do you think it is different from what happens in real life?

first person to the last person.

△ Guide

Have the class read the *Reflect* questions and record their answers. Then, have students discuss their answers with the class.

Help students develop an understanding of the importance of keeping accurate records during a scientific investigation, by asking students to describe any difficulties they had in completing the data table and to consider how confident they are in the accuracy of what they recorded.

TEACHER TALK

"What were some of the problems you encountered while completing the data table?

How sure are you that all of the information you recorded in the data table is completely accurate?

What information are you unsure about?"

Challenge the class to identify some of the reasons why they are having difficulty completing the data table and identifying the initial carrier.

Ask some students to share with the class who interacted with whom, and write those interactions down. Students may disagree with the order of events in their interactions. This is useful because it will help students recognize the need for improving data collection.

Ask students how they could improve their data collection. Students should come up with the idea that they really should have kept track as they were going along and that perhaps they should repeat the activity, this time keeping a running list of everyone they interacted with during the activity.

Ask students why the information recorded by just one person would not provide enough evidence to trace the infection to its source. Guide students to appreciate that collaboration will play a key role in determining the identity of the initial carrier.

Have the class design a way of organizing the information they record during interactions so that it can be easily shared with the class.

⬡ Get Going

Have students construct a data table for recording information about their interactions using the ideas they just discussed. You may want to have students do this on note cards that can easily be carried from one interaction to the next.

Procedure: Repeat the Simulation

15 min.

Students repeat the simulation, this time recording their data during their interactions.

3. Can you identify the initial carrier now?
4. If you were to repeat the activity, what could you do to improve your data collection?

Procedure: Repeat the Simulation

You may have decided that it would have been easier to record your data while you were having your interactions. Repeat the simulation. This time record your data as you do the activity.

Compile your class data. As a class, try to figure out who was the initial carrier.

Stop and Think

⬡ Get Going

Distribute a new set of materials. Again, make sure you know which cup contains the basic solution.

It is again important to caution students not to drink the liquids.

Have students run the procedure, this time recording their data while doing the activity.

After the allotted time has elapsed, put a drop of phenolphthalein pH indicator into each student's cup. Have students record who was and was not infected.

□ Assess

- During the simulation, make sure both students in an interaction are sharing their liquid. Also check to make sure that students are exchanging the proper amounts of liquid and are stirring after each exchange.

NOTES

Compile y... ...ata. As a class, try to figu... ...was the initial carrier.

Stop and Think

Answer the following questions. Be prepared to discuss your answers with the class.

1. It is possible to begin finding the initial carrier by looking at whose cups are not infected. How can you eliminate others?
2. Who do you think was the initial carrier? Why? What made the person easier to find this time?
3. Which cup(s) can you rule out as being the one that first had the disease? How do you know?
4. Now that you have finished the activity, how would you explain how a disease spreads through a community?
5. This investigation was a simulation. How do you think this investigation is like spreading a real disease? How do you think it is different from what happens in real life?

Stop and Think

15 min.

Hold a class discussion in which the data is shared and analyzed to identify the initial carrier.

△ Guide

Have students answer the questions in small groups and write down their answers.

Lead a class discussion on their responses to each question. During the discussion, elicit some of the following ideas:

1. You can eliminate any cups that were not infected and assume that any cup that interacted with one of these uninfected cups was also uninfected at the time of interaction.
2. The number of infections approximately doubles with each additional interaction.
3. It is not possible to trace the infection back to one person, only the two persons who were infected after the first interaction. The initial carrier could be identified if everyone had kept a sample of their original liquid for later testing.
4. Disease spreads through a community by human interactions, sometimes without warning or signs.
5. This simulation is similar to the spreading of a real disease because it shows how disease can spread rapidly and unknowingly during interactions and infect a community. It is different from what happens in real life because this simulation makes the appearance of germs more apparent and shows a quicker transmisson of the germs.

Create a tree of infections to help students figure out who infected them and

META NOTES

Students should notice that the number of infections approximately doubles with each interaction. At first, this doubling increases exponentially (2, 4, 8, ...), but after several interactions it begins to level out (e.g. 2, 4, 7, 12..). If students graph the number of new infections after each interaction, they will observe this decrease in the rate of new infections. Why doesn't the number of infections continue to grow exponentially? As the number of infected student increases, there is an increased likelihood that an infected student will interact with another student that has already been infected. As a result, the number of new infections slows down after several interactions.

Models and Simulations

5 min.

Students read about the distinction between a model and a simulation.

Be a Scientist

Models and Simulations

Simulations use a **model** to imitate, or act out, real-life situations.

A model is a **representation** of something in the world. One model that you know is a globe. The parts of the globe represent parts of the Earth. Scientists use models to investigate things that are too difficult or too dangerous to examine in real life. The models are at a size that people can easily examine. To use a model to investigate, the model needs to be similar to the real world in ways that are important for what the scientist is investigating.

Sometimes what you want to model is a situation or an event. To do this, you create a model that includes the things that are part of an event and then you use that to act out a situation in a simulation.

In the glow-powder activity, you simulated how disease spreads by using a model of germs and things in the world that might have germs on them. The glow dust corresponded to germs. The beanbag corresponded to things in the world that can have germs on them. As you passed around the beanbag with glow powder on it and touched other things in the classroom, you simulated how germs can be spread from place to place. That simulation allowed you to experience how easily germs spread.

You also participated in an activity that simulated how diseases are transmitted as you meet friends. Each cup with water represented a person. The chemical in the water of one of the cups represented a disease. Each time you interacted with somebody, you passed some of your water to him or her. That person passed some water to you. Your water became "infected" after you interacted with someone with "infected" water. After your water was "infected," you "infected" the water of everybody you interacted with. You simulated how diseases are transmitted using a model that included people with a disease and without a disease.

Models and simulations help scientists learn. The simulations you did helped you learn about how germs spread and how diseases are transmitted. You will experience several other models in this Unit and throughout PBIS.

model: a way of representing something in the world to learn more about it.

representation: a likeness or image of something.

Molecular model.

who they subsequently infected.

△ Guide

Ask students to give examples of models. Make a list of student responses. Point out that most of the things on the list are objects of one type or another. Help students to recognize that what they did when they participated in the glow-powder and water-sharing activities was not an object. What they did was make up of a series of events. Therefore, a distinction needs to be made between objects and events.

Models and Simulations may be read any time after the simulation that the term model comes up in discussion, or that the term model is incorrectly used interchangeably with the term simulation by students. This selection clarifies the distinction between these two terms—a model is a representation of something in the world, and a simulation uses a model to imitate, or act out real-life situations. The selection clearly identifies the

NOTES

Epidemiologists

5 min.

Students read about the role of an epidemiologist.

Epidemiology

Epidemiology is the study of what affects health and illness. (The word comes from Greek words that mean "among the people.") **Epidemiologists** are scientists who study diseases and how they spread. Their work is very similar to what you did. They look at who has a disease and who does not and try to find connections. When they are studying the spread of disease, they look for the person who begins the spread of the disease in a group. They call this person the "initial carrier" or "**sentinel case**." The first person then gives the disease to someone else through an interaction. Epidemiologists would call this first interaction the "initial step."

epidemiology: the study of how and why diseases occur.

epidemiologist: a scientist who studies diseases and how they spread.

sentinel case: a person who begins the spread of disease in a group.

What's the Point?

This activity, like the one you did earlier, was a simulation. Simulations are important in science when you cannot use or do the real thing. You cannot make people sick to test how diseases spread.

When you repeated the simulation, you recorded your interactions as they happened. You found it much easier to track the disease. This shows you how important it is to keep good records.

As you interacted with more and more people, you may have caught and then spread the disease. The disease may have infected each member of your class. This simulation demonstrates some important points about how diseases spread. It showed that sometimes you cannot know who is sick. Often, sick people do not appear sick and yet they can spread their disease. The investigation also demonstrates that spreading diseases can be very easy. However, it is not always easy to figure out who the initial carrier was.

models and the simulations in the glow-powder and water-sharing activities

△ Guide

Ask students what they think should be done if there is an epidemic. Make a class list of their suggestions. Students' typically focus on the need to determine the cause of the epidemic and to figure out a way to stop the disease from spreading any further. Help students make the connection between their suggestions and the need for someone to take on that role, i.e. an epidemiologist.

This selection may be read any time after the simulation when any student uses the word, or you may introduce it during a discussion. The point here is that epidemiologists perform the kinds of investigations students were just doing. They try to find out how different diseases spread, and they do detective work to try to find who got it from whom and also make and run

first ... initial step."

...ead of disease in a group.

What's the Point?

This activity, like the one you did earlier, was a simulation. Simulations are important in science when you cannot use or do the real thing. You cannot make people sick to test how diseases spread.

When you repeated the simulation, you recorded your interactions as they happened. You found it much easier to track the disease. This shows you how important it is to keep good records.

As you interacted with more and more people, you may have caught and then spread the disease. The disease may have infected each member of your class. This simulation demonstrates some important points about how diseases spread. It showed that sometimes you cannot know who is sick. Often, sick people do not appear sick and yet they can spread their disease. The investigation also demonstrates that spreading diseases can be very easy. However, it is not always easy to figure out who the initial carrier was.

What's the Point?

models to test their ideas.

Teacher Reflection Questions

- What was the value of engaging students in two iterations of the water-sharing activity? What important understandings were facilitated by not instructing students to record data during the first iteration?
- What components of this *Learning Set* engaged students in social and interactive scientific inquiry through evidence-based discussions and reflection on learning?
- What classroom management issues did you face in carrying out this investigation? How could you improve these when doing this investigation with other classes?

NOTES

SECTION 1.3 INTRODUCTION

1.3 Investigate

What Are Communicable Diseases?

◀ *1 class period*

A class period is considered to be one 40 to 50 minute class.

Overview

Students read about the distinctions between communicable and non-communicable diseases. Examples of both types of disease are identified.

Targeted Concepts, Skills, and Nature of Science	Performance Expectations
Scientists often collaborate and share their findings. Sharing findings makes new information available and helps scientists refine their ideas and build on others' ideas. When another person's or group's idea is used, credit needs to be given.	Students should share their responses to questions with the class.
Communicable diseases are diseases that spread from one person to another. Non-communicable diseases cannot be passed from the person who is sick to other people.	Students should be able to distinguish between communicable and non-communicable diseases and give several examples of each type.

Materials	
1 per class	Class *Project Board*

Homework Options

Reflection

- **Science Content:** What are some communicable diseases that you or a person you know has had? What were some of the symptoms of that illness?

- **Science Content:** Many people suffer from hay fever, an allergic reaction to pollen in the air. Would hay fever be considered a communicable disease? *(No, allergies such as hay fever are not communicable.)* Defend your answer with evidence from your reading or with information you obtain from independent research. *(Allergic reactions are the result of an immune response to a substance perceived by the body as foreign.)*
- **Science Process:** Imagine you are a doctor and a person comes to see you because they have a fever and a rash of red spots. How could you determine whether this condition was communicable? *(One could monitor those who come into contact with the person to see if they develop the same symptoms.)* What might be some ethical dilemmas faced by doctors trying to determine whether a disease is or is not communicable? *(Experimentation on human subjects is unethical. Should the person be quarantined or allowed to continue to come into contact with others?)*

Preparation for Back to the Big Question

- **Science Process:** In what ways is keeping track of our ideas on the *Project Board* similar to keeping track of data as an investigation is carried out? *(In both cases, it is important to record information as it is being obtained because it is difficult to recall details if they have not been written down. In both cases, organizing information helps make it easier to share and to recognize relationships in the information.)*

SECTION 1.3 IMPLEMENTATION

◀ *1 class period**

1.3 Read

What Are Communicable Diseases?

communicable disease: a disease that spreads very readily from person to person.

non-communicable disease: a disease that cannot be passed on by the person who is sick to other people.

Diseases that spread from one person to another are called **communicable diseases**. You probably know a lot of communicable diseases, like colds and the flu. But only some diseases are communicable.

Imagine a visitor comes into your house and is sneezing and coughing. You might think the visitor has a cold. If he or she does, you can get the cold from them. The simulation you just did showed you that. A cold is a disease one person can give to another person. As you learn more about diseases, you will want to think about which ones are communicable and which ones are not. Only the communicable ones can spread from one person to another. Diseases that are not communicable cannot do that.

Cancer is one example of a **noncommunicable disease**. If a visitor to your house has cancer, you cannot get cancer from him or her. It does not matter if both of you drink from the same glass or if that person sneezes into the air. You cannot "catch" cancer. You may know someone who has a noncommunicable disease like cancer. These diseases make you sick, but they cannot spread from one person to another. Some other diseases that are not communicable are diabetes, heart disease, high blood pressure, and asthma.

Communicable diseases, such as colds, can spread from one person to another.

Project-Based Inquiry Science   GF 16

1.3 Read

What are Communicable Diseases?

Students learn the differents between communicable and noncommunicable diseases.

○ Engage

Ask students if they can think of any diseases that are not "catching". Ask them how they know the disease is not "catching".

△ Guide

Have students read the selection. You may consider having students take turns reading to the class, and discuss the content as you go along. Or, you might present the content in class and have them do the reading as part of a homework assignment.

*A class period is considered to be one 40 to 50 minute class.

Stop and Think

20 min.

Have students read the questions and record their answers.

Stop and Think

1. How is a communicable disease different from a noncommunicable disease?
2. What are two communicable diseases?
3. What are two noncommunicable diseases?

Explain

Make your first attempt at explaining how you can get sick with communicable diseases. Develop a claim about how you can get sick with communicable diseases, and record it on a *Create Your Explanation* page. Then use observations from the simulations you've run as evidence to support your claim. Write a statement using your evidence and science knowledge that supports your claim. This will be your explanation. You may find it hard to develop an explanation right now. Do your best with what you know now. You will have chances later in the Unit to revise your claims and explanations.

Recommend

Now, make your first recommendations about staying healthy. Develop recommendations that answer two questions:

- How can I keep from getting sick with a communicable disease?
- How can I prevent my friends from getting sick with a communicable disease?

Record each of your recommendations as a claim on a *Create your Explanation* page. Record evidence and science knowledge that support your recommendation. Then write a statement that shows why your recommendation is trustworthy. You will have chances later in the Unit to revise your recommendations and to better explain them.

Lead a class discussion on students' ideas about communicable and non-communicable diseases and encourage students to use their life experiences as well as the reading to help them answer the questions. Focus students' attention on the distinctions between communicable and noncommunicable diseases. Guide students to see the connection between germs and communicable diseases and to recognize that noncommunicable diseases are not caused by germs.

Teacher Reflection Questions

- In this section, students were introduced to new information. What formative assessments did you use with the students to identify what they have learned?
- How does reading about communicable and noncommunicable diseases set the stage for the focus of the next Learning Set?

NOTES

NOTES

BACK TO THE BIG QUESTION INTRODUCTION

Learning Set 1

Back to the Big Question

◀ ***1 class period***

A class period is considered to be one 40 to 50 minute class.

Overview

Students read a summary of the key ideas in *Learning Set 1* and reflect on what they have learned. Students then update the *Project Board*, focusing on Columns 3 and 4: *What are we learning?* and *What is our evidence?*

Homework Options

Targeted Concepts, Skills, and Nature of Science	Performance Expectations
Scientists often collaborate and share their findings. Sharing findings makes new information available and helps scientists refine their ideas and build on others' ideas. When another person's or group's idea is used, credit needs to be given.	Students share what they have learned, what evidence supports that learning, and what they need to investigate as they fill out the class *Project Board*.
Scientific questions are directed toward objects and events that can be described, explained, or predicted by scientific investigations.	Students should be able to generate appropriate questions for the *What do we need to investigate?* column of the *Project Board*.
Explanations are claims supported by evidence, accepted ideas, and facts.	As students identify ideas and explanations that should be added to the What are we learning? column of the *Project Board*, they should be able to provide corresponding evidence to support their claim in the *What is our evidence?* column.

Materials	
1 per class	Class *Project Board*

Reflection

- **Practice/Review:** How were the glow-powder and water-sharing activities similar to a real disease? *(In both activities, a substance was passed from person to person just as germs are passed from person to person.)* How were these activities different from a real disease? *(In both activities, the substance that was passed around was not a living thing and caused no harm to the body.)*

Preparation for Learning Set 2

- Ask students what exactly it is that they think is passed from one person to another when a disease spreads. (*Students typically state that it is germs.*) Ask students what they think germs are. *(Students may know that they are microscopic living things.)*

NOTES

BACK TO THE BIG QUESTION IMPLEMENTATION

◀ *1 class period**

Learning Set 1

Back to the Big Question

How can you prevent your good friends from getting sick?

The *Big Question* for this Unit is *How can you prevent your good friends from getting sick?* You began this Unit by investigating with your class how diseases spread. You used simulations to find out how you can get sick. You found that even one of your good friends can make you sick.

Epidemiologists spend a lot of time studying how diseases spread. By repeating a simulation, you learned how important it is for scientists and for you to accurately record data. Keeping accurate records of your work will be important as you work toward answering the *Big Question*.

You also read about the difference between communicable and non-communicable diseases. Communicable diseases spread from person to person. There are many communicable diseases you already know about and others you will learn about. For the next few weeks, you will study different diseases and learn more about how they can spread from one person to another.

Learning Set 1: Back to the Big Question

How can you prevent your good friends from getting sick?

20 min.

Students reflect upon what they have learned, update the Project Board, *and focus their attention on the* Big Question *and its two challenges.*

○ Engage

Begin by asking students why it was important to record data while they were having their interactions in the water-sharing activity. (*Because it is easy to forget details or lose track of information when you are involved in something.*) Connect this with the importance of updating the *Project Board* throughout the Unit and maintaining focus on the *Big Question.*

△ Guide

Have students read the selection. You might consider having students take turns reading to the class, and discussing the content as you go along.

*A class period is considered to be one 40 to 50 minute class.

Reflect

10 min.

Students answer questions that guide them to consider what they are learning and to support their claims with evidence about what they have learned.

Reflect

Think about and answer the following questions. Be prepared to discuss your answers with your class.

1. What did you learn about the importance of keeping accurate records? Use an example from this *Learning Set* to justify your answer.
2. What is the difference between a communicable and a noncommunicable disease?
3. How do you get sick? Use evidence from this Learning Set to support your answer.
4. What is a model? Why are models important in science?
5. How did you use models to run simulations in this *Learning Set*?

Update the *Project Board*

You started a *Project Board* centered on the idea of learning more about how you can prevent your good friends from getting sick. Now you have done a simulation. You know more about the factors that affect how a disease spreads. You are now ready to fill in the *Project Board* more completely.

Up to now, you only recorded information in the first two columns of the *Project Board.* You will now focus on the next two columns. These are the *What are we learning?* and *What is our evidence?* columns. When you record what you are learning in the third column, you will be answering some questions in the *What do we need to investigate?* column. You will describe what you learned from the investigation you just did. But you cannot just write what you learned without providing evidence for your conclusions. Evidence is necessary to answer scientific questions. You will fill in the evidence column based on data and trends you found in your investigations.

You will also include your understanding of the science readings and your discussions with each other. You may use the text in this book to help you write about the science you have learned. However, make sure to put it into your own words. The class will fill in the large *Project Board*. Make sure to record the same information on your own *Project Board* page.

△ Guide and Evaluate

Lead a class discussion on student responses and evaluate whether their responses contain the information below.

1. It is important for scientists to keep accurate records to refer back to because they are more reliable and complete than when recalled. In the water-sharing activity, the students first tried to

record their findings at the end of the activity, and it was hard to remember all their interactions. The students could not figure out which cup was "infected." When they ran the simulation the second time, they kept accurate records and from them, they could trace the "infection" to its source.

2. A *communicable disease* is one that can spread from person to person. A *noncommunicable disease* cannot be spread from person to person.
3. You get sick by the spread of germs from an infected person. You get the germs through contact with the infected person or by contact with something they touched. Evidence for this is the glow-powder simulation that showed that the powder could get on other surfaces as well as people and how it spread throughout the classroom through your interactions. Further evidence is the water-sharing activity, which simulated how one infected person could infect many other people (*cups of water*).
4. A *model* is a representation of something in the world. Models allow you to investigate something when the real thing is too big, too small, too dangerous, too difficult, or impossible to investigate.
5. The students used models to run two simulations. The first was the glow-powder simulation. In it, the beanbag was a model of a person and the glow powder was a model of germs. The model was used to simulate the spread of germs by human contact. The second was the water-sharing simulation. In this activity, the cup represented a person, the water represented body fluids, and the basic solution represented body fluid infected with germs. The model was used to simulate the spread of germs through a population by exchange of body fluids.

Update the *Project Board*

15 min.

4. Wha... ...y are models imp... ...ce?

5. How did you use models to run simulations in this *Learning Set*?

Update the *Project Board*

You started a *Project Board* centered on the idea of learning more about how you can prevent your good friends from getting sick. Now you have done a simulation. You know more about the factors that affect how a disease spreads. You are now ready to fill in the *Project Board* more completely.

Up to now, you only recorded information in the first two columns of the *Project Board*. You will now focus on the next two columns. These are the *What are we learning?* and *What is our evidence?* columns. When you record what you are learning in the third column, you will be answering some questions in the *What do we need to investigate?* column. You will describe what you learned from the investigation you just did. But you cannot just write what you learned without providing evidence for your conclusions. Evidence is necessary to answer scientific questions. You will fill in the evidence column based on data and trends you found in your investigations.

You will also include your understanding of the science readings and your discussions with each other. You may use the text in this book to help you write about the science you have learned. However, make sure to put it into your own words. The class will fill in the large *Project Board*. Make sure to record the same information on your own *Project Board* page.

GF 19 GOOD FRIENDS and GERMS

Earlier in this *Learning Set*, students focused mainly on the first and second columns – *What do we think we know?* and *What do we need to investigate?*. Now that they have completed several investigations, it is time for students to reflect on what they have learned from these investigations and add that information to the *Project Board*.

TEACHER TALK

"At the beginning of this Learning Set, you identified the need to investigate how germs spread. Now that you have completed several investigations, let's reflect on what you have learned about the spread of germs from these investigations."

Tell students that they should be concentrating here on the third and fourth Columns – *What are we learning?*, and What is our evidence? Remind students about what goes in these columns.

Ask the class what they feel needs to be updated in these two columns. During the discussion, help students identify and clarify what they are learning about how people get sick.

How can you prevent good friends from getting sick?

What do we think we know?	What do we need to investigate?	What are we learning?	What is our evidence?	What does it mean for the challenge or question?

Guide students by asking:

- Where should you put the information about communicable and noncommunicable diseases?
- What can you list about what you think you know about communicable and noncommunicable diseases?

META NOTES

It is likely that students will bring up specific types of communicable and non-communicable diseases. They may bring up what makes a disease easy or not so easy to spread. They may bring up sickness in a specific body system. Some of these things will belong in the third and fourth columns (*if they have good evidence for them*). Some will belong in the first two columns. While this discussion will focus mostly on the third and fourth columns, do continue to fill in the first two columns as new things come up. Be selective in what you choose to record on the class' Project Board. Exhaustive lists of diseases, symptoms or other details can clutter the board so that students "can't see the forest for the trees" and lose focus on core ideas. Consider using umbrella statements such as "Our class knows the names of many diseases and their symptoms, for example..." (*and then record one or two examples*) or "Many types of interactions can result in the spread of disease, for example: shaking hands, sharing drinks, and kissing to name a few."

Facilitate a class discussion of how people get sick and guide students to specifically associate evidence with their claims. For example:

> **Claim:** Germs spread when there are germs on your hands and you touch things.
>
> **Evidence:** Glow-powder simulation.
>
> **Claim:** People can get sick from other people.
>
> **Evidence:** In the disease activity, you poured water into each other's cups as if you were passing a disease to each other. Only some people got sick, but if they did, they spread it to other people. You saw this whenever a person whose water was not originally infected with the germ turned pink after they had shared water with an infected person.

Hold a discussion about what students have learned that would help them to prevent their friends from getting sick (the *Big Question*). If they feel they have begun to learn anything about this, write it in the fifth column of the *Project Board*. Students may have some things they want to add, but also do this in the form of claims and evidence. For example:

> **Claim:** I should wash my hands when I sneeze or blow my nose.
>
> **Evidence:** I know that hands spread germs, and when I sneeze, I get germs on my hands. If I wash my hands, those germs won't spread.
>
> **Claim:** I should stay home if I am sneezing a lot.
>
> **Evidence:** When I sneeze, I spread germs around—on my hands and in the air. The glow-powder activity showed us how easy it is for germs to spread to unexpected places. A cold is a communicable disease. If I stay home, I will not spread the germs, but if I go to school, I could spread the germs all over.

The fifth column of the *Project Board* will be new to students. This is where they will record what they've learned that will help them answer the *Big Question: How can I prevent my good friends from getting sick?* The ideas they put in the column during this discussion might not be very detailed or accurate. They will have a chance to come back and revisit those ideas at the end of each *Learning Set.*

How can you prevent good friends from getting sick?

What do we think we know?	What do we need to investigate?	What are we learning?	What is our evidence?	What does it mean for the challenge or question?
Illnesses are caused by germs.	What makes me sick? What are germs? What kinds of germs are there?			
Germs can spread easily.	How do germs spread? How can we prevent spreading germs? How can we get rid of germs?	Germs can be spread through a population quickly by human interactions. The spread of disease is hard to detect and trace to its source.	Spreading of glow powder when beanbag was passed from hand to hand. Disease spread by sharing liquids. The spread of the disease went undetected in both simulations until students specifically looked for the glow powder or the basic solution in the liquid. The liquid-sharing activity provided evidence of how hard it is to track where a disease (*basic solution*) started.	Limiting contact with others when you are sick can help prevent the spread of disease.

Washing hands helps prevent the spread of germs.	How effective is hand washing at getting rid of germs?	Hands must be carefully and completely washed to remove germs. Even then, some germs may remain.	Many students still had glow powder on their hands after washing the first time. In some cases, even a second washing did not remove all traces of the powder.	Good personal hygiene helps prevent the spread of germs.
A list of many different illnesses.	What kinds of illnesses are there? Are all illnesses caused by germs?	Diseases that spread from one person to another are called communicable diseases. Only some diseases are communicable. Cancer, heart disease, diabetes, high blood pressure, and asthma are noncommunicable diseases.	Students read this in *Section 1.3*.	Some diseases are caused by germs, others are not.
Lots of statements about the symptoms of specific illnesses.	How do germs affect my body? What happens to you when you get sick?			
Statements about bacteria and viruses may come up.	What are bacteria and viruses?			
There are immunizations that can prevent you from getting certain diseases.	What is in an immunization? How does an immunization prevent you from getting a disease?			

LEARNING SET 2 INTRODUCTION

Learning Set 2

What Kinds of Things Make You Sick?

◀ *18 class periods**

Students examine the symptom/disease relationship. They investigate cells and cellular structures using the microscope, read about the nature of bacteria and viruses, design investigations involving the culturing of bacteria, and examine case studies of a variety of diseases caused by bacteria and viruses.

Overview

In *Learning Set 1*, students investigated communicable disease and how it spreads. In *Learning Set 2*, students will develop the understanding that bacteria and viruses cause communicable diseases. They also will investigate cells and read about the structure of bacteria and viruses.

Section 2.1 launches the central question for this Learning Set — *What kinds of things make you sick?* Students formulate questions that can be answered through scientific inquiry. Working in small groups, students evaluate one another's questions to be sure that they meet the criteria for an appropriate question. Each group formulates two questions that will help guide their investigation into what kinds of things make you sick.

In *Section 2.2,* through the example of gastrointestinal illness, students consider the characteristics of disease — what causes it, how it spreads, the body systems it affects, its symptoms and how it can be treated. This reinforces the ideas introduced in *Learning Set 1*, and provides a good reason for developing a *Communicable-Disease Information Table* to keep a running record of what they are learning about diseases. This table will be updated throughout the Unit, and will serve as an important resource when developing students' final presentations and recommendations.

LOOKING AHEAD

- **Prepare microscopes and stains for class use.**
- **Obtain prepared slides for animal cell activity and onions for plant cell activity.**
- **Prepare Petri dishes with agar for culturing bacteria.**
- **This will take a considerable amount of time and planning. Do not prepare the dishes too far in advance (4 or 5 days maximum). Prepare enough dishes for both *Learning Sets 2.6* and *2.7.***

*A class period is considered to be one 40 to 50 minute class.

In *Section 2.3,* students read about cells and cellular structure. They are shown microscope techniques such as preparing wet-mount slides, staining specimens, and estimating a cell's size. Students then use the microscope to observe cells, identify cell structures, and differentiate between plant and animal cells. Students create drawings of the cells they observe and share their drawings with group members. The group then chooses one cell and creates a poster of it to present to the class.

In *Section 2.4,* students extend their understanding of cells by reading about bacteria, single-celled organisms. As they read, students differentiate between eukaryotic and prokaryotic cells, the structure of bacterial cells, and how they carry out various life functions. They also consider ways in which bacteria may be beneficial or harmful, and how some bacteria produce toxins that adversely affect body systems. Students then apply their knowledge in *Section 2.5* by examining a case study of bacterial food poisoning. Working as an epidemiologist would, they consider how temperature and food-storage methods affect the growth of bacteria, and they analyze data from the case study to determine the source of the food poisoning. Students complete the investigation by presenting their conclusions and supporting evidence to the class in an *Investigation Expo.*

In *Sections 2.6* and *2.7,* students investigate where bacteria might live in their classroom. Students are shown how to safely culture bacteria and use this technique to design and carry out investigations to identify places in their classroom that harbor bacteria. In *Section 2.6,* groups suggest ideas for a possible investigation. The ideas are discussed and the class selects one as the basis for a class investigation. The class is guided through the process of designing and carrying out an investigation in which samples collected from various places in the classroom are cultured to determine where bacteria can be found. Based on a surprising result of this investigation, students develop an independent research question that can be answered by gathering, growing, and judging the amount of bacteria that has grown. In *Section 2.7,* students go through the process of designing and carrying out an investigation modeled in *Section 2.6* on their own. Working collaboratively within their groups, students design an experiment to answer the research question, present their experimental design to the class, and hear about other groups' designs during class presentations. Then they revise their plan, carry out their independent experiment, and report their claims and supporting evidence to the class in an *Investigation Expo.*

Now that students have become familiar with bacteria, in *Section 2.8* they read about viruses, another microbe that causes communicable diseases. Students are given an overview of the structure of a virus, how viruses function and reproduce, and how they can be transmitted. This background knowledge helps students in *Section 2.9* as they examine a case study of a disease caused by a virus. The case study describes both the characteristics of the smallpox disease and the historical role it played in the development of immunization by vaccination. In *Section 2.10,* small groups of students examine case studies of a variety of diseases caused by bacteria or viruses. As groups present information about different diseases to the class, the information is added to the *Communicable Disease Information Table.* In *Back to the Big Question*, students reflect upon what they have learned and construct explanations for how they get sick using *Create Your Explanation* pages. After sharing their explanations with the class, they revise them based on peer feedback. Students use their explanations to create recommendations for keeping themselves and their friends from getting sick with communicable diseases. This section concludes with students updating the *Project Board*, focusing their attention on how what they have learned will help them address the *Big Question* and its two challenges.

Targeted Concepts, Skills, and Nature of Science	Section
Scientists often collaborate and share their findings. Sharing findings makes new information available and helps scientists refine their ideas and build on others' ideas. When another person's or group's idea is used, credit needs to be given.	*All sections*
Scientists must keep clear, accurate and descriptive records of what they do so they can share their work with others; consider what they did, why they did it, and what they want to do next.	2.3, 2.5, 2.6, 2.7
Scientific questions are directed toward objects and events that can be described, explained, or predicted by scientific investigations.	2.1, 2.6, 2.7
Identifying factors that lead to variation is an important part of a scientific investigation.	2.5, 2.6, 2.7
Explanations are claims supported by evidence, accepted ideas, and facts.	BBQ
The cell is the basic unit of structure and function in all living things. Cells are usually microscopic in size.	2.3, 2.8
Cells have specialized structures that perform specific functions. Most cells have a cell membrane, cytoplasm and genetic material. Many cells have a nucleus. Some cells have cell walls and/or chloroplasts.	2.3, 2.4
Some living things are single cells; others, including humans, are multicellular.	2.4
Disease is a breakdown in the structures or functions of an organism. Some diseases are the result of failures of a body system. Other diseases are the result of damage by infection from other organisms (germs).	2.2, 2.4, 2.8, 2.10

Targeted Concepts, Skills, and Nature of Science	Section
Many diseases are caused by bacteria or viruses.	2.2, 2.4, 2.5, 2.8, 2.9, 2.10
All organisms require energy to survive. Cells obtain energy in a variety of ways.	2.3, 2.4
Metabolism is the sum of all chemical reactions in an organism.	2.4
Homeostasis is the process by which an organism keeps its internal environment within the range required for survival despite changes in the external environment.	2.4
Immunization is a medical treatment that helps protect you from disease.	2.8, 2.9
Vaccination is the process by which a person is given a vaccine – a substance that protects a person from a disease.	2.9, 2.10
Antibiotics kill or prevent the growth of bacteria.	2.4, 2.10
Students recognize and analyze patterns and trends.	2.5, 2.6, 2.7
Students manipulate a compound microscope to view microscopic objects.	2.3
Students prepare a wet-mount slide, use appropriate staining techniques, and determine the size of a microscopic object, using a compound microscope.	2.3

NOTES

LEARNING SET 2 IMPLEMENTATION

◀ *18 class periods**

Learning Set 2

What Kinds of Things Make You Sick?

The next question you might think about is *What kinds of things make you sick?* You know germs can make you sick. They cause many diseases. You may have already added the question, "What are germs?" to the *Project Board*. There are many types of germs. You may know about germs, but you may not know the different types of germs that exist. You are now going to learn about two groups of germs: bacteria and viruses.

You know what a communicable disease is and you know that colds are one type of communicable disease. You will learn more about specific communicable diseases and how they are spread. In this *Learning Set*, you will be introduced to different diseases. You will learn about the symptoms of each disease, who can get sick, how you can avoid getting sick, and how to cure the disease.

All this information will become part of a *Communicable-Disease Information Table* you will complete. As you fill in diseases on this chart, you will be on your way to answering the *Big Question*: *How can you prevent your good friends from getting sick?*

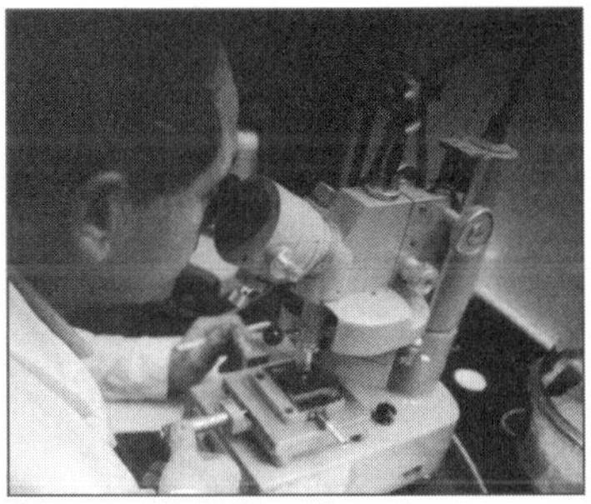

Germs are only visible under a microscope.

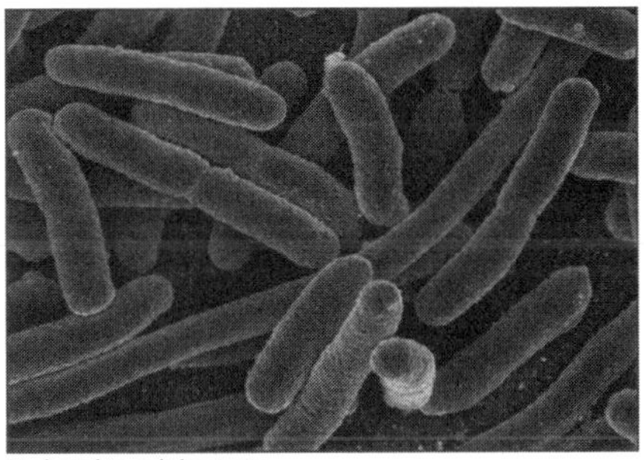

Escherichia coli *bacteria.*

GF 21 GOOD FRIENDS and GERMS

Learning Set 2

What kinds of things make you sick?

15 min.

Students consider the smaller question for Learning Set 2 — *What kinds of things make you sick?*

○ Engage

TEACHER TALK

"In Learning Set 1, you learned about communicable disease and how it spreads. You already know that "germs" make you sick and we recorded this on the *Project Board*. In this *Learning Set*, we are going to focus on the germs that cause communicable diseases."

*A class period is considered to be one 40 to 50 minute class.

Engage students by asking how they have seen germs characterized in television commercials, or by showing a videotape of a television commercial that anthropomorphizes germs.

Challenge students to compare and contrast the germs shown in television commercials with what they think actual germs are like.

Have students draw a sketch of the germs shown in the television commercial and describe what they did in the commercial.

TEACHER TALK

"How do you think real germs are similar or different from those shown in television commercials? Do you think germs have arms, legs, and eyes like those in the commercials? Do you think real germs talk to one another and try to run away from "germ-killing" products?

Why can't you see germs? How could you tell if there were germs on an object?

Are germs alive? Why do you think this?

Are all diseases caused by germs? What supporting evidence for your answer can you give?"

Remind students that the *Big Question* of the Unit is *How can you prevent your good friends from getting sick?* Ask students to think about the *Big Question* as they read the introduction to *Learning Set 2.*

Read through the Introduction with the students. Consider having students read aloud. Emphasize that the smaller question they will be focusing on in this *Learning Set* is *What Kinds of Things Make You Sick?* Also let them know that as they pursue answers to this question, they will investigate the microscopic world of cells, bacteria, and viruses.

NOTES

SECTION 2.1 INTRODUCTION

2.1 Understand the Question

Thinking about What Kinds of Things Make You Sick

◀ *1 class period*

A class period is considered to be one 40 to 50 minute class.

Overview

This section is designed to engage students in active consideration of the question: *What Kinds of Things Make You Sick?* Working collaboratively, students identify what they think they know about what makes you sick, what they are unsure of, and what needs to be investigated. Students formulate questions that might help them answer the question of *Learning Set 2* and share them with their group. Groups evaluate whether their questions meet the criteria that questions must have: interesting to them personally; require several resources to answer; relate to the *Big Question*; require collecting and using data; and not have a yes/no answer. Each group then selects their two most interesting questions to share with the class, and the *Project Board* is updated.

Targeted Concepts, Skills, and Nature of Science	Performance Expectations
Scientists often collaborate and share their findings. Sharing findings makes new information available and helps scientists refine their ideas and build on others' ideas. When another person's or group's idea is used, credit needs to be given.	Students should share their individual questions with their group and their group's questions with the class as they consider what they think they need to learn and what they need to investigate as they fill out the class *Project Board*.
Scientists must keep clear, accurate, and descriptive records of what they do so they can share their work with others; consider what they did, why they did it, and what they want to do next.	Students participate in a class discussion while updating their *Project Board*, which is an organized record of their ideas and what they are learning.

Materials	
1 per class	*Project Board*

Activity Setup and Preparation

This activity begins with students working individually to formulate questions and ends with a class discussion that leads to updating the *Project Board.* Decide how you would like students to keep a record of individual work, whether it will be in the form of a science notebook, a learning journal, or some other format. Make sure the class *Project Board* is posted where everyone can see it.

Homework Options

Reflection

- **Science Content:** What challenges do you think you will face investigating germs? *(Germs are microscopic in size. Germs are potentially dangerous. I do not really know what a germ looks like, so how will I know one when I see it?)*
- **Science Content:** When you discussed the question, *What makes you sick?* with your group, which ideas shared by other students were new to you? *(Answers will vary, but often include references to a disease that they were not familiar with, or issues relating to behaviors that lead to sickness.)*
- **Science Process:** Of the questions that were added to the *Project Board*, which did you think was the best? What made this question such a good one? *(The question should match the criteria for good questions. The question was interesting; it is something I personally want to learn more about. It is possible to find answers to this question by doing an investigation, and I can think of some ways in which this might be done.)*

Preparation for 2.2

- **Science Process:** Why is it difficult to tell if the illness you have is a cold or whether it is the flu? *(In general, the flu is worse than the common cold, and symptoms such as fever, body aches, extreme tiredness, and dry cough are more common and intense. People with colds are more likely to have a runny or stuffy nose.)*

SECTION 2.1 IMPLEMENTATION

2.1 Understand the Question

Thinking about What Kinds of Things Make You Sick

The question for this *Learning Set* is *What kinds of things make you sick?* It is a good idea to think about what you already know about what makes you sick. It is also important to think about what you are unsure about and what you would like to investigate.

Get Started

Think about the question. Share your ideas about what makes you sick. What do you already know about germs? What are the different types of germs? How are they different? Listen carefully to all the ideas presented. You may want to write down some of the ideas you hear. Think of several questions that might help you answer the question for this *Learning Set*. On your own, develop two questions that might help you understand how you get sick. When you write your questions, keep in mind that your questions should

- be interesting to you;
- require several resources to answer;
- relate to the *Big Question*; and
- require collecting and using data.

Your questions should not have a yes/no or a one-word answer.

When you have completed your two questions, take the questions back to your small group. Share all the questions with each other. Carefully consider each question and decide if it meets the criteria for a good question. With your group, refine the questions that do not meet the criteria. Choose the two most interesting questions to share with the class. Give your teacher the rest of the questions so they might be used later.

Update the *Project Board*

You will now share your questions with your class. Be prepared to support your questions with the above criteria. Your teacher will help you with the criteria if needed. Your teacher will then add your questions to the *Project Board*. In this *Learning Set*, you will work to answer some of these questions.

GF 22

Project-Based Inquiry Science

Engage

Challenge students to identify what they need to know about germs.

2.1 Understand the Question

Thinking about What Kinds of Things Make You Sick

10 mins.

Students share their ideas about germs and identify what they are unsure of and would like to investigate.

META NOTES

Small-group work provides an opportunity for high levels of participation and learning with low risk for the participants. It also provides opportunities for peer learning and support, recognition, and team building.

META NOTES

To be effective, small group work requires attention to issues such as group composition, task assignments, leadership, functional roles, and monitoring.

Effective small-group assignments are clear, thought-provoking, and directed toward producing a product within a given time frame.

TEACHER TALK

"In the previous activity, you drew sketches of germs shown in television commercials and shared some of your ideas about how real germs are similar or different from those shown in the commercial. Now it is time to consider what you actually do or do not know about real germs and what you think needs to be investigated.

What do you already know about germs? How do you know this?

What are the different types of germs? How are they different?

What ideas about germs are you unsure of? What are some important ideas about germs that need to be investigated?"

META NOTES

During small group work, circulate unobtrusively from group to group. By moving around the room instead of standing aloofly up front, you are more likely to get questions or requests for help. Try to visit all groups so none feel singled out and stay only long enough to get an idea of the trend of the discussion.

The purpose of your visit should be to:

- Answer questions about the assigned task.
- Determine the progress being made on the assigned task. Check on groups that finish too quickly to see if they are on track.
- Pick up some ideas first-hand that can be used in the class discussion following the small group work. For example: "I heard Group 2 make the point that ..."
- Show that you are interested in their work and value it.

△ Guide

Lead a class discussion of ideas that students find confusing or are unsure of and of other ideas that they would like to investigate. For example, students are often confused about where germs can be found, not realizing that they are ubiquitous on and in the human body. They often think that all microbes are "bad" or that factors pertaining to personal health (like "catching" a disease) are beyond their control. Students often have no idea what germs actually look like, how they reproduce and grow, how they affect human body systems, or what preventative measures they can take to stay healthy. Guide students to identify specific ideas about germs that they are unsure of and should therefore be investigated.

NOTES

you sick. It is also important to think about what you are unsure about and what you would like to investigate.

Get Started

Think about the question. Share your ideas about what makes you sick. What do you already know about germs? What are the different types of germs? How are they different? Listen carefully to all the ideas presented. You may want to write down some of the ideas you hear. Think of several questions that might help you answer the question for this *Learning Set*. On your own, develop two questions that might help you understand how you get sick. When you write your questions, keep in mind that your questions should

- be interesting to you;
- require several resources to answer;
- relate to the *Big Question*; and
- require collecting and using data.

Your questions should not have a yes/no or a one-word answer.

When you have completed your two questions, take the questions back to your small group. Share all the questions with each other. Carefully consider each question and decide if it meets the criteria for a good question. With your group, refine the questions that do not meet the criteria. Choose the two most interesting questions to share with the class. Give your teacher the rest of the questions so they might be used later.

Update the *Project Board*

Get Started

10 min.

Students formulate questions, about what makes you sick, that can be answered through scientific inquiry. Working in small groups, students evaluate one another's questions to be sure that they meet the criteria for a good question. Each group completely develops two questions that meet the criteria and will help guide their investigation into what can make you sick.

△ Guide

Read the *Get Started* section together with the class and discuss the criteria for the questions to be written. Check that students understand each of the criteria. Then, have students work individually to formulate their questions. Next, have students share their questions with their group. As questions are shared, the group should evaluate whether each question meets the stated criteria.

Have groups create a chart listing each question shared by a group member in one column. The other columns should each display one of the criteria. As the group evaluates each question, have them place a check in the column for each criteria it meets. When all questions have been evaluated, have students select the two questions that best meet all of the criteria to be presented to the class. Have groups prepare to defend their choices when they are presented to the class.

META NOTES

Effective monitoring involves awareness, listening and intervention. Try to be aware of participation and interest levels, anxiety or frustration, conflict and incompatibility. Listen for expression of key ideas, progress toward completing the task, attitudes, creativity and winding down. Intervene only if groups have a poor grasp of the instructions, are floundering, or ask for help.

Update the Project Board

15 min.

Briefly review the Project Board with the class. Add students' ideas and the questions developed during this activity to the Project Board.

our questions should n... /no or a one-w...

When you have completed your two questions, take the questions back to your small group. Share all the questions with each other. Carefully consider each question and decide if it meets the criteria for a good question. With your group, refine the questions that do not meet the criteria. Choose the two most interesting questions to share with the class. Give your teacher the rest of the questions so they might be used later.

Update the *Project Board*

You will now share your questions with your class. Be prepared to support your questions with the above criteria. Your teacher will help you with the criteria if needed. Your teacher will then add your questions to the *Project Board*. In this *Learning Set,* you will work to answer some of these questions.

GF 22

Project-Based Inquiry Science

△ Guide and Assess

Lead a discussion in which groups share their two most interesting questions with the class. The goal of this discussion is to engage the class in evaluating whether the questions meet the stated criteria. It also provides the basis for connecting the readings and investigations in *Learning Set 2* to questions raised by the students themselves. Challenge students to support the questions they developed by describing how they meet the criteria.

TEACHER TALK

"What are the five criteria for questions that were described in the *Get Started* section for this activity? *(Elicit the criteria and record this on the board for the class to reference.)*

Let's check to see if they meet the five criteria listed on the board. If you think a question does not meet the criteria, share your reasoning, and then allow the group to respond to your analysis."

Assess students' ability to generate questions that meet the criteria and are suitable for scientific investigation. Students often need help generating questions from what they are confused about. An important part of managing this discussion is to provide that help. There is an opportunity here to model questions that are suitable for scientific investigation and to point out what makes a question suitable.

Assessment Options

Targeted Concepts, Skills, and Nature of Science	How do I know if students got it?
Scientists often collaborate and share their findings. Sharing findings makes new information available and helps scientists refine their ideas and build on others' ideas. When another person's or group's idea is used, credit needs to be given.	**ASK:** When groups shared their questions and the class discussed whether they met the criteria, what comments by other students did you find most helpful in revising your own questions? **LISTEN:** Students should cite examples of valid comments related to meeting the criteria for questions.
Scientists must keep clear, accurate, and descriptive records of what they do so they can share their work with others, consider what they did, why they did it, and what they want to do next.	**ASK:** In what ways is the *Project Board* helpful to our search for answers to the *Big Question*? **LISTEN:** Students should cite examples of how the *Project Board* serves as a vehicle for sharing ideas, keeping track of what they have learned, and seeing what they are doing in a larger context.

Teacher Reflection Questions

- Research into students' understanding of the concepts in this Unit indicates that students often have difficulty understanding that not all diseases are caused by germs, and not all microbes are "bad." How well do you think your students understand these ideas? What did they say in the class discussion or within their small groups that makes you think they understand these ideas? How might you tailor the investigations in this *Learning Set* to better attend to these misunderstandings?
- What naïve or pre-scientific conceptions about germs have come up in class discussions? Where in the Unit might you challenge some of these conceptions?

- Students often pose questions that are not subject to scientific investigation. For example, ethical questions such as *"Should people with a disease be quarantined?"* or broad, ill-defined questions such as *"Why do some people get sick when exposed to germs and not others?"* Questions appropriate for scientific inquiry must be directed toward objects or events that can be described, explained, or predicted by a scientific investigation. Ideally, they should identify a proposed relationship between an independent and a dependent variable. How did you facilitate students' thinking about questions so that they are directed toward objects and events that can be described, explained, or predicted by scientific investigations?
- What exemplars did you use that were particularly effective in helping students to develop the ability to formulate questions appropriate for scientific inquiry?

NOTES

SECTION 2.2 INTRODUCTION

2.2 Read

Learn about Other Communicable Diseases

◀ 1 class period

A class period is considered to be one 40 to 50 minute class.

Overview

Students read about the causes, symptoms, spread, prevention, and treatment of gastrointestinal illnesses (a common communicable disease). This serves as an introduction to the *Communicable Disease Information Table,* in which students keep track of many aspects of diseases in order to assist them in answering the *Big Question.* Within this activity, students read that germs could be parasites, bacteria, or viruses. Bacteria are introduced in the next activity, after the students learn about cells. Students will read about viruses in the latter part of this *Learning Set.*

Targeted Concepts, Skills, and Nature of Science	Performance Expectations
Scientists often collaborate and share their findings. Sharing findings makes new information available and helps scientists refine their ideas and build on others' ideas. When another person's or group's idea is used, credit needs to be given.	Students should share their individual questions with their group and their group's questions with the class as they consider what they think they need to learn and what they need to investigate as they fill out the class *Project Board.* Students will also collaborate as they add information to the *Communicable Disease Information Table.*
Explanations are claims supported by evidence, accepted ideas, and facts.	When proposing methods by which gastrointestinal illnesses spread and for preventing their spread, students should be able to support their claims using examples from *Learning Set 1* and their readings.

Targeted Concepts, Skills, and Nature of Science	Performance Expectations
Disease is a breakdown in the structures or functions of an organism. Some diseases are the result of failures of a body system. Other diseases are the result of damage by infection from other organisms (germs).	Students should be able to identify an infection of the body's intestines by parasites, bacteria, or viruses as the cause of gastrointestinal illness.
Communicable diseases may be transmitted by direct contact with an infected person or by indirect contact with an infected object or substance.	Students should be able to describe several methods by which disease germs can spread, ranging from contact with an infected organism (e.g. humans, pets) to contact with a contaminated object or substance (e.g. toilets, uncooked meat.)
Communicable diseases are commonly spread through hand-to-hand contact. Hand washing is a simple habit that can help keep you healthy.	Students should be able to describe several ways in which germs on unwashed hands can enter the body and how washing hands can help prevent the spread of gastrointestinal illnesses.

Materials	
1 per class	class Communicable Disease Information Table (e.g. projection or chart)

NOTES

Homework Options

Reflection

- **Science Content:** How would a gastrointestinal illness cause you to become dehydrated? *(Your body loses more water than usual through vomiting and the watery bowel movements of diarrhea.)* What are two ways in which dehydration can be treated? *(In a mild case, by drinking plenty of fluids; in a severe case by adding fluids to the bloodstream intravenously.)*
- **Science Process:** Do a survey of your home environment and identify places that you would be particularly careful to clean if someone in your family had a gastrointestinal illness. *(Students should identify bathroom and kitchen surfaces at the least, but should also mention other surfaces frequently touched by multiple persons, such as remote controls, computer keyboards and mice, doorknobs, and light switches.)*

Preparation for 2.3

- **Science Content:** Why can't we see the germs on our hands? *(They are microscopic in size, too small to be seen with the unaided eye.)* What scientific tool would enable us to observe germs? *(The microscope.)*

NOTES

NOTES

SECTION 2.2 IMPLEMENTATION

2.2 Read

Learn about Other Communicable Diseases

When you are sick, your body responds to the sickness. When you have a cold, you get a stuffy or runny nose and maybe a cough. The stuffy or runny nose and cough are called **symptoms** of the cold. All diseases have symptoms. Some diseases are easy to detect because their symptoms are very obvious. A cold is easy to detect. However, some diseases have symptoms that are less obvious or are just like the symptoms of other diseases.

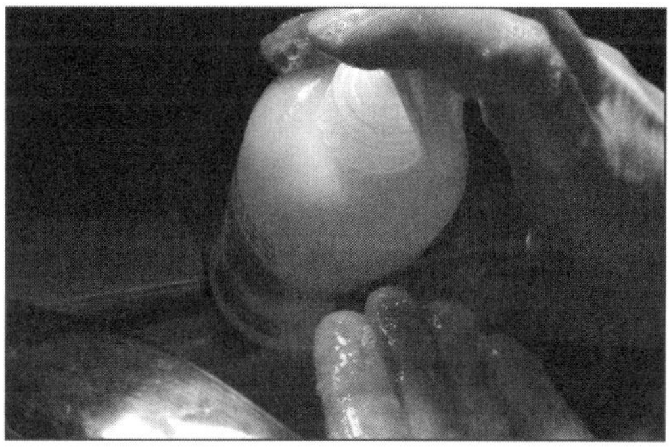

Washing your hands can help to prevent the spread of disease.

Other parts of your body that get sick are your stomach and your intestines. **Gastrointestinal** illnesses are common illnesses that happen in your stomach or intestines. These illnesses have very obvious symptoms. The most common is **diarrhea**, but they can also make you vomit and give you abdominal (belly) cramps.

Both colds and gastrointestinal illnesses are communicable diseases. They are spread from one person to another. Think back to the glow-powder activity. Think of the powder as germs that can cause an illness. What made the "germs" move from one person to another? Germs that cause many illnesses can move from another person to you because they get on your hands. Then you may put your hands near your face. This is a way colds, gastrointestinal illnesses, and other sicknesses move from one person to another.

Read about diarrhea on the next page. As you are reading, pay attention to words you might not understand. Record these words as you are reading. You will find out what they mean as the Unit goes on.

symptom: an indication in your body that you have a disease.

gastrointestinal: related to the stomach and intestines.

diarrhea: frequent and watery bowel movements.

GF 23 GOOD FRIENDS and GERMS

◀ *2 class periods**

2.2 Read

Learn about Other Communicable Diseases

5 min.

Students read an introductory selection about symptoms and how they relate to gastrointestinal illnesses.

△ Guide

After reading the first paragraph about symptoms, ask the class what other symptoms they have experienced when they were sick, or have observed in others that were sick. Students should be able to supply common examples such as pink eyes for conjunctivitis, and a red, sore throat for strep throat. Ask students to think of symptoms that are less obvious. Some less obvious symptoms might be fatigue or lack of appetite. As students share their experiences, guide them in distinguishing between the disease and its symptoms. Also, call attention to the similarities in the symptoms of many different illnesses.

*A class period is considered to be one 40 to 50 minute class.

TEACHER TALK

"Which of the symptoms listed by our class are present in many different illnesses?

You will now read about a common type of communicable disease - gastrointestinal illness. While you are reading, make notes of any new vocabulary words and important new ideas."

Finish the introductory reading on this page and emphasize the connection to the glow powder in *Learning Set 1*. Ask students if they can make a connection with the water-sharing simulation they did.

NOTES

Gastrointestinal Infections

What Are Gastrointestinal Infections?

Diarrhea is usually caused by an **infection** in your body's intestines. Infections in the intestines are called gastrointestinal infections. Germs like **parasites**, viruses, or bacteria can all cause gastrointestinal infection. Diarrhea can also be caused by other illnesses, medicine you are taking, or changes in diet. Diarrhea is communicable only when it is caused by germs.

infection: the growth of germs in your body.

parasite: organism that lives and feeds either inside of or attached to another organism and does harm to that organism.

sanitation: the disposal of sewage and waste.

hygiene: things people do to stay healthy.

dehydration: (medical) a condition in which the body does not have enough fluid to function properly.

Which germs are responsible for diarrhea depends upon where a person lives. It also depends upon **sanitation** and **hygiene** standards. Countries that have poor sanitation have frequent outbreaks of diarrhea. In developed countries like the United States, outbreaks of diarrhea are most often caused by food poisoning. Sometimes food poisoning happens when bacteria in food that is not handled, stored, or cooked properly gets into a person's body.

What Is Diarrhea?

When you hear the word diarrhea, you probably think about discomfort and embarrassment. Diarrhea is no fun. But nearly everybody gets it once in a while. When you have diarrhea, you have frequent and watery bowel movements.

How Long Are Gastrointestinal Infections Contagious?

Infections that cause diarrhea are highly contagious. They can spread from person to person through dirty hands, contaminated food or water, and some pets. Most cases are contagious for as long as a person has diarrhea. However, some infections can be contagious for even longer.

Drinking untreated water can cause food poisoning.

Can You Prevent These Infections?

The most effective way to prevent the spread of contagious diarrheal infections is to wash your hands often. Dirty hands carry germs into the body. This can happen when you do things like bite your nails or use your hands when eating. It is important to always wash your hands well with soap and water after using the bathroom and

Gastrointestinal Infections

15 min.

Students read about the symptoms, causes, contagiousness, prevention, and treatment of gastrointestinal infections.

○ Get Going

Have students read this entire section. Make it clear that their focus should be on the symptoms, causes, and prevention/treatment of gastrointestinal illness. When students have completed the reading, ask them to record their answers to the *Stop and Think* questions.

Stop and Think

10 min.

Discuss students' understanding of symptoms, treatment, and spread of gastrointestinal illness using the Stop and Think *questions.*

before eating. Washing your hands is especially important if you know there is a disease going around. Keeping bathroom surfaces clean can also help to prevent infections.

Food and water can also spread germs that cause diarrhea. To help protect yourself, cook foods thoroughly and wash raw fruits and vegetables well before eating them. Avoid eating undercooked hamburgers and poultry or raw eggs, and always refrigerate leftover food quickly. Make sure to clean your kitchen counters and cooking utensils, especially after they have been in contact with raw meat, eggs, and poultry.

Pets, particularly reptiles, can also spread germs. (Reptiles include animals like lizards, snakes, and turtles.) Pets should be kept away from family eating areas. Never wash pet cages or bowls in the same sink that your family uses to prepare meals. And always wash your hands after handling your pet!

Pets such as this bearded dragon can spread germs.

How Are These Infections Treated?

Most infections that cause diarrhea will go away without treatment. Resting at home and drinking plenty of fluids to avoid **dehydration** are the best ways to get over the illness. Dehydration means your body does not have all the fluids (water) it needs. Sometimes doctors tell people to treat the symptoms of diarrhea by taking medicines to make their **stools** less watery. Other times diarrhea needs to be treated with **antibiotics**, or drugs that kill some germs.

Stop and Think

1. The effects of an illness in your body are called symptoms. One symptom of a cold is a runny nose. What are the symptoms of gastrointestinal infections?
2. How do gastrointestinal infections spread?
3. How can you keep from getting a gastrointestinal infection?
4. You have learned a little about one gastrointestinal infection. Based on the reading, how can you make a good friend sick with a gastrointestinal infection?
5. What is one way you can prevent your good friend from getting sick with a gastrointestinal infection?

stool: the solid waste that is produced by the body.

antibiotic: a type of medicine that kills germs or prevents them from growing

△ Guide and Assess

When students have completed the *Stop and Think* questions, hold a brief class discussion in which students share their answers.

Assess students' responses to the *Stop and Think* questions by listening for the following information in their responses:

1. Diarrhea, vomiting, and stomach cramps are symptoms of gastrointestinal illness.

2. They can spread by unwashed hands, contact with contaminated food or water, pets, and surfaces.
3. Infection can be prevented by washing your hands often, especially after going to the bathroom and before eating, after touching a family pet, by cleaning bathroom surfaces, and by cooking foods thoroughly.
4. Response should contain not doing any of the items in the response for question 4.
5. Response should contain doing any of the items in the response to question 4.

At the end of the discussion, point out that they have just examined a specific example of how not to make our friends sick with one type of communicable disease – gastrointestinal infection. Let the students know that, by answering the last question in the section, they have begun to answer and understand the Big Question.

NOTES

Record Your Ideas

5 min.

Students set up a Communicable Disease Information Table to keep track of what they learn about diseases throughout the Unit.

Record Your Ideas

Throughout this Unit, you will record what you are learning about communicable diseases by entering the information in a table like the one shown. On your copy of the table, record what you now know about gastrointestinal diseases. Complete as many columns as you can. Some answers you may not know yet. You will be able to come back to the table later.

Communicable Disease Information Table 2.1

Name: _______________ Date: _______________

Name of disease	Cause (bacterial or viral)	How does it spread?	Which body systems does it affect?	What are the symptoms?	How is it treated? Is there a vaccine?	Other notes
Write the scientific name and the common name. People usually say that they have "the flu," which is a shorter way to say that they have "influenza."	Many diseases are caused by either a virus or a bacterium.	Some diseases are spread through blood, body fluids (like saliva), air, contaminated food, and animal bites.	Some diseases affect several systems. List the main one(s).	The symptoms are what help you to determine that you have a particular disease.	"Treat the symptoms" means doing something so the symptoms will not be as bad. Sometimes, you can treat a disease; sometimes you can only treat the symptoms.	Use this column to write anything you think is interesting or important to remember about a particular disease.

What's the Point?

You read about gastrointestinal diseases. Gastrointestinal diseases that are caused by germs are communicable. This means they can be passed from one person to another person. The reading described some of the symptoms of this disease. One of the symptoms is diarrhea.

You will enter information about each disease you learn about in a *Communicable-Disease Information Table*. The information will help you prepare to answer the Big Question: *How can you prevent your good friends from getting sick?*

△ Guide

Together with students read the section describing the *Communicable Disease Information Table.* Convey the purpose of the table to the students and clarify what each of the columns represents.

TEACHER TALK

"The *Communicable-Disease Information Table* is a place for us to keep track of what we have learned about communicable diseases. You will be adding information to this table throughout the Unit. You will use this information in several *Learning Sets* and when you address the two challenges and answer the *Big Question* at the end of the Unit. Sometimes we may learn some things about a disease in one investigation and other things about the disease in later investigations. So don't be concerned if you cannot complete every column on the table now, you will have many opportunities to add to the table in later *Learning Sets*. It is now time for you to construct your table and add information about your first communicable disease, the common cold."

⬡ Get Going

Have students construct a *Communicable-Disease Information Table* using the table printed in the student text as a template. To get started, tell students to add the information given in the textbook for the common cold and ask if they can add anything.

Next, ask students to add information from their reading about gastrointestinal illness to the table. You might consider having students do this in pairs or groups, but make it clear to students that each person is responsible for filling in their own table and keeping it up-to-date throughout the Unit.

□ Assess

While students are working on gastrointestinal illness, check that all students have set up their own *Communicable-Disease Information Table* properly, and are recording the appropriate information in each column. This also provides the opportunity to address individual issues or questions that may have arisen, and to reinforce the importance of this table to their overall goal.

△ Guide

Create a class table, and have a discussion centered on what information students think should be added to each column for the disease gastrointestinal illness. The table below shows some typical student responses.

Name of disease	Cause (bacterial or viral)	How does it spread?	Which body system does it affect?	What are the symptoms?	How is it treated? Is there avaccine?
Common cold	Bacteria or virus	Bodily fluids in the air or on surfaces; sneezing or coughing; blowing your nose; touching your hands to other people or objects	Respiratory	Cough; body aches; congestion; runny nose	No cure, only treatment for symptoms.
Gastrointestinal illness	Parasite, bacteria, or virus.	Dirty hands, contaminated food or water, pets, and dirty surfaces (like the bathroom).	Stomach/intestines	Diarrhea; vomiting; cramps	Rest; drinking plenty of fluids; if severely dehydrated, IV fluids

◇ Evaluate

Before moving on, make sure that everyone's ideas are represented. Check to make sure that the information provided by students is correct and is recorded in the appropriate column. Also make sure students understand that the *Communicable Disease Information Table* will be used many more times during the unit and should therefore not be lost or discarded.

Assessment Options

Targeted Concepts, Skills, and Nature of Science	How do I know if students got it?
Scientists often collaborate and share their findings. Sharing findings makes new information available and helps scientists refine their ideas and build on others' ideas. When another person's or group's idea is used, credit needs to be given.	**ASK:** What information about gastrointestinal disease that you added to your *Communicable-Disease Information Table* was based on others' ideas? **LISTEN:** Students should cite examples of information that they had not originally identified, but became aware of as a result of the class discussions.

Targeted Concepts, Skills, and Nature of Science	How do I know if students got it?
Explanations are claims supported by evidence, accepted ideas, and facts.	**ASK:** Why were some of the claims in the class discussions about gastrointestinal illness not added to the *Project Board*? **LISTEN:** Students should state that claims for which there is no supporting evidence are not accepted by the scientific community or by the class.
Disease is a breakdown in the structures or functions of an organism. Some diseases are the result of failures of a body system. Other diseases are the result of damage by infection from other organisms (germs).	**ASK:** What body structures or functions have broken down when a person has diarrhea? **LISTEN:** Students should identify the digestive system as the affected body structure and excretion as the affected body function.
Communicable diseases may be transmitted by direct contact with an infected person or by indirect contact with an infected object or substance.	**ASK:** What claims can you make about the spread of gastrointestinal disease and what evidence do you have to support your claims? **LISTEN:** Students should recognize that gastrointestinal disease can spread from person to person through dirty hands, contaminated food or water, and some pets. Students should cite statements from the reading selection for each claim.
Communicable diseases are commonly spread through hand-to-hand contact. Hand washing is a simple habit that can help keep you healthy.	**ASK:** What is the most effective way to prevent infections that cause diarrhea? **LISTEN:** Students should specify frequent hand-washing. Students should also describe mechanisms by which germs can get into their body from their hands, i.e. nail biting and using hands when eating.

Teacher Reflection Questions

- In this section, students read about one example of a communicable disease. What aspects of the disease were most interesting to students? Which ones were sources of confusion?
- How much understanding of the distinction between a disease and its symptoms was evident in class discussions? How could you make this distinction clearer to students?
- How were you able to connect the information about gastrointestinal illness to the investigations carried out in *Learning Set 1?*
- What are the benefits of vocabulary in this Unit only being introduced within the context of activities students are doing, usually after they have been discussing the ideas that are part of the definition of the word?

NOTES

SECTION 2.3 INTRODUCTION

2.3 Explore

Observing Cells

◀ $3\frac{1}{2}$ *class periods*

A class period is considered to be one 40 to 50 minute class.

Overview

In this section, students explore animal cells, plant cells, and their cellular structures. Students read about the structure and functions of specific organelles (specialized cell structures) and compare and contrast the organelles found in plant and animal cells. Students then read about the compound light microscope and magnification before using this instrument to observe prepared slides of animal cells and plant cells, including a wet-mount slide of plant cells that they prepare. Students examine the cells under the microscope at different magnifications and make detailed drawings of what they observe. Students then work in small groups sharing the drawings they made and discussing the parts of the cells they observed and which ones were difficult or impossible to see. After labeling all the cellular structures in their drawings that they are able to identify, they pick one cell and make a poster of it, labeling the organelles and their functions. Groups present their posters to the class and discuss their observations. The class then updates their *Project Board*, connecting what they have just learned about cells to understanding the nature of germs.

Targeted Concepts, Skills, and Nature of Science	Performance Expectations
Scientists often collaborate and share their findings. Sharing findings makes new information available and helps scientists refine their ideas and build on others' ideas. When another person's or group's idea is used, credit needs to be given.	Students should share their observations of the cells they examined under the microscope with their group. Their group's observations will be shared when they present their cell posters to the class.
Scientists must keep clear, accurate, and descriptive records of what they do so they can share their work with others, consider what they did, why they did it, and what they want to do next.	Students should make detailed drawings with accompanying descriptive notes of cells observed with the microscope so their work can be shared during the poster session.

Targeted Concepts, Skills, and Nature of Science	Performance Expectations
Students should be able to manipulate a compound microscope to view microscopic objects.	Students successfully bring a prepared slide into focus at various magnifications.
Students should be able to prepare a wet-mount slide, use appropriate staining techniques, and determine the size of a microscopic object, using a compound microscope.	Students should be able to prepare a wet-mount slide of onion skin and stain the specimen to reveal details of cellular structure. Students should also successfully estimate the size of a single cell.
The cell is the basic unit of structure and function in all living things. Cells are usually microscopic in size.	Students should be able to identify a single cell when viewing a specimen with the microscope and describe how cells carry out life functions. They should also be able to identify cells as the basic structural component of living things.
Cells have specialized structures that perform specific functions. Most cells have a cell membrane, cytoplasm and genetic material. Many cells have a nucleus. Some cells have cell walls and/or chloroplasts.	Students should be able to describe the appearance and function of various organelles. They should also be able to distinguish between plant and animal cells based on the presence or absence of certain organelles.
All organisms require energy to survive. Cells obtain energy in a variety of ways.	Students should be able to identify mitochondria as the cell organelle involved in converting food energy into a form of energy the cell can use. Students should also be able to distinguish between how animal and plant cells obtain food.

Materials	
1 per class	Laminated *Project Board*
1 per class	Overhead transparencies of cells in text
1 each per group	Dropper bottle of iodine
1 each per group	Transparent millimeter ruler
1 per student	*Observing Cells* page

Materials	
1 per student	Compound microscope
1 per student	Set of prepared slides of animal cells
1 per student	Set of prepared slides of plant cells
1 per student	Sample of fresh onion skin
1 per student	Plastic microscope slide
1 per student	Microscope slide cover slip
1 per student	Thin-stem pipette
1 per student	Tweezers
1 per student	Safety scissors
1 per student	Goggles
1 per student	Apron
1 per student	Paper towels
1 per student	Water access
1 per student	Poster material

Activity Setup and Preparation

Check for any structural differences between the microscope shown in the text and those you will be using and make any necessary adjustments to procedures based on these differences.

- Determine classroom rules for handling microscopes and establish procedures for distributing and returning microscopes.
- If you do not have enough microscopes for students to work individually, consider alternatives such as sharing in pairs or groups, or setting up stations which students can visit.

CAUTION:

DO NOT have students prepare wet-mounts of cheek cells — the danger of cross-infection through shared contact with live human tissues is significant. Therefore, many school districts proscribe this type of activity.

- Pre-cutting the onion into roughly square sections helps students more easily obtain the specimen of onion skin and placing the sections in small plastic bags speeds up distribution.
- If students are having difficulty conceptualizing the size of a cell, consider doing an internet search to find examples and explanations to show the class.

Homework Options

Reflection

- **Science Content:** Suppose you were given a sample of cells to observe but you were not told whether the sample was taken from a plant or an animal. What evidence would you look for that would help you decide whether the sample came from a plant or an animal? *(The presence of a cell wall or chloroplasts would be evidence that it is a plant cell. Their absence would be evidence that it was an animal cell.)*
- **Science Process:** A student determines that the field of view of a microscope is 2.5 mm under low power. The student then observes some onion cells and estimates that 10 cells will fit lengthwise across the field of view. What is the estimated length of an onion cell? *(2.5mm/10 cells = 0.25 mm/cell)*
- **Science Process:** Create a resume listing the skills you developed during this Learning Set. *(Answers will vary, but typically include references to skills related to use of the microscope.)*

Preparation for 2.4

- **Science Content:** Now that you have read about cells and have observed both plant and animal cells, why is it logical to conclude that bacteria are made of cells? *(Living things are all composed of cells, and bacteria are alive; therefore, bacteria are made of cells. However, while this is true, students may not realize that bacteria are unicellular.)*
- **Science Process:** Draw a picture of what you think a bacteria looks like. *(Their drawings of bacteria indicate whether students think bacteria are unicellular or multicellular and the organelles indicate whether they think bacteria are animals or plants.)*

SECTION 2.3 IMPLEMENTATION

◀ *4 class periods**

2.3 Explore

Observing Cells

What structure do all living things, whether they are germs or humans, have in common? The **cell** is the basic unit common to all living things. Every living thing is made of cells. Most germs are made of a single cell. You are made of about 100 trillion cells. To understand how germs spread disease and how they affect the cells in your body, you need to know the structure and the functions of a cell.

Cells are the "building blocks" of all plants and animals. They are very small. Most cells are too small to see with the human eye. To actually observe them, you have to make them appear larger. To observe cells, you will look at them through a microscope. A microscope **magnifies** objects. It makes things appear much larger than they are.

Before you begin your observations, read about what a cell is and how it functions. Also, read about the parts of a microscope and their function.

cell: the structural and functional unit of all living organisms. (What a living thing is made of and what makes it work.) It is sometimes called the "building block" of life.

magnifies: makes something look larger, but does not actually enlarge the physical size of the object.

cell theory: a theory about the relationship between cells and living things.

The Cell Theory

After scientists were able to study cells for some time, they put together a theory, called the **cell theory**, about cells and living things. The cell theory explains the relationship between cells and all living things, large and small. The theory states that

- all living things are composed of cells.
- cells are the basic units of structure and function in living things.
- all cells are produced from other cells.

The cell theory helps scientists learn more about living things by studying cells. Because all cells come from other cells, scientists can study cells to learn about reproduction and growth.

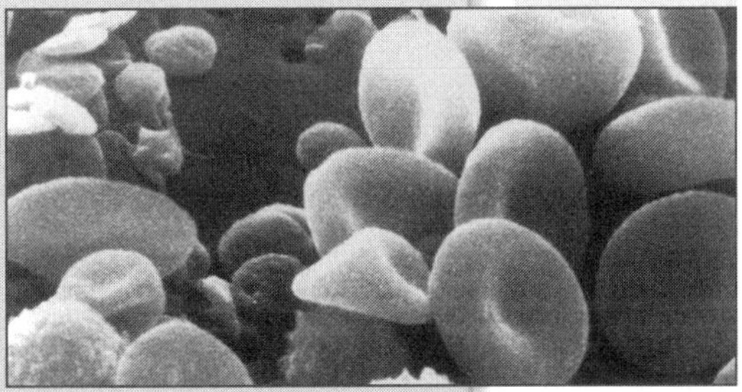

Human blood cells under a microscope.

GF 27 GOOD FRIENDS and GERMS

2.3 Explore

Observing Cells

5 min.

Introduce the concept that living things are made of cells and that cells contain specialized structures, or organelles.

○ Engage

Before students begin reading the introduction on page GF 27, ask students to close their eyes and visualize a cell. Invite several students to come to the board and sketch their mental image of a cell. Ask the class to compare the sketches on the board to their own ideas and suggest any additions that might be made to the sketches.

Create a class drawing of an animal cell that incorporates students' ideas.

*A class period is considered to be one 40 to 50 minute class.

TEACHER TALK

"Close your eyes and create a mental image of a cell. Try to include as many details as you can think of in your image. Now open your eyes and draw a sketch of your mental image."

Elicit several volunteers to draw their sketch on the board. Then ask the class:

TEACHER TALK

"Which of these sketches looks most like your own mental image? How is it similar and how is it different from your image? What changes or additions to these sketches could you suggest?"

⬡ Get Going

Have students read the introduction. A common misconception held by middle-school students is that living things contain cells, rather than that they are made of cells. Lead a short discussion about what is meant by the statement "Cells are the 'building blocks' of all plants and animals." Emphasize that cells are too small to be seen with the unaided eye. Make it clear that visible structures, such as pores in the skin, or the flaky texture of dry skin, are not cells.

△ Guide

Have students examine the photograph of cells at the bottom of page GF 27. Ask students to speculate about what living thing is made up of these cells. After students have shared some of their ideas, tell them that this is what blood looks like under the microscope. Students are often surprised blood is made of cells. It also provides an opportunity to reinforce the idea that cells are also the functional components of living things.

organelle: a specialized structure in a cell.

nucleus: the control center of the cell.

chromosomes: contain the genetic material of the cell.

genetic material: contains the information that determines the traits of an organism; hereditary material.

cell membrane: surrounds the cell; controls the movement of materials into and out of the cell.

Cells – The Basic Units of Life

Scientists have spent many years observing cells. They have found that there are many different sizes and shapes of cells. Some different types of cells are shown in the photographs. Although cells come in different sizes and shapes, they have some structures in common.

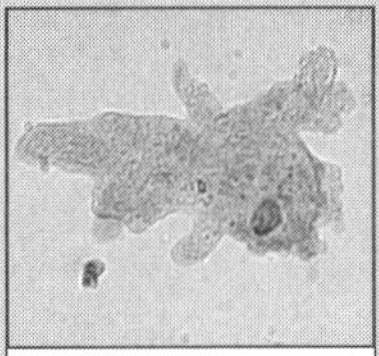

Single-celled organism
Amoeba proteus

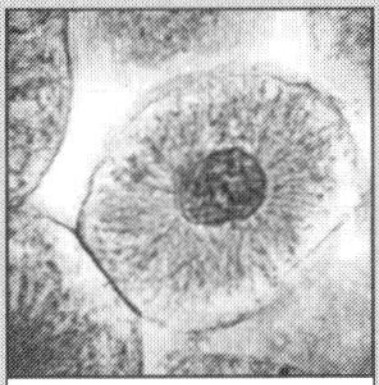

Animal cell

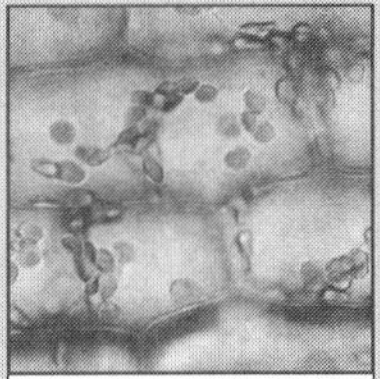

Plant cells

Each different structure in a cell has a specific function (role). Each structure acts as if it were a special organ. That is why cell structures are known as **organelles**.

The diagram shows some of the organelles that are found in animal cells. All these structures are also found in plant cells.

Animal Cell

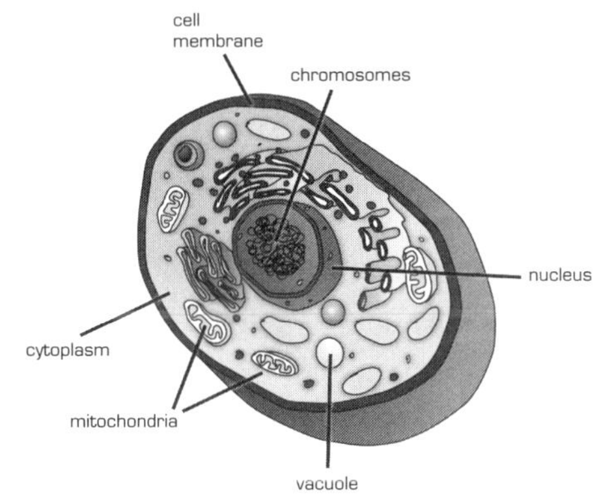

The **nucleus** is the control center of the cell. It holds the information that directs what the cell does. The nucleus contains **chromosomes**. Chromosomes store the **genetic material**. Genetic material determines the specific traits of an organism. The **cell membrane** surrounds the cell. It controls the movement of materials into and out of the cell.

The **cytoplasm** is the watery fluid inside the cell. It contains all the cell's organelles. **Mitochondria** (singular, mitochondrion) provide the cell with energy. In the mitochondria, food and oxygen are converted to energy, carbon dioxide, and water by the process of **cell respiration**. The **vacuole** is a storage area. It is used to store water and nutrients.

Cells – The Basic Units of Life

15 min.

Students read about cellular structures and the functions of various cell organelles.

Before reading this selection, ask students to consider what they would change about their drawing if it was supposed to represent an animal cell. Make adjustments to the class drawing as needed.

TEACHER TALK

"If you were told that your drawing was supposed to show a cell from an animal, what would you change about your drawing?"

Next, read through the *Cells – The Basic Units of Life* selection with students.

Consider projecting the diagram of the animal cell and using it as a focal point for the discussion of the different types of organelles in the cell. Discuss the appearance of each organelle and its role in the cell.

Emphasize that each organelle has a specific function and help students make connections with organs in their own body that perform similar functions (e.g. nucleus/brain, cell membrane/skin). An important point to make here is that each individual cell is alive and carries out all of the life processes. For example, cells take in food, obtain energy from food, get rid of wastes, grow and reproduce.

Ask students to reassess their cell drawings based on what they have read, and to make any necessary additions or alterations. Make any necessary adjustments to the class drawing.

Challenge students to predict how they might have to change their drawing if it was supposed to represent a plant cell instead of an animal cell.

TEACHER TALK

"How would you change your drawing if it was supposed to show a plant cell instead of an animal cell?"

NOTES

2.3 Explore

Plant Cells

Plant cells have all the same organelles found in animal cells. They also contain a few structures not found in animal cells.

In plants, the organelles in which food is made are called **chloroplasts**. The chloroplasts contain a green pigment called **chlorophyll**. Chlorophyll is used in **photosynthesis**. You may recall from previous science classes that photosynthesis is the process by which plants use light energy, water, and carbon dioxide to make sugar and oxygen. Plant cells also have a **cell wall**. The cell wall supports and protects the plant cell.

Plant Cell

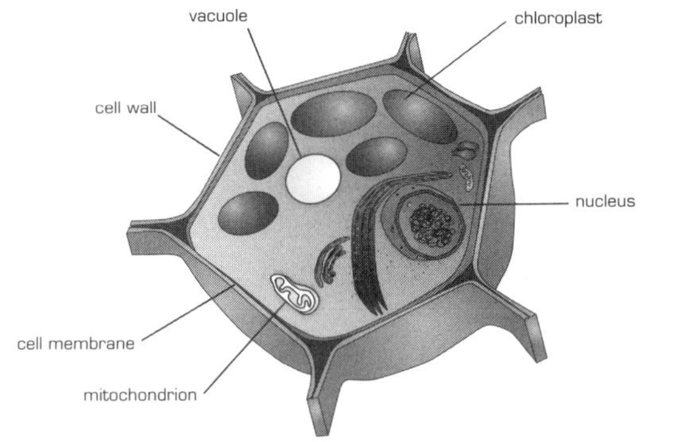

cytoplasm: the watery fluid that contains the organelles of the cell.

mitochondrion (plural, mitochondria): provides the cell with energy.

cell respiration: the process by which food and oxygen are converted to energy, carbon dioxide, and water.
vacuole: a storage area for food and water.

chloroplast: organelle that contains the green pigment chlorophyll used in photosynthesis.

chlorophyll: a green pigment.

photosynthesis: the process by which plants make sugar and oxygen using light, water, and carbon dioxide.

Stop and Think

1. Name four structures found in both animal and plant cells. What is the function of each?
2. What structures are found in plant cells that are not found in animal cells?
3. What can a plant cell do that an animal cell cannot? What structure carries out this function?

cell wall: protects and supports the plant cell.

GF 29

GOOD FRIENDS and GERMS

△ Guide

Finally, have students read the selection on Plant Cells on page GF 29. Consider projecting the diagram of the plant cell and using it as a focal point for discussing of the different types of organelles in the plant cell. Here again, discuss the appearance of each organelle and its role in the cell.

Ask students to compare and contrast the organelles found in animal cells and plant cells.

Have students record their answers to the three *Stop and Think* questions.

Stop and Think

5 min.

Guide students to think about the functions of cell organelles and how they relate to life functions carried out by an organism as a whole.

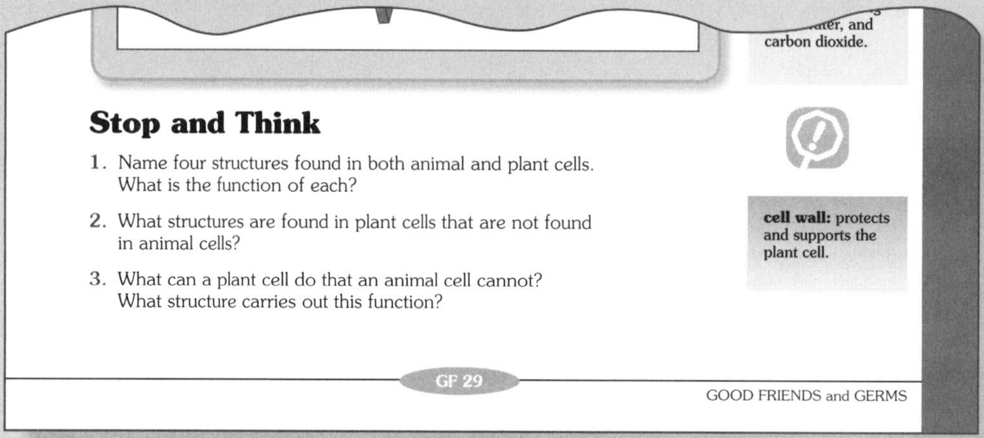

ater, and carbon dioxide.

Stop and Think

1. Name four structures found in both animal and plant cells. What is the function of each?
2. What structures are found in plant cells that are not found in animal cells?
3. What can a plant cell do that an animal cell cannot? What structure carries out this function?

cell wall: protects and supports the plant cell.

GF 29

GOOD FRIENDS and GERMS

△ Guide and Assess

When students have finished answering the *Stop and Think* questions, hold a brief class discussion in which students share their answers.

Asses whether students' responses to the three questions contain the following information:

1. Animal and plant cells both contain the following structures:
   - **Cell membrane:** surrounds the cell and controls movement of material into and out of the cell
   - **Cytoplasm:** watery fluid that contains all the cell's organelles
   - **Nucleus:** control center of cell, holds information that directs cell activities
   - **Chromosomes:** store genetic material
   - **Mitochondria:** converts energy stored in food to a form the cell can use
   - **Vacuole:** storage area within cell, usually contains water and nutrients
2. Plant cells contain the following structures not found in animal cells:
   - **Cell wall:** supports and protects cell
   - **Chloroplast:** contains the green pigment chlorophyll used in photosynthesis
3. Plant cells can use light energy to convert water and carbon dioxide into sugar and oxygen. This process, called photosynthesis, is carried out in the chloroplast.

compound light microscope: a microscope that has more than one lens and uses light transmitted to your eye to form an image.

The Microscope and Magnification

You cannot see most cells with the unaided eye. You need a microscope to observe them. There are different types of microscopes. The type you will be using is called a **compound light microscope**. The diagram shows the parts of this type of microscope.

The *stage* holds the microscope slide. The *clips* hold the slide in place. An opening in the stage lets light shine through. The *ocular lens* magnifies the object. The *objective lenses* found on a *revolving nosepiece* also magnify the object. There are usually three objective lenses. The shortest, or low-power lens, usually magnifies the object four times (written as 4×). The medium-power lens usually magnifies the object 10×. The longest lens, the high-power lens, usually magnifies the object 40×. The *coarse-adjustment knob* moves the *body tube* up and down. It is used only with the low-power lens. The *fine-adjustment knob* also moves the body tube up and down. It is used with the medium- and high-power lenses.

You have probably used a handheld magnifying glass to make objects appear larger. A magnifying glass, also called a hand lens, is a single lens that magnifies objects. An important development of the microscope was to use two lenses. Each lens multiplies the magnification of the other. If a lens magnifies an object 10×, the object appears 10 times larger. If you then add another lens that also magnifies the object 10×, the object now appears 100 times larger (10 × 10)!

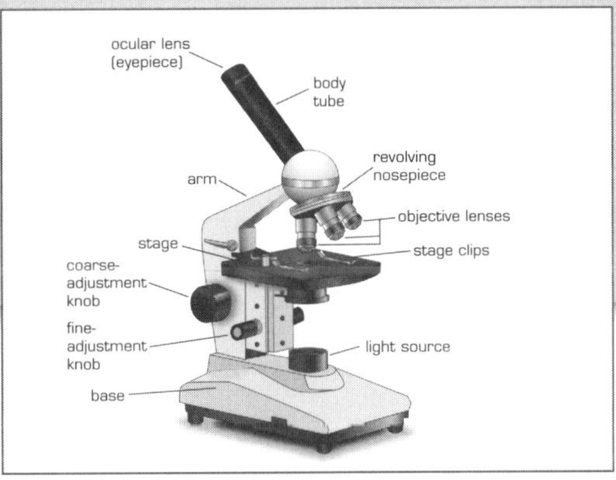

The Microscope and Magnification

15 min.

Students read about the parts of the compound light microscope and how to determine its magnification.

△ Guide

Tell students that they now have the opportunity to observe actual animal and plant cells using the microscope. However, before using the microscope they are going to review the parts of the microscope and how it works.

Have students read the selection *The Microscope and Magnification.* Review the parts of the microscope and their functions with students. Demonstrate the proper use of the microscope, showing how the various parts operate. Demonstrate how to bring a slide into focus and how to make any necessary adjustments.

Inform students of any classroom rules you have about using the microscope, and remind them of any safety issues such as wearing goggles and apron, or tying their hair back.

Describe the procedures that will be used in your classroom to pick up and return microscopes.

NOTE: This is an opportune time to conclude the first class session of this section.

NOTES

Procedure: Observing Animal Cells

You will be observing animal cells under the microscope. Scientists work hard to make accurate observations. Careful observations help scientists make predictions, draw conclusions, and find comparisons. It is very important that you take time to make careful observations when looking at the cells. Do not rush through your observations. Another important scientific skill you will need to use in making your observations is sketching.

1. Make sure that the nosepiece of the microscope is set so the low-power lens is in place. Put the slide on the stage. Hold it in place with the clips.

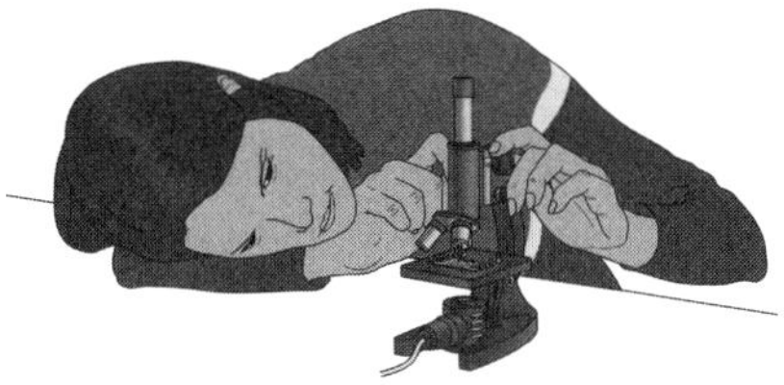

2. Look at the microscope from the side. Find the coarse-adjustment knob. Use it to lower the low-power lens. Bring it as close to the slide as possible. Do not let the lens touch the slide.
3. Look at the slide through the eyepiece. Very slowly move the coarse-adjustment knob away from the slide. Turn the knob until the object on the slide comes into focus (becomes clearer).
4. On the *Observing Cells* page, make a sketch of what you see as you look through the low-power lens. You should try to sketch exactly what you see. Make the object fill the same amount of space in your sketch as it does in the microscope. Be careful to include as many details as you can see. Suppose you see a big circle in the middle of the cell. In your diagram, sketch the big round circle and put it in the middle of the cell. If there are many smaller things in the middle of the cell, you should sketch them all. Try sketching them in the same places where you see them.
5. Look through the microscope again. Adjust the slide so the object you are looking at is in the center. Rotate the nosepiece to the medium-power lens. Use the fine-adjustment knob this time. Bring the slide into focus.

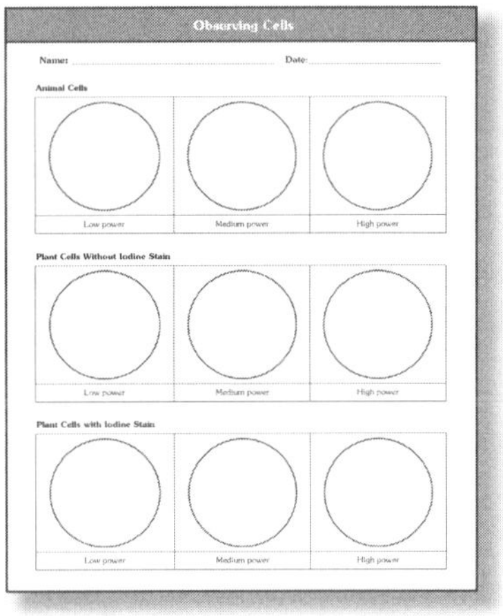
Observing Cells

Name: Date:

Animal Cells

Low power | Medium power | High power

Plant Cells Without Iodine Stain

Low power | Medium power | High power

Plant Cells with Iodine Stain

Low power | Medium power | High power

Procedure: Observing Animal Cells

15 min.

Students use the microscope to observe a prepared slide of animal cells.

△ Guide

Have students read through the procedure before distributing materials. Emphasize the importance of making detailed drawings of what they observe with the microscope. Remind students that they will make three drawings of animal cells on their *Observing Cells* page. They will make drawings of the animal cells at low, medium, and high power.

⬡ Get Going

Distribute *Observing Cells* pages, microscopes, and prepared slides of animal cells (squamous epithelium).

Have students observe the animal cells and record what they observe as detailed sketches on their *Observing Cells* page. Remind students that the sketch should show exactly what they see, and that the sketch should fill the same amount of space in the circle as do the actual cells in the microscope's field of view. Encourage students to label parts they recognize based on what they have read about animal cells, and to make written notes of their observations to accompany their drawings.

□ Assess

Assess students' skill in manipulating the microscope and bringing a slide into focus. Provide individual assistance where needed.

As students are making their observations, assess the quality of their sketches. Are structures misplaced or poorly represented? Are students including written descriptions and labeling structures?

△ Guide

Briefly discuss students' observations of the animal cells. Ask students which organelles they were able to identify in the cells they observed. Encourage students not to be discouraged if they could not see all of the organelles they read about. Point out that some organelles are not distinguishable at the magnifications of classroom microscopes, and some are transparent, jelly-like structures that are not easily visible unless stained.

Collect the prepared slides of animal cells.

NOTES

6. Sketch what you see under medium power.
7. Repeat Steps 5 and 6 using the high-power lens.

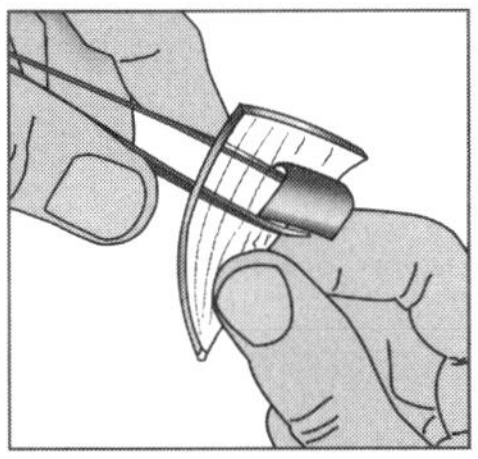

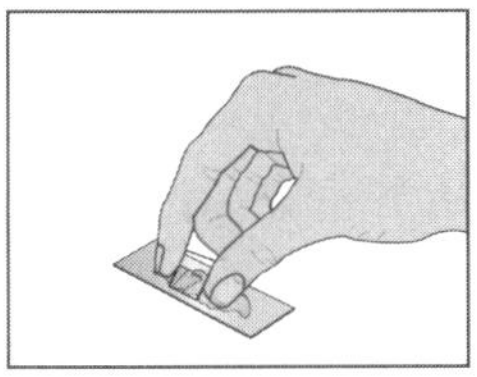

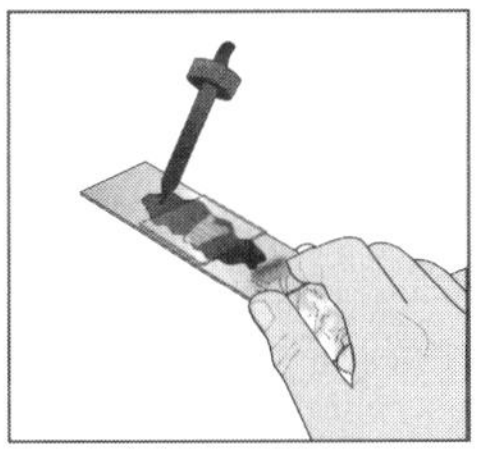

Procedure: Observing Plant Cells

You will now look at a type of plant cell, an onion cell. You will first prepare a wet-mount slide of the onion skin cell and then look at it under the microscope. Your teacher may also give you prepared slides of other plant cells to observe.

1. Get a piece of onion skin. Using tweezers, remove a thin, transparent layer from the inside surface. Transparent means that you should be able to see through the layer that you remove.
2. Place the piece flat in the center of the slide. Be careful that the piece does not fold over.
3. Cover the onion skin with one or two drops of water. Gently lower a cover slip over the slide, as shown in the diagram.
4. Observe the slide under the microscope. Sketch what you see under each magnification on the *Observing Cells* page.
5. Switch to low power and remove the slide.
6. Put on rubber gloves and safety goggles. Place a drop of iodine stain at one edge of the cover slip. Hold a piece of paper towel at the other end of the cover slip. The paper towel will draw the iodine stain under the cover slip and across the onion skin.
7. Observe the slide under the microscope again. Sketch what you see under each magnification on the *Observing Cells* page.
8. You may observe other prepared slides of plant cells under the microscope.

Stop and Think

1. Describe in words the animal cells you looked at.
2. Describe in words the plant cells you looked at:
   a) without the iodine stain.
   b) with the iodine stain.

Procedure: Observing Plant Cells

20 min.

Students prepare a wet-mount slide of onion skin cells, stain the cells for better viewing, and observe the cells using the microscope.

△ Guide

Tell students that they are now going to prepare their own slide of onion skin cells.

Have students read through the procedure before distributing materials.

Demonstrate the proper technique for preparing a wet-mount slide and for staining the onion skin cells. Show students how to use the tweezers to remove the thin, transparent layer of cells from between the thicker layers of the onion.

Remind students that they will be making sketches on their *Observing Cells* page of how the onion cells appear before and after staining at low, medium, and high power.

⬡ Get Going

Distribute materials and have students prepare a wet-mount slide of onion skin cells.

Have students observe the unstained onion skin under low, medium, and high power and make a sketch of what the cells look like at the different magnifications on their *Observing Cells* page.

Have students remove the slide from the microscope stage and stain the cells with iodine solution using the technique you demonstrated earlier. Then have students observe the stained onion skin under low, medium, and high power and make detailed sketches of what the stained cells look like at the different magnifications on their *Observing Cells* page.

If time permits, you may want to distribute prepared slides of other plant cells for students to view under the microscope.

□ Assess

Assess students' skill in preparing and staining their wet-mount slides. Provide individual assistance where needed.

As students are making their observations, assess the quality of their drawings.

△ Guide

Ask students to review the image of a plant cell on page GF 28 in the student text. You may want to project the plant cell image for the class. Ask each student to label the organelles in their drawings.

NOTES

you see unde... ...gnification on the ... page.

8. You may observe other prepared slides of plant cells under the microscope.

Stop and Think

1. Describe in words the animal cells you looked at.
2. Describe in words the plant cells you looked at:
   a) without the iodine stain.
   b) with the iodine stain.

3. Why do you think you used the iodine stain? What effect did it have on the cells?
4. Which cell parts were easy to see and which cells parts were difficult or impossible to see in
   a) the animal cell?
   b) the plant cell?
5. Which cell parts looked similar in all the cells you observed?
6. Which cell parts looked different in all the cells you observed?
7. What parts did you observe in both animal and plant cells?
8. What differences between the animal and the plant cells did you observe?

Conference

In your small group, share all the sketches you made. Identify the parts of the cells and label them with the proper names. Check with the members of your group to see if you have labeled the cells the same way.

Create a poster for a presentation. Choose one cell picture from those in your group and draw it on your poster. Make the picture as similar to the original sketch as possible. Label the organelles that you could see with the micro... ...ite the job or function of each organelle next to the

Stop and Think

5 min.

Discuss students' observations of cells using the Stop and Think *questions.*

○ Get Going

Have students read the *Stop and Think* questions and record their answers.

When students have completed recording their answers to the questions, hold a brief class discussion in which students share their observations and ideas about cells.

□ Assess

Assess whether students' responses to the *Stop and Think* questions contain the following information:

1. Students' descriptions of the animal cells typically cite:
   - a cell membrane forming the outer boundary of the cell

- a large, dark, ball-shaped nucleus
- cytoplasm with flecks of material in it

2. Students' descriptions of plant cells typically cite:

   a) Without stain:

   Cell wall

   Cytoplasm

   Vacuole

   b With stain:

   Nucleus

   More detail in cell wall

3. Iodine turns blue-black in the presence of starch. Many cell structures are colorless; those that contain starch will change color in the presence of iodine and will therefore be more visible. Even those that will do not change color will be stained a light brownish yellow, and will show more contrast so that details are easier to see.

4. a) In the animal cell, the cell membrane, nucleus, and cytoplasm were easy to see. It was difficult or impossible to see vacuoles, individual chromosomes or mitochondria.

   b In the plant cell, it was easy to see the cell wall, chloroplasts (only in very fresh onion skin), nucleus and vacuoles. It was difficult or impossible to see the cell membrane, mitochondria, or individual chromosomes.

5. The nucleus and cytoplasm looked similar in both plant and animal cells.

6. The cell wall and chloroplasts were not present in the animal cell. The vacuoles were larger in the plant cell than in the animal cell. The shape of the cell membrane and/or cell wall varied from cell to cell.

7. Cytoplasm, a nucleus, and vacuoles were present in both plant and animal cells. (Although both contain a cell membrane, it is difficult for students to distinguish between the cell wall and the cell membrane in plant cells.)

8. Plant cells contained a cell wall and chloroplasts; animal cells did not. Plant cells also tended to have larger vacuoles and a more rectangular shape than animal cells.

NOTE: This is an opportune time to conclude the second class session of this section.

3. Why do you think you used the iodine stain? What effect did it have on the cells?
4. Which cell parts were easy to see and which cells parts were difficult or impossible to see in
   a) the animal cell?
   b) the plant cell?
5. Which cell parts looked similar in all the cells you observed?
6. Which cell parts looked different in all the cells you observed?
7. What parts did you observe in both animal and plant cells?
8. What differences between the animal and the plant cells did you observe?

Conference

In your small group, share all the sketches you made. Identify the parts of the cells and label them with the proper names. Check with the members of your group to see if you have labeled the cells the same way.

Create a poster for a presentation. Choose one cell picture from those in your group and draw it on your poster. Make the picture as similar to the original sketch as possible. Label the organelles that you could see with the microscope. Write the job or function of each organelle next to the label on the poster.

Be prepared to present your poster to the class.

Communicate Your Results

Investigation Expo

Place your poster on the wall. As you observe the other posters, notice any differences in the ways your classmates sketched their cells. After everyone has had a chance to see all the posters, discuss any differences you noticed.

Conference

5 min.

△ Guide

Ask students to share their sketches of animal and plant cells within their small group and to check that the members of their group have all labeled cells the same way.

Briefly discuss the criteria that groups might want to use for selecting the sketch that will be made into a poster, such as clarity, correct scale, level of detail, and organelles.

Then have the students select one cell sketch from their group to make into a poster for a class presentation. Let students know that during their presentation they will be describing their cell, naming the organelles and describing the function of each organelle.

original sketch as possible. Label the organelles that you could see with the microscope. Write the job or function of each organelle next to the label on the poster.

Be prepared to present your poster to the class.

Communicate Your Results

Investigation Expo

Place your poster on the wall. As you observe the other posters, notice any differences in the ways your classmates sketched their cells. After everyone has had a chance to see all the posters, discuss any differences you noticed.

Communicate Your Results

30 min.

Have groups create posters and present their observations of cells to the class.

Have each group place their cell poster on a wall and present their cell poster to the class, showing the organelles they observed in the cell and describing the function of each organelle.

Lead a discussion in which students assess what they have been doing. Point out what they have been doing, making observations and describing what they see, is something that scientists do every day. Emphasize that taking careful notes and making detailed sketches allows for a more detailed understanding of what was observed. Scientists try to take clear and detailed notes so that anyone reading their notebook could follow what they have done and understand it.

Have a discussion emphasizing that cells function like independent small systems. Guide students in making connections between the functions of organelles in a cell and the functions of organs in the body. Help students to recognize that just as the organs/organ systems in the body work together to keep the body alive and healthy, organelles keep the cell alive and healthy.

Update the *Project Board*

You have just observed animal and plant cells. You know that it is important to learn about cells in order to answer the question: *What kinds of things make you sick?* Focus on the two columns on the *Project Board*: *What are we learning?* and *What is our evidence?* Record what you have learned about cells. Make sure to record the same information on your own *Project Board* page.

What's the Point?

It is important to know about the cells in your body. Cells move disease around and cells keep you healthy. All the different cells in your body have specific jobs, called functions.

Most cells are too small to see with only your eyes. You need a microscope. The parts inside cells are even smaller. The parts, called organelles, each have a function. They work together to keep the cell functioning properly.

It is important to know about the structure and functions of cells to understand what makes you sick and how you can keep from getting sick. Most germs are cells. All your body parts are made of cells. You learned that there are parts that are common to all cells. That means that cells that make you sick and cells that keep you healthy are very similar to each other.

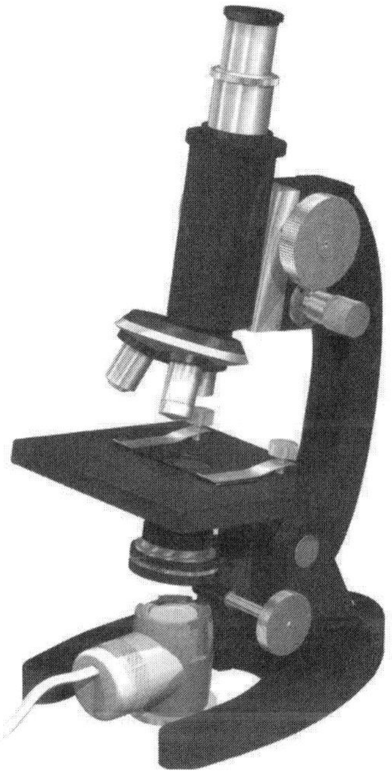

Technology Connection

Science and technology work together. Until a microscope was made, no one could see a cell. As better microscopes were developed, scientists were able to see more. They learned more about the parts of cells and their functions. This led the scientists who made microscopes to make even more powerful ones.

Update the *Project Board*

10 min.

Have students add new information and supporting evidence to the Project Board.

Have students read the *Update the Project Board* section in their text. Point out that the focus is on adding information about cells to the *Project Board.*

Remind students that for every idea added to the *What are we learning?* column, there should be corresponding supporting evidence in the *What is our evidence?* column.

Ask students if there is any information they want to add to the *Project Board* from what they learned in the previous activity on gastrointestinal illness.

During the discussion, help students recognize how what they are learning is related to the smaller question, *What Kinds of Things Make You Sick?* Help students recognize the connections between the presence of cells everywhere. Germs are cells and all of their body parts are cells. Therefore, the more students learn about cells will help them understand what happens to make you sick.

Have students read the *Technology Connection* in the margin and emphasize that the microscope opened up a whole new world to scientists, enabling them to see more and thus have more to question and investigate.

NOTE: This is an opportune time to conclude the third class session of this section.

NOTES

More to Learn

More to Learn

How Big Is Very Small?

Cells are very small. You cannot see most cells without using a microscope. You looked at plant and animal cells under a microscope. Exactly how small were the cells you looked at? Can you use a microscope to estimate the size of the cells?

If you can calculate the size of the microscope's **field of view** using a ruler and estimate the number of cells that you can fit across the field of view, you can estimate the size of each cell. The field of view is the circle of light you see through the microscope.

The first procedure below guides you through the steps to find the **diameter** of the field of view of your microscope under low and medium power using a ruler. Most high-power lenses have a field of view that is less than one millimeter, the smallest marking on your ruler. You will not be able to use a ruler for the high-power lens. You will use a **ratio**. You will then estimate how many objects could fit into each field to determine the size of the object.

field of view: the circle of light you see when you look through a microscope.

diameter: the length of the line through the middle of a circle from one side to the other.

ratio: a comparison of two numbers.

Procedure: Determining the Field of View

1. With the low-power lens in place, put a transparent ruler on the microscope's stage. Position the millimeter marks on the ruler below the lens. Center the ruler in the field of view.
2. Using the coarse-adjustment knob, focus on the marks on the ruler.
3. Move the ruler so that one of the millimeter markings is just at the edge of the field of view. Record the diameter of the field of view in millimeters. This is the size of the field of view of the low-power lens.
4. Use the same procedure to measure the field of view of the medium-power lens.

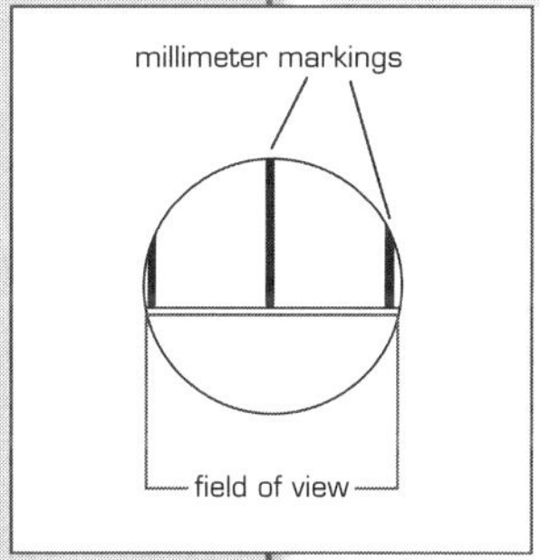

More to Learn: How Big Is Very Small?

20 min.

Students are shown a technique that can be used to estimate the size of objects viewed under the microscope.

△ Guide

Draw a circle on the board and ask for a student volunteer to draw a diameter of the circle. Review the meaning of the term diameter with students.

Have students read the introductory paragraphs before the procedure.

By middle school, most students have already studied estimating size using non-standard units of measurement. As an example, show students an index card and tell them it is 3 inches wide by 5 inches long. Ask them to suggest a way in which they could estimate the diameter of a penny using only a blank index card. Students will typically suggest marking off the number of pennies that fit across an index card and then dividing that number into the known dimension of the card. Have one student try this and note that 4 pennies just fit across the 3 inch side of the card. Have students calculate the diameter of a penny – $\frac{3}{4}$ inch (3 inches √ 4 pennies = $\frac{3}{4}$ inch per penny.)

Tell students that they are going to use a similar technique to estimate the size of a cell.

Procedure: Determining the Field of View

20 min.

Students measure the diameter of the microscope's field of view at different magnifications.

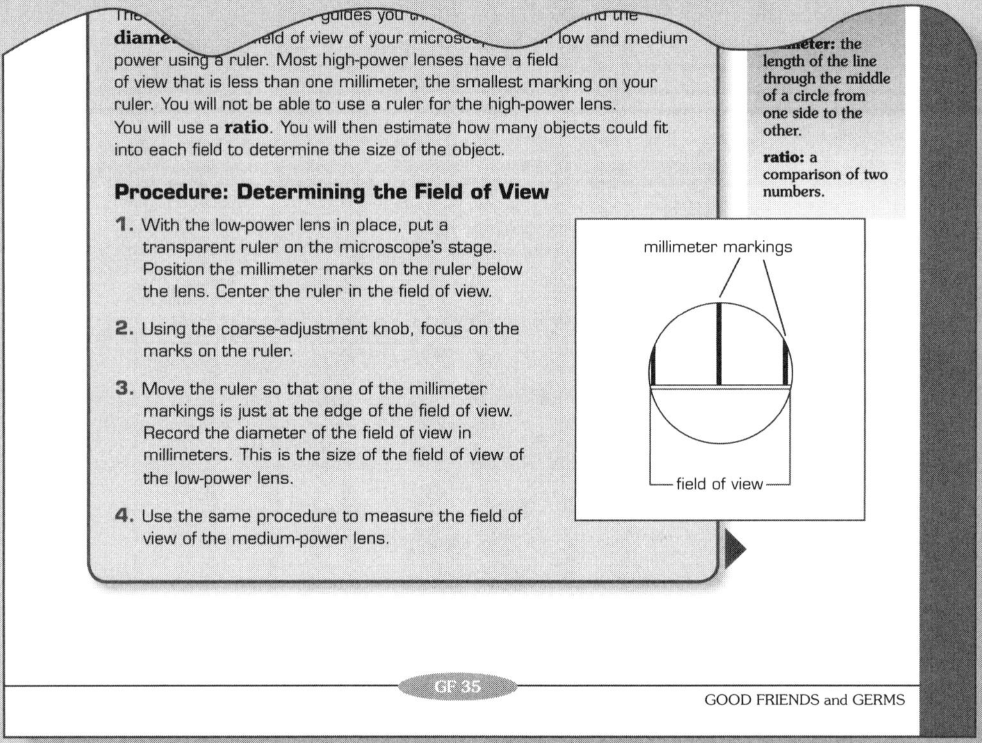
The ... guides you ... the **diame**... field of view of your microsc... low and medium power using a ruler. Most high-power lenses have a field of view that is less than one millimeter, the smallest marking on your ruler. You will not be able to use a ruler for the high-power lens. You will use a **ratio**. You will then estimate how many objects could fit into each field to determine the size of the object.

...eter: the length of the line through the middle of a circle from one side to the other.

ratio: a comparison of two numbers.

Procedure: Determining the Field of View

1. With the low-power lens in place, put a transparent ruler on the microscope's stage. Position the millimeter marks on the ruler below the lens. Center the ruler in the field of view.
2. Using the coarse-adjustment knob, focus on the marks on the ruler.
3. Move the ruler so that one of the millimeter markings is just at the edge of the field of view. Record the diameter of the field of view in millimeters. This is the size of the field of view of the low-power lens.
4. Use the same procedure to measure the field of view of the medium-power lens.

GF 35 GOOD FRIENDS and GERMS

Have students read through the procedure before handing out any materials.

Have students describe the method for measuring the size of the field of view in their own words. Guide students to recognize that even though the image of the ruler may be magnified under the microscope, the actual distance between one marking and the next in the image is still the same as the actual distance between the markings on the ruler – 1 millimeter. Help students make the connection between estimating the size of a penny by comparing it to the known size of the index card and estimating the size of a cell by comparing it to the known size of the field of view.

5. The high-power objective lens on your microscope probably has a field of view that is less than one millimeter. You cannot measure this field of view using a ruler. You need to do mathematical calculations.

First, calculate the ratio of the magnification of the high-power objective lens to the magnification of the low-power objective lens.

$$\text{ratio} = \frac{\text{magnification of high-power lens}}{\text{magnification of low-power lens}}$$

Use this ratio to determine the diameter of the field of view under high-power magnification.

$$\text{field diameter (high power)} = \frac{\text{field diameter (low power)}}{\text{ratio}}$$

Procedure: Estimating Size

1. You now know the diameter of the field of view under the low-, medium-, and high-power lenses. You can use this information to estimate the size of an object you see under the microscope.
2. Look at the object under the low-power lens. Estimate how many of these objects would fit across the diameter of the field of view.
3. Calculate the width of the object.

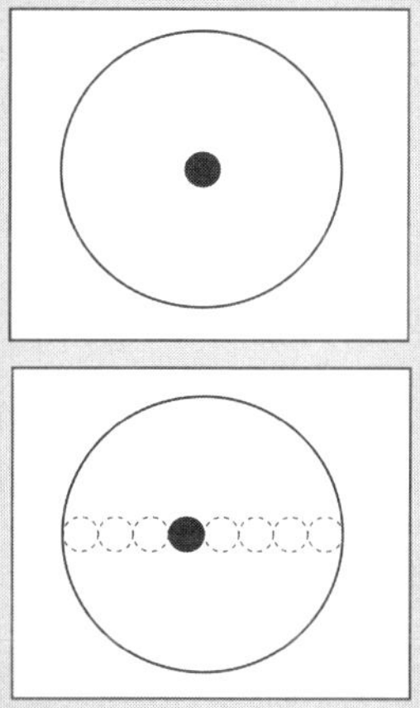

$$\text{width of object} = \frac{\text{width of field of view}}{\text{number of objects across the diameter of the field of view}}$$

⬡ Get Going

Distribute the microscopes and transparent rulers. Have students look into the microscope with nothing mounted on it. Identify the white circle of light that they see as the field of view. Then have students follow the steps of the procedure to measure the size of the field of view under low power and under medium power. Next, have students calculate the size of the field of view under high power using the equations in the text.

Assess

Have students share their measurements of the field of view under low power and high power. Encourage students to report anomalous results and to repeat their measurements. Guide students to understand that some variation is to be expected since the edges of the field of view rarely align exactly with markings on the ruler, so every measurement involves a degree of estimation.

Ask for student volunteers to write their calculations on the board and check them together with the class.

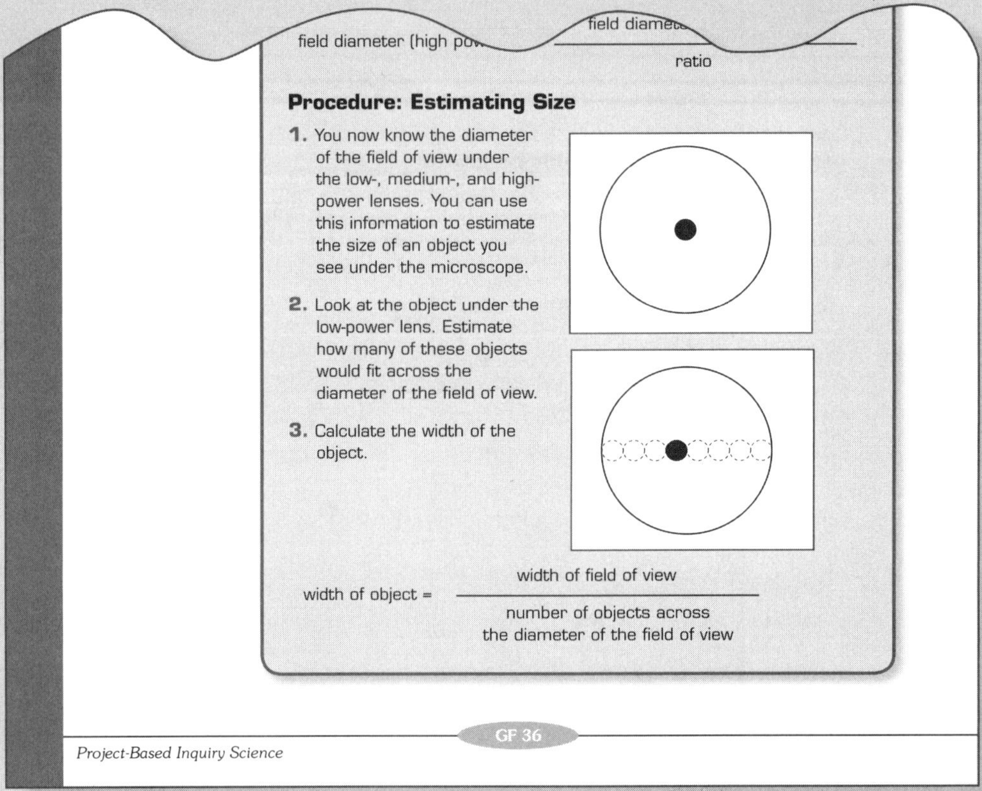
field diameter (high po

field diame

ratio

Procedure: Estimating Size

1. You now know the diameter of the field of view under the low-, medium-, and high-power lenses. You can use this information to estimate the size of an object you see under the microscope.
2. Look at the object under the low-power lens. Estimate how many of these objects would fit across the diameter of the field of view.
3. Calculate the width of the object.

$$\text{width of object} = \frac{\text{width of field of view}}{\text{number of objects across the diameter of the field of view}}$$

Project-Based Inquiry Science

GF 36

Procedure: Estimating Size

20 min.

Students estimate size by comparing an object's diameter with the diameter of the field of view.

Get Going

Have students read through the procedure. Ask students to look at the student text and estimate the diameter of the black dot as if this picture represented what they would see under low power. After students have had the opportunity to estimate the diameter of the black dot, ask for a volunteer to demonstrate on the board how he or she arrived at their answer. Have the class check these calculations, and make any necessary corrections. Next, ask students to estimate the diameter of the black dot if this picture represented what they would see under medium power.

Again, ask for a student volunteer to demonstrate on the board how he or she arrived at their answer and have the class review the calculations and suggest any needed corrections. Finally, ask students to estimate the diameter of the black dot if this picture represented what they would see under high power, have a volunteer share his or her calculations and have the class check the results. If time permits, you may want to consider distributing prepared slides and having students estimate the size of actual cells.

Assessment Options

Targeted Concepts, Skills, and Nature of Science	How do I know if students got it?
Scientists often collaborate and share their findings. Sharing findings makes new information available and helps scientists refine their ideas and build on others' ideas. When another person's or group's idea is used, credit needs to be given.	**ASK:** Which features of cell posters made by other groups were particularly effective in communicating information? **LISTEN:** Students should identify features such as detail in drawings, clear labeling, descriptive notes and use of graphic organizers.
Scientists must keep clear, accurate, and descriptive records of what they do so they can share their work with others, consider what they did, why they did it, and what they want to do next.	**ASK:** Look back at your notes and diagrams of the first prepared slide of an animal cell that you observed. If someone else looked at your notes and diagrams, how closely would their mental image of the cells you observed match what you actually observed? **LISTEN:** Students should be able to identify details missing in their notes and diagrams and recognize ways in which their recording of observations could be improved.

Targeted Concepts, Skills, and Nature of Science	How do I know if students got it?
Students should be able to manipulate a compound microscope to view microscopic objects.	**ASK:** Here is a prepared slide of some plant cells. Bring it into focus using your microscope. **LISTEN:** Watch the students for proper placement of slide on stage, the careful lowering of body tube, the adjustment of light, and the appropriate use of coarse and fine adjustment knobs. Observe the slides yourself to verify that the student has successfully brought it into focus.
Students should be able to prepare a wet-mount slide, use appropriate staining techniques, and determine the size of a microscopic object, using a compound microscope.	**ASK:** Your friend gets a microscope as a birthday gift and calls you on the telephone asking for your help in making a wet-mount slide of onion skin and staining it. Describe how you would guide your friend through the process. **LISTEN:** Students should be able to articulate the correct sequence of steps for making and staining a wet-mount slide. **ASK:** You observe one of your hairs under the microscope at medium power and estimate that you could fit 8 hairs across the field of view. The field of view of this microscope at medium power is 1.0 mm. Describe how you would calculate the approximate width of your hair. **LISTEN:** Students should state that they would divide the width of the field of view by the number of hairs that would fit across it (1mm ÷ 8 = 0.125mm).

Targeted Concepts, Skills, and Nature of Science	How do I know if students got it?
The cell is the basic unit of structure and function in all living things. Cells are usually microscopic in size.	**ASK:** Here is a photograph of some onion skin as seen under the microscope. Draw an outline around a single cell. **LISTEN:** Students should be able to successfully outline a single cell. **ASK:** Draw a diagram of what you think a thin slice of carrot would look like under the microscope. **LISTEN:** Student diagrams should reflect a cellular structure.
Cells have specialized structures that perform specific functions. Most cells have a cell membrane, cytoplasm and genetic material. Many cells have a nucleus. Some cells have cell walls and/or chloroplasts.	**ASK:** You are viewing a sample of some cells under the microscope. How could you tell if the sample came from a plant or an animal? **LISTEN:** Students should identify cell organelles that distinguish plant cells from animal cells. They should mention looking for the presence or absence of cell walls and chloroplasts.
All organisms require energy to survive. Cells obtain energy in a variety of ways.	**ASK:** Which cell organelle is involved in converting food energy into a form of energy the cell can use? **LISTEN:** Students should identify mitochondria in their response.

Teacher Reflection Questions

- How did the activities in this section contribute toward students' understanding that cells are systems with many parts that work together?
- Which microscope skills did students have the most difficulty with? How might you help them to improve those skills?

- In what ways did interspersing the reading selections in this section amid the activities make for a more effective approach toward learning?
- What difficulties, if any, did students have when updating the *Project Board* with information about cells? If they did have any difficulties, what other guiding questions could you ask them to assist them in making a connection about cells and what kinds of things make you sick?

NOTES

SECTION 2.4 INTRODUCTION

2.4 Read

Some Germs Are Single-Celled Organisms Called Bacteria

◀ *1 class period*

A class period is considered to be one 40 to 50 minute class.

Overview

In this section, students extend what they have learned about cells to one type of germ, bacteria. Students read about the cellular structure of bacteria and the distinction between prokaryotic and eukaryotic cells. Students then consider ways in which bacteria may be beneficial or harmful. Students read about why bacteria are considered to be alive and how their ability to adjust to changes in their environment enables them to survive under a wide range of conditions. Finally, students read about the ways in which bacteria can affect the body.

Targeted Concepts, Skills, and Nature of Science	Performance Expectations
Scientists often collaborate and share their findings. Sharing findings makes new information available and helps scientists refine their ideas and build on others' ideas. When another person's or group's idea is used, credit needs to be given.	Students should share their individual questions with their group and their group's questions with the class as they consider what they think they need to learn and what they need to investigate as they fill out the class Project Board. Students will also collaborate as they add information to the *Communicable-Disease Information* Table.
Some living things are single cells; others, including humans, are multicellular.	Students should be able to identify bacteria as unicellular organisms and distinguish between prokaryotes and eukaryotes.
Cells have specialized structures that perform specific functions. Most cells have a cell membrane, cytoplasm and genetic material. Many cells have a nucleus. Some cells have cell walls and/or chloroplasts.	Students should be able to describe the organelles found in bacteria.

Targeted Concepts, Skills, and Nature of Science	Performance Expectations
All organisms require energy to survive. Cells obtain energy in a variety of ways.	Students should be able to describe how bacteria obtain the energy they need to survive.
Homeostasis is the process by which an organism keeps its internal environment within the range required for survival despite changes in the external environment.	Students should be able to define metabolism and give several examples of how bacteria change their physiological activities to maintain homeostasis.
Many diseases are caused by bacteria or viruses.	Students should be able to identify bacteria as one type of germ and describe some of the negative effects that bacteria can have on the body.
Disease is a breakdown in the structures or functions of an organism. Some diseases are the result of failures of a body system. Other diseases are the result of damage by infection from other organisms (germs).	Students should be able to describe ways in which toxins produced by certain bacteria damage the body and can cause failures of a body system.
An antibiotic is a drug that kills or prevents the growth of bacteria.	Students should be able to identify how antibiotics can be used to treat illnesses caused by a microorganism.

Materials	
1 per class	Optional: projections of bacterial cell structure shown in text, images of bacteria showing the three shapes: spherical, rod-shaped, and spiral.

Homework Options

Reflection

- **Science Content:** How is a bacterial cell different from the cells of plants and animals? *Bacteria are prokaryotes. The genetic material in a prokaryotic cell is not contained in a nucleus.)*
- **Science Content:** Research a disease-causing bacterium and create a "Wanted Poster" for it on a single sheet of paper. Include a drawing of the bacterium, a description of the bacterium, how

it attacks and spreads, its hideout (where it is most likely found), most common injury to its victims, and how it is armed and dangerous (degree of damage caused).

- **Science Process:** A single bacterium enters a student's body with food eaten at 12 P.M. and begins to grow in the student's intestines. This particular type of bacteria can reproduce by dividing every 20 minutes. How many bacteria could that one bacterium have multiplied to by 5P.M.? *(32,768 – 3 divisions/hour x 5 hours = 15 divisions: 2, 4, 8, 16, 32, 64, 128, 256, 512, 1024, 2048, 4096, 8192, 16,384, 32,768)*

Preparation for 2.5

- **Science Content:** Based on what you now know about bacteria, what conditions would be favorable to their growth? *(Abundant food, moisture, warm temperatures, and an absence of chemical disinfectants or antibiotics – conditions opposite to the unfavorable ones that trigger spore development.)*

NOTES

NOTES

SECTION 2.4 IMPLEMENTATION

◀ *1 class period**

2.4 Read

Some Germs Are Single-Celled Organisms Called Bacteria

bacteria (singular, bacterium): the most common form of one-celled organisms.

eukaryote: an organism whose cells have nuclei.

prokaryote: a single-celled organism that does not have a nucleus.

microbe (or microorganism): an organism that can be seen only through a microscope.

The question you are investigating is *What kinds of things make you sick?* You know germs can make you sick. Some germs are **bacteria**. In this section, you will read about bacteria.

One-Celled Organisms: Bacteria

In the last section, you looked at several plant and animal cells. All the cells you looked at had a nucleus. Scientists call an organism made of cells that have nuclei a **eukaryote**. A eukaryote can be made of one cell or many cells.

There is another type of organism called a **prokaryote**. All prokaryotes are made of a single cell. However, these cells do not have nuclei. Bacteria (singular, bacterium) are examples of prokaryotes.

Look at the picture. It looks like the cells that you were observing through the microscope. However, notice the difference. There is no nucleus.

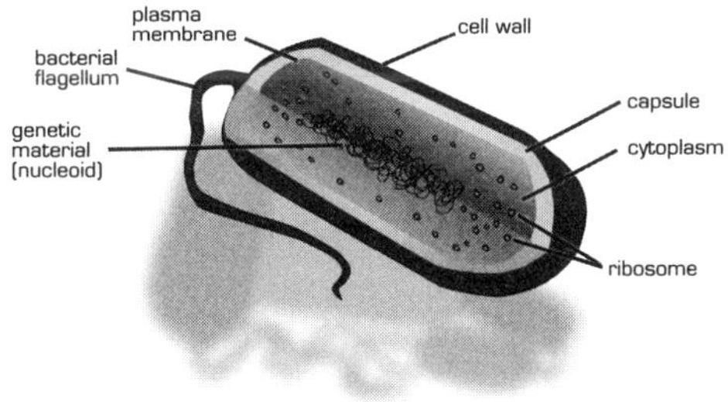

Bacteria and other very small organisms are also called **microorganisms** or **microbes**. These are organisms that can only be seen through the microscope.

Bacteria are found everywhere, and they can survive almost anything. Some kinds of bacteria live in your stomach, which has a lot of acid. Bacteria live on kitchen counters that are cleaned often. Bacteria can live in the harshest places in the world. They have been found in polar icecaps. They also live in hot-water vents at the bottom of the ocean.

There are many different kinds of bacteria. There are over 300 kinds of bacteria in your mouth alone. In your body, there are 10 times more bacteria than there are cells in your body.

2.4 Read

Some Germs Are Single-Celled Organisms Called Bacteria

5 min.

Students read about bacteria, single-celled organisms. Students are made aware of the differences between eukaryotes and prokaryotes, the structure of prokaryote bacteria cells and how they carry out various life functions. They also consider evidence that bacteria are alive, ways in which bacteria may be beneficial or harmful, and how some bacteria produce toxins that adversely affect the body.

○ Engage

Before students begin reading, ask them to recall the earlier activity in which they studied gastrointestinal illnesses. Remind students that bacteria are one type of germ that causes gastrointestinal illness. Ask students what they think bacteria look like on the cellular level. Have several students draw what they think bacteria look like for the class. Have the class select one drawing and edit it according to their suggestions.

*A class period is considered to be one 40 to 50 minute class.

META NOTES

This type of activity engages students in the topic by relating to their own life experiences and background knowledge. It also organizes the content by focusing on the major concepts they will be reading about.

TEACHER TALK

"In our earlier activity on gastrointestinal infections, we read that bacteria are one type of germ that causes such illnesses. Why is it reasonable to think that bacteria are made of cells? What do you think bacteria look like at the cellular level?"

Ask several volunteers to draw what they think bacteria look like on the board, and then have the class select one to discuss.

TEACHER TALK

"What changes to this drawing can you suggest, based on what we know about cells?"

Have students read the introduction and the definition of bacteria in the marginal box.

One-Celled Organisms: Bacteria

5 min.

Students read about prokaryotic bacteria cells and how they differ from eukaryotic cells.

The questio... investigating is *What k... make you sic...* You know germs can make you sick. Some germs are **bacteria**. In this section, you will read about bacteria.

One-Celled Organisms: Bacteria

In the last section, you looked at several plant and animal cells. All the cells you looked at had a nucleus. Scientists call an organism made of cells that have nuclei a **eukaryote**. A eukaryote can be made of one cell or many cells.

There is another type of organism called a **prokaryote**. All prokaryotes are made of a single cell. However, these cells do not have nuclei. Bacteria (singular, bacterium) are examples of prokaryotes.

Look at the picture. It looks like the cells that you were observing through the microscope. However, notice the difference. There is no nucleus.

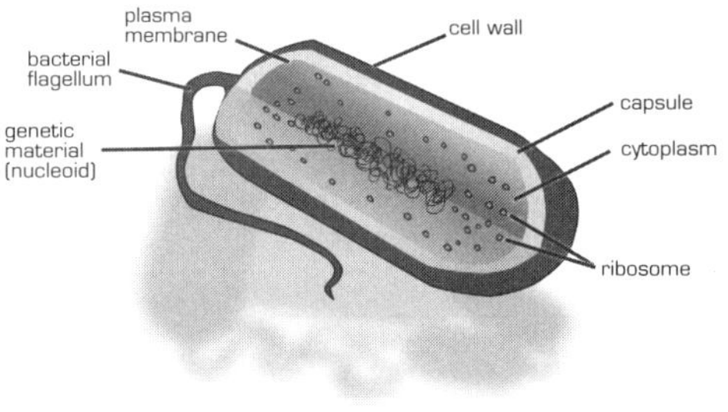

Bacteria and other very small organisms are also called **microorganisms** or **microbes**. These are organisms that can only be seen through the microscope.

Bacteria are found everywhere, and they can survive almost anything. Some kinds of bacteria live in your stomach, which has a lot of acid. Bacteria live on kitchen counters that are cleaned often. Bacteria can live in the harshest places in the world. They have been found in polar icecaps. They also live in hot-water vents at the bottom of the ocean.

There are many different kinds of bacteria. There are over 300 kinds of bacteria in your mouth alone. In your body, there are 10 times more bacteria than there are cells in your body.

...: an organism whose cells have nuclei.

prokaryote: a single-celled organism that does not have a nucleus.

microbe (or microorganism): an organism that can be seen only through a microscope.

GF 37 GOOD FRIENDS and GERMS

△ Guide
Before students read this section, project an image of the bacterium shown in the text. Ask students to compare and contrast its structure and organelles with that of the animal and plant cells they observed in the last activity. Create a class list of similarities and differences.

⬡ Get Going
Have students read the first three paragraphs about eukaryotes and prokaryotes.

□ Assess
Assess whether students understand the distinction between eukaryotes and prokaryotes by asking them whether an onion would be classified as a eukaryote or a prokaryote and to provide evidence to support their answer. Point out that even though onion cells and bacteria both have a cell wall, onion cells have a nucleus and bacteria do not.

△ Guide
Ask students where they think bacteria might live. Make a class list of their ideas.

⬡ Get Going
Have students read to the end of this section and emphasize that bacteria can live under a wide variety of environmental conditions. Update the class list and ask students which of the places that bacteria can live are most surprising.

NOTES

Bacteria: Good or Bad Guys?

5 min.

Students read about beneficial and harmful bacteria.

Bacteria: Good or Bad Guys?

You know that some bacteria can make you sick. However, some bacteria in your body actually keep you healthy. Some bacteria in your digestive system keep you healthy by breaking down food materials. Other bacteria in your digestive system produce vitamins that your body needs to stay healthy.

People also use bacteria for a wide variety of purposes. The dairy industry uses bacteria when making cheese, yogurt, sour cream, and other products. Bacteria are also used to break down waste in sewage-treatment plants, clean up oil spills, and develop antibiotics.

Bacteria play important roles in the natural world. Bacteria live in the soil. They help plants grow successfully. Bacteria help make gardens and farms more fertile. They make it possible for elements such as carbon, nitrogen, and oxygen to be returned to the atmosphere by decomposing waste and dead organisms. If there were no bacteria, the world would not look the way it does.

Did You Know?

Parasites are organisms that benefit by living on or in another organism. Some bacteria and **fungi** are parasites. Fungi are eukaryotes that have cell walls, reproduce by **spores**, and get food by absorbing it from their surroundings. One type of fungus causes athlete's foot, a disease that affects the outer layer of human skin.

Are Bacteria Alive?

Bacteria are single-celled organisms. Just like all other organisms, bacteria need energy and materials to stay alive, grow, develop, and reproduce. They obtain the energy and the nutrients they need from the food they eat. Most bacteria obtain their food from other living organisms. This is how many of the disease-causing bacteria obtain their food. You are one of the organisms from which they obtain food. Some bacteria obtain their food from dead organisms, and a few can make their own food.

Bacteria use **enzymes** to break down this food. They take up nutrients through their membranes. Bacteria then convert nutrients into energy. The process by which animals break down their food to obtain energy and then use the energy in their life processes is called **metabolism**.

Bacteria get larger by using energy to grow. Each cell grows larger. Then it divides into new cells. One bacterium will grow and divide into two bacteria. This is a way bacteria reproduce. Then each of the two bacteria divides, making four bacteria cells. Each of these four bacteria cells divides to make eight. These divisions can happen very quickly in a favorable environment. What a favorable environment is varies among the different bacteria. After many divisions, a colony of bacteria has developed. There are many individual bacteria cells in a colony. Bacteria continue to divide until they have used up

fungi (singular, fungus): eukaryotes that have cell walls, reproduce by spores, and get food by absorbing it from their surroundings

enzyme: a substance that causes a chemical change in another substance.

△ Guide

Before students read this section, ask them whether they think that bacteria are good guys or bad guys and create a class list of their ideas.

⬡ Get Going

Have students read through this section in their groups noting updates for the class list or have them make a new list of ways in which bacteria are helpful and harmful.

Hold a class discussion of ways in which bacteria are useful. Emphasize that only the bad bacteria are considered germs, and that bacteria play an important role in the living environment.

Are Bacteria Alive?

5 min.

Students read about the life functions of bacteria.

△ Guide

Before students read this section, take a poll of how many students think bacteria are alive and how many do not think bacteria are alive. Ask the class how they would decide if something was alive. Ask for examples of criteria something would have to satisfy in order to be considered alive. Create a class list of their ideas.

NOTES

all their food or created too much waste to survive.

Like all other organisms, bacteria need to adjust to a changing external environment. They need ways to keep their internal environment —the amount of fluid, the materials, and the temperature—inside the cell fairly constant. The process of keeping a constant internal environment is called **homeostasis**. Bacteria have developed various ways of doing this.

One important way that many bacteria respond to their environment is **spore** formation. When conditions for bacterial growth are not favorable, these bacteria form a tough capsule around their genetic material and some cytoplasm, forming a spore. The spore can remain **dormant** for years. When conditions return to favorable, the bacteria will begin to grow and reproduce again. Spore formation allows some bacteria to survive extremely harsh conditions—lack of food, extreme temperatures, dryness, and chemical **disinfectants**.

Look at the pictures of different types of bacteria. Most bacteria have one of three shapes: rod-shaped, spherical, or spiral. The string-like parts in the bacteria in the bottom picture are called **flagella** (singular, flagellum). The bacterium uses those to move.

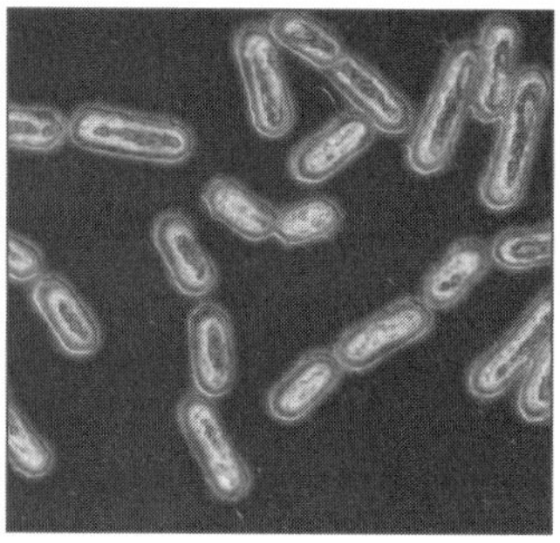

Rod-shaped bacteria (bacillus).

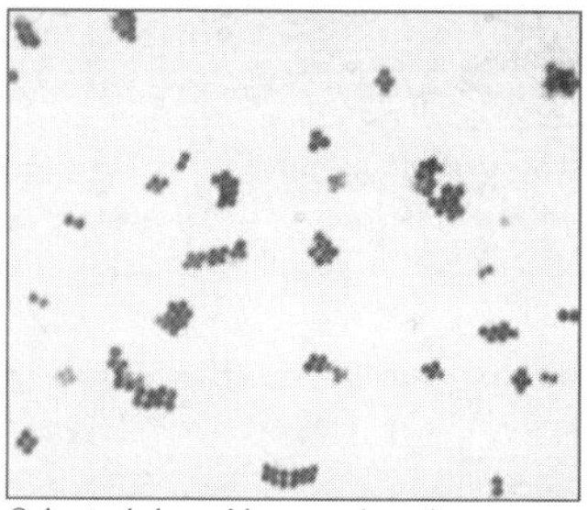

Spherical-shaped bacteria (cocci).

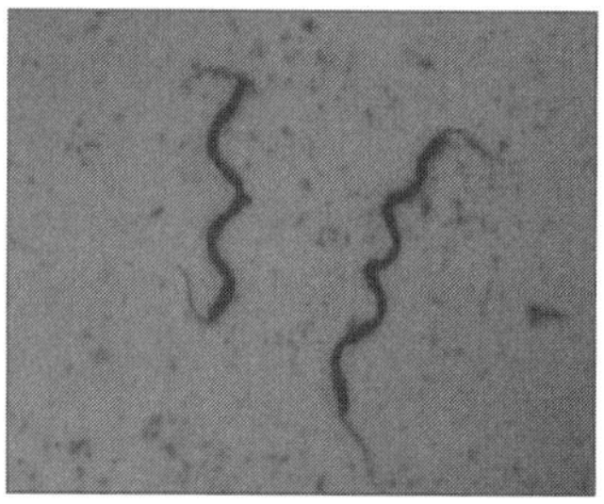

Spiral-shaped bacteria (spirilla).

metabolism: the combination of chemical reactions that takes place in an organism; food is converted into energy that the organism uses to carry out its life processes.

colony: a group of only one kind of bacteria that grows from a single, original bacterium.

homeostasis: the process by which an organism keeps its internal environment in a constant condition despite changes in its external environment.

spore: in bacteria, a dormant structure that allows the bacterial cell to survive unfavorable conditions. In fungi, a cell that develops into a new organism.

dormant: temporarily not active.

disinfectant: a substance that can kill microorganisms.

flagellum (plural, flagella): a string-like part on some cells that helps the cell move.

⬡ Get Going

Read through this section by having students read aloud or in pairs. After they have completed the reading, ask students to refer back to their criteria for being alive, and ask them to cite evidence from what they have read that would support a claim that bacteria are alive. Ask students to describe how bacteria reproduce and form colonies. Then discuss how bacteria have adapted to survive when the environment changes through spore formation. Review the three common shapes of bacteria and how bacteria use flagella to move.

How Do Bacteria Make You Sick?

toxin: a poisonous substance.

toxic: poisonous.

protein: a substance that is found in all cells; it is necessary for life.

What happens after bacteria are in your body depends on the type of bacteria. Some cannot survive inside you, so they cannot harm you. But other bacteria can survive. The bacteria that survive begin to reproduce inside you. In larger numbers, they can affect your body. Some of these bacteria produce **toxins** (**toxic** substances). Toxins can have different effects on your body. Some can make you vomit or cause diarrhea. Others can affect nerve cells and prevent signals from passing through your nerves. This can cause paralysis. Other bacteria can actually invade and destroy your cells. One type of bacteria that attacks and destroys the cells of your intestines also causes severe diarrhea. Some bacteria can produce specific **proteins** to which your body reacts. Your body's reaction to the protein makes you sick. As you investigate the disease you will select, you will find out how the type of bacteria that causes the disease affects the human body.

Stop and Think

1. Why are some organisms called microorganisms or microbes?
2. What is the difference between a eukaryote and a prokaryote?
3. What type of organism is bacteria, a eukaryote or a prokaryote?
4. Describe three ways that bacteria benefit humans.
5. Why is it important to know that some bacteria can form spores when you are thinking about how to prevent others from getting sick?
6. Explain two ways that bacteria can make you sick.
7. Describe a situation in which a bacteria-free environment would be
   a) desirable.
   b) not desirable.

What's the Point?

Bacteria are single-celled organisms. Like all living things, bacteria need to maintain a constant internal environment. This process is called homeostasis. Bacteria have three basic shapes — rod, spherical, or spiral. Bacteria can be found in most places in the world. Without bacteria, the world would not work as it does. The bacteria that live in your body do important jobs to keep you healthy. Some bacteria can be very dangerous and can make people sick by releasing toxins or destroying tissues.

How Do Bacteria Make You Sick?

5 min.

Students read about the harmful effects of certain bacteria on cells and body systems.

△ Guide

Have students read through this section and ask them to update their list of how bacteria are harmful. Guide students to make the connection between the damage to cells caused by bacteria and the malfunctioning of a body system.

You may want to briefly change the tone and pace of the classroom discussion by giving the students a background of the science that they have been and will be experimenting with. Tell students that the Petri dish was invented in 1887 by the German bacteriologist Julius Petri, while he was working as an assistant to Robert Koch. Robert Koch became famous for isolating the bacteria that cause anthrax, tuberculosis, and cholera. He also developed Koch's Postulates – four criteria for establishing the causal relationship between a microbe and a disease. This will give students insight to how the process of science and discovery has evolved, along with showing how some things affect one another in a variety of ways.

☐ Assess

Assess students' understanding of the mechanisms by which bacteria can cause sickness. For example, some toxins produced by bacteria can prevent signals from passing through nerve cells. Ask students to predict some of the ways in which blocked nerve signals might affect the body. Some bacteria invade and destroy cells. Ask students to predict how destruction of skin or muscle cells might affect the body.

Stop and Think

10 min.

Discuss the distinctions between eukaryotes and prokaryotes and the ways in which bacteria can be harmful or beneficial using the Stop and Think questions.

your cells. One type of bac... attacks and destroy... of your intestines also causes severe diarrhea. Some bacteria can produce specific **proteins** to which your body reacts. Your body's reaction to the protein makes you sick. As you investigate the disease you will select, you will find out how the type of bacteria that causes the disease affects the human body.

Stop and Think

1. Why are some organisms called microorganisms or microbes?
2. What is the difference between a eukaryote and a prokaryote?
3. What type of organism is bacteria, a eukaryote or a prokaryote?
4. Describe three ways that bacteria benefit humans.
5. Why is it important to know that some bacteria can form spores when you are thinking about how to prevent others from getting sick?
6. Explain two ways that bacteria can make you sick.
7. Describe a situation in which a bacteria-free environment would be
   a) desirable.
   b) not desirable.

What's the Point?

Bacteria are single-celled organisms. Like all living things, bacteria need to maintain a constant internal environment. This process is called homeostasis. Bacteria have three basic shapes — rod, spherical, or spiral. Bacteria can be found in most places in the world. Without bacteria, the world would not ...

⬡ Get Going

Have students record their answers to the seven Stop and Think questions. When students have finished answering the seven questions, hold a brief class discussion in which students share their answers.

☐ Assess

Assess whether students' responses to the *Stop and Think* questions contain the following information:

1. Microorganisms or microbes can only be seen through the microscope.
2. A eukaryote has cells that have their genetic material enclosed in a nucleus; a prokaryote has cells that contain genetic material, but it is not enclosed in a nucleus.
3. Bacteria are considered prokaryotes because their genetic material is not enclosed in a nucleus.
4. Ways in which bacteria benefit humans include:
   - helping break down food materials;
   - producing vitamins;
   - being involved in making foods such as yogurt, cheese, and sour cream;
   - breaking down wastes in sewage treatment plants;
   - helping clean up oil spills;
   - being used to develop antibiotics;
   - fix nitrogen in the soil which helps plants grow more successfully;
   - decomposing waste and dead organisms.
5. Spores can remain dormant and survive under extreme conditions. Spores may cling to objects, be carried by wind or water, and contaminate air that is breathed in or food that is ingested. Once inside the body, the spore may begin to grow and cause a bacterial infection.
6. Bacteria can make you sick by invading and destroying cells, and by producing toxins that have adverse affects on the body.
7. a) A bacterial-free environment would be desirable in situations where they have a harmful effect, such as when trying to prevent the spread of a disease-causing bacteria.

   b) It is not desirable in situations where bacteria have a beneficial effect, such as when trying to produce certain foods, or decompose wastes of dead organisms.

Assessment Options

Targeted Concepts, Skills, and Nature of Science	How do I know if students got it?
Scientists often collaborate and share their findings. Sharing findings makes new information available and helps scientists refine their ideas and build on others' ideas. When another person's or group's idea is used, credit needs to be given.	**ASK:** What responses from other students to your questions were most helpful in revising your own questions? **LISTEN:** Students should cite examples of valid comments relating to meeting the criteria for questions.
Some living things are single cells; others, including humans, are multicellular.	**ASK:** How do prokaryotes and eukaryotes differ? Give one example of each. **LISTEN: S**tudents should describe prokaryotes as unicellular organisms that do not have a nucleus. Students should be able to identify bacteria as prokaryotes and plants or animals as eukaryotes**.**
Cells have specialized structures that perform specific functions. Most cells have a cell membrane, cytoplasm and genetic material. Many cells have a nucleus. Some cells have cell walls and/or chloroplasts.	**ASK:** Which cell organelles are present in single-celled bacteria? **LISTEN:** Students should be able to identify the organelles found in bacteria such as a cell wall, flagellum, cell membrane, cytoplasm, ribosomes, and genetic material.
All organisms require energy to survive. Cells obtain energy in a variety of ways.	**ASK:** How do bacteria obtain energy? **LISTEN:** Students should be able to describe how bacteria obtain the energy they need to survive by absorbing food through their membranes and then breaking it down using enzymes.

Targeted Concepts, Skills, and Nature of Science	How do I know if students got it?
Metabolism is the sum of all chemical reactions in an organism.	**ASK:** What is meant by metabolism? **LISTEN:** Students should be able to define metabolism as the chemical reactions that take place in an organism.
Homeostasis is the process by which an organism keeps its internal environment within the range required for survival despite changes in the external environment.	**ASK:** Describe one way in which bacteria maintain homeostasis. **LISTEN:** Students should be able to cite examples such as spore formation when environmental conditions become unfavorable and using flagella to move in response to changes in the environment.
Many diseases are caused by bacteria or viruses.	**ASK:** How do bacteria make you sick? **LISTEN:** Students' responses should include invasion and destruction of cells and the production of toxins.
Disease is a breakdown in the structures or functions of an organism. Some diseases are the result of failures of a body system. Other diseases are the result of damage by infection from other organisms (germs).	**ASK:** What are some of the ways in which the toxins produced by bacteria affect the body? **LISTEN:** Students should be able to describe effects of the toxins such as vomiting, diarrhea, paralysis and destruction of cells.
An antibiotic is a drug that kills or prevents the growth of bacteria.	**ASK:** How does treating a person with antibiotics help fight a bacterial infection? **LISTEN:** Students should recognize that antibiotics kill bacteria or prevent them from growing.

Teacher Reflection Questions

- What insights into your students' conceptions of life were revealed as you discussed the question *Are bacteria alive?* What misconceptions do they hold that will need to be addressed in future activities? How might these misconceptions be addressed?
- How might you determine whether students understood the ways in which some bacteria are harmful while others are beneficial?
- Assessments become formative when the information obtained from them is used to adapt teaching and learning to meet student needs. In what ways were your assessments during this activity formative?

NOTES

SECTION 2.5 INTRODUCTION

2.5 Case Study

Analyzing Catering-Service Sickness Data

◀ *2 class periods*

A class period is considered to be one 40 to 50 minute class.

Overview

In this section, students apply what they have learned about bacteria to a real-life situation. They examine a case study of a restaurant catering service whose customers complain that they are getting sick after eating the restaurant's food. Students analyze data collected about the conditions under which food is stored and the number of customers that became sick after eating food kept in each of the storage conditions in order to determine the cause of the sickness. Students then present their conclusions and supporting evidence to the class in an *Investigation Expo.*

Targeted Concepts, Skills, and Nature of Science	Performance Expectations
Scientists often collaborate and share their findings. Sharing findings makes new information available and helps scientists refine their ideas and build on others' ideas. When another person's or group's idea is used, credit needs to be given.	Students should share their data analysis with their group and their group's analyses with the class as they try to determine the cause of the gastrointestinal illness.
Scientists must keep clear, accurate, and descriptive records of what they do so they can share their work with others, consider what they did, why they did it, and what they want to do next.	Students reorganize data to more clearly show relationships among variables and record their conclusions in order to share their ideas with the class.
Identifying factors that lead to variation is an important part of a scientific investigation.	Students should be able to identify the variables represented by the data collected in the case study.
Explanations are claims supported by evidence, accepted ideas, and facts.	Students should be able to support their conclusions about the cause of the illness with evidence drawn from the data.

Targeted Concepts, Skills, and Nature of Science	Performance Expectations
Many diseases are caused by bacteria or viruses.	Students should be able to make connections between illness and bacteria and identify conditions that are favorable and unfavorable to their growth.
Students should be able to recognize and analyze patterns and trends.	Students should be able to analyze the data provided in the case study and reorganize it to more clearly show trends they have identified.

Materials	
1 per class	projection of *Uncle Joe's Catering-Service Sickness* Data

Homework Options

Reflection

- **Science Content:** What information from your previous reading about bacteria played a role in your analysis of the data in this case study? *(Warm temperatures are favorable to the growth of bacteria. Some bacteria can remain dormant as spores until conditions are favorable to their growth.)*
- **Science Process:** When Darrell set up the data table in the book, he organized it by date. Why is the date not a good basis for organizing the data in this case study? *(There is no cause-and-effect relationship between bacteria and the date. There is, however, a plausible cause-and-effect relationship between temperature and bacterial growth, and a cause-and-effect relationship between storage location and temperature.)*
- **Science Process:** What other variables could Darrell have tracked for his investigation? How would that information have been useful to the investigation? *(Additional variables include, but are not limited to the following examples: Which foods were eaten by those who became sick could help identify which food was infected. Persons who did not eat the food that became sick after coming into contact with a sick person could help determine whether the illness is a communicable disease.)*

Preparation for 2.6

- **Science Content:** If you wanted to grow bacteria in a laboratory, what kind of environment would you provide for the bacteria? *(Warm, moist, abundant food, absence of chemical disinfectants or antibiotics)*

SECTION 2.5 IMPLEMENTATION

◄ *2 class periods**

2.5 Case Study

Analyzing Catering-Service Sickness Data

In this section, you are going to read about Darrell. Darrell and his uncle are trying to find out why customers of his uncle's catering business are getting sick. Read the case carefully. As you are reading, be sure to use the *Communicable-Disease Information Table* as a resource to understand what could be making the people who buy food in Darrell's uncle's restaurant sick.

Examine a Case Study

One summer, Darrell got a job in his Uncle Joe's restaurant. His uncle makes great chicken and coleslaw. Near the restaurant is a city park. Each summer, families meet at the park for reunions. They often purchase chicken and coleslaw from Darrell's uncle for their reunions. Darrell's uncle was concerned because several people had called the restaurant the previous summer to complain about getting sick after eating his restaurant's food.

2.5 Case Study

Analyzing Catering-Service Sickness Data

5 min.

Students connect prior knowledge of gastrointestinal illness with a real-life situation.

○ Engage

Before reading this section, have students take out their *Communicable-Disease Information Table* and review information about gastrointestinal illnesses that they recorded during previous activities. Have students share what they know about gastrointestinal illnesses. Have students recall that bacteria are one of the causes of gastrointestinal illness.

*A class period is considered to be one 40 to 50 minute class.

Darrell's uncle uses the freshest products to make his chicken and coleslaw. He is not sure why people are getting sick. Darrell wonders, "What is causing people to get sick at the family reunions?"

Darrell knows that bacteria on food can cause gastrointestinal illness. He knows that bacteria reproduce quickly in a warm environment. Darrell's uncle asks people to keep their coleslaw and chicken in a cooler during the reunion. The cooler helps keep the food at the right temperature for safety. But maybe all the people do not follow his instructions. Darrell decides to collect data on how the food is stored and at what temperature it is kept. He thinks it is also important to know the outside temperature.

Darrell has noticed that sometimes people put the food out on the table rather than leaving it in the cooler. Every time he makes a delivery to a reunion, Darrell watches to see if the food is stored in a cooler or placed on the table. He records the daily outside temperature and where the food is stored. In addition, if people call to complain, he politely asks them how many people got sick at the party. After collecting data for three weeks, he shared it with his uncle. Darrell thought a table would be a good way to display his information.

Uncle Joe's Catering Service Sickness Data

Date	Observations about where food is stored	Outside temperature	Number of sick people
July 1	In cooler	77°	0
July 3	On table	76°	1
July 5	On table	80°	2
July 7	On table	82°	4
July 9	In cooler	82°	1
July 11	In cooler	83°	0
July 13	On table	81°	1
July 15	On table	86°	4
July 17	On table	84°	2
July 19	In cooler	84°	0
July 21	On table	87°	6

△ Guide

Have students read the introduction and encourage them to refer to the *Communicable Disease Information Table* throughout their examination of the case study, and also to update it with any new information they come across in their reading.

are trying [illegible] at why customers of his uncle's [illegible]ring business are getting sick. Read the case carefully. As you are reading, be sure to use the *Communicable-Disease Information Table* as a resource to understand what could be making the people who buy food in Darrell's uncle's restaurant sick.

Examine a Case Study

One summer, Darrell got a job in his Uncle Joe's restaurant. His uncle makes great chicken and coleslaw. Near the restaurant is a city park. Each summer, families meet at the park for reunions. They often purchase chicken and coleslaw from Darrell's uncle for their reunions. Darrell's uncle was concerned because several people had called the restaurant the previous summer to complain about getting sick after eating his restaurant's food.

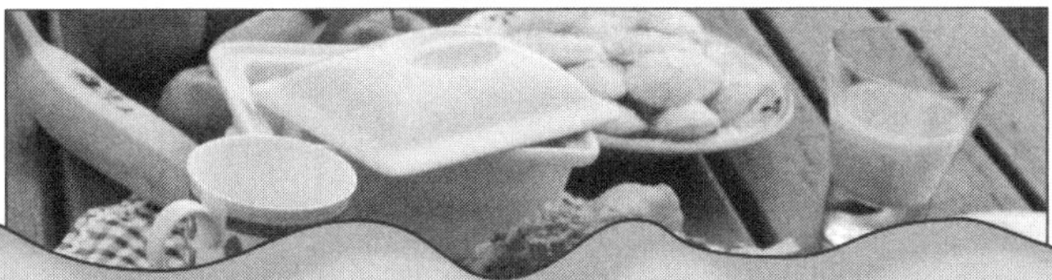

Examine a Case Study

15 min.

Students examine a case study of people getting ill after eating food from a restaurant.

⬡ Get Going

Have students read the case study. You might want students to read individually or take turns reading within their groups. After students have read the case study, ask them to discuss with their group whether Darrell's question could be answered by scientific investigation. Ask students to consider whether there is a clear connection between Darrell's question and the data he collected. Have students recall that questions that can be answered through scientific investigations are directed toward objects and events that can be described, explained or predicted by the investigation.

Lead a class discussion focusing on the weaknesses of the question *"What is causing people to get sick at the family reunions?"* Help students to develop the understanding that Darrell's question is too broad and ill defined to direct a scientific investigation. The question does not clearly indicate how observations or measurements and the collecting and recording of data could lead to an answer.

Ask students to suggest how the question could be refocused or clarified. Guide students in the development of alternatives such as: *"What effect does the method of food storage have on the number of people who get sick?"* or *"How do outside temperatures affect the number of people who get sick?"* Help students to recognize that these examples clearly identify the variables that can be observed or measured to answer the question.

Stop and Think

5 min.

Students identify the variables that were tracked in the investigation and tell why they were chosen.

Stop and Think

1. What variables did Darrell track for his investigation?
2. Why did Darrell record the outside temperature?

Analyze Your Data

The table shows the data Darrell collected. The data should be organized so Darrell's uncle can make sense of it to answer the question. He needs to be able to see differences or make comparisons between different variables. Organize the data so that you can make comparisons or see differences.

Then make three statements about what the data tells you about why the people were getting sick. What will you tell Darrell's uncle about the food to keep other people from getting sick? Record your recommendation and be prepared to share it with others.

Reflect

1. Why do you think your way of organizing information would help Darrell's uncle understand the data so he could answer the question?
2. What does this investigation have to do with bacteria and spreading disease?
3. Can you tell from the case study if the illness people got might be a communicable disease?

Communicate Your Results

Investigation Expo

In your small group, discuss what conclusions you can draw from the data. Record your conclusions and be ready to present them to the class in an *Investigation Expo*.

As a class, review and discuss the conclusions drawn by each group. What data did they use to support their conclusions? Come up with conclusions the whole class agrees upon.

Get Going

Have students record their answers to the *Stop and Think* questions. When students have finished answering the two questions, hold a brief class discussion in which students share their answers. Ask students to support their answers with evidence from the reading or the data table.

☐ Assess

Assess whether students' responses to the questions contain the following information:

1. Food storage location, outside temperature and number of sick people.
2. Darrell knows that bacteria on food can cause gastrointestinal illness and that bacteria reproduce quickly in a warm environment. The food out on the table is at the outside temperature.

⬡ Get Going

Analyze the Data
10 min.

Students analyze the data to identify relationships between variables.

Project the *Uncle Joe's Catering Service Sickness Data* table that Darrell constructed. Have students analyze the data, trying to identify any relationships that exist between the variables in the data table. Encourage students to reorganize the data table so that the relationship between variables that they have identified is clearly shown. Then ask students to work individually to write their three statements about what the data tells them about why the people were getting sick. Make it clear to students that these statements should not be opinions or predictions — just three factual statements of relationships shown by the data. Tell students that they should be able to back up each of their statements with evidence drawn from the data. Ask students to make a recommendation to Darrell's uncle about the food that will help keep other people from getting sick. Tell students to record their recommendation and to be prepared to share it with others later.

△ Guide and Assess

Reflect
5 min.

Students consider how reorganizing the information in the data table helps them identify relationships and connect this investigation with the spread of disease.

Have students record their answers to the *Reflect* questions. Facilitate a brief class discussion checking for students' comprehension of how data can be manipulated to reveal relationships among variables. Guide students in connecting the data back to the original question: *What is causing people to get sick at the family reunions?*

Assess students' understanding of both the data and communicable disease by listening to their responses to the question *"Can you tell from the case study if the illness people got might be a communicable disease?"* Students' responses should indicate that since no data was collected about contact between affected and unaffected persons, there is no basis for concluding that the illness was spread from person to person.

NOTE: This is an opportune time to conclude the first class session of this activity.

Communicate Your Results

20 min.

Students discuss conclusions that can be drawn from the data presented to the class.

Reflect

1. Why do you think your way of organizing information would help Darrell's uncle understand the data so he could answer the question?
2. What does this investigation have to do with bacteria and spreading disease?
3. Can you tell from the case study if the illness people got might be a communicable disease?

Communicate Your Results

Investigation Expo

In your small group, discuss what conclusions you can draw from the data. Record your conclusions and be ready to present them to the class in an *Investigation Expo*.

As a class, review and discuss the conclusions drawn by each group. What data did they use to support their conclusions? Come up with conclusions the whole class agrees upon.

GF 43 GOOD FRIENDS and GERMS

△ Guide

Have students work in small groups to discuss what conclusions can be reached based on the data. Remind students that the conclusions should answer the question, and should be supported by evidence. Groups should write down their conclusions and prepare to present them to the class during an *Investigation Expo.*

⬡ Get Going

Ask groups to present their conclusions and supporting evidence to the class. Ask the class to determine whether the data presented by the group supported their conclusions.

After all groups have presented their conclusions, facilitate a class discussion that leads to a consensus about which conclusions are best supported by the data. Have students agree upon a set of recommendations that could be made to Uncle Joe about how to keep his customers from getting sick.

△ Guide

Ask the class to step back from the data analysis and reflect on the process in which they have engaged. Help students to recognize that they have reviewed an investigation carried out by someone else, and have collaborated in analyzing the data, drawing conclusions, discussing the results, and reaching a consensus about what conclusions can be reached. Point out to students that this is the way in which scientists work.

Conference

The *Project Board* is beginning to take shape. You need to add the information you found about bacteria and cells to the *Project Board*. What information is most important to add to the board? Write three things you think are important. Next to each of your ideas, write in which column of the *Project Board* you think your information belongs.

Also, now that you have learned more about cells and bacteria, write two new questions. What questions are you interested in answering?

Update the *Project Board*

Share the three things you think are important to add to the *Project Board* with the class. As a class, decide which information about cells and bacteria is important to add to the *Project Board*. If you add a fact to the *What are we learning?* column, be sure you have evidence to support it. Record this in the *What is our evidence?* column.

Then discuss the questions that you and your class think need to be answered about what makes you sick. Record these in the *What do we need to investigate?* column. Keep a record of what was added to the *Project Board* on your own *Project Board* page.

What's the Point?

In the last section, you learned that under favorable conditions, bacteria reproduce and multiply very quickly. Bacteria that cause disease multiply quickly in warm, moist conditions. The food that Darrell's uncle sold probably contained very few bacteria. However, if the customers did not keep the food cold, a few bacteria would very quickly have grown to become many bacteria in warm temperatures. When the customers ate the food containing all these bacteria, they got sick.

Conference

5 min.

Students consider what information should be added to the Project Board.

△ Guide

Remind students that they have just completed two activities that involved bacteria in which they have been presented with new information. Point out that before moving on, they need to add this information to the *Project Board.* Have students think about what they have learned and identify information that they think is important.

Get Going

Ask students to write down the three things they have learned that are most important to add to the *Project Board*. Make clear to students that next to each of their ideas, they should identify the column on the *Project Board* in which it should be recorded. Then ask students to formulate two new questions that need to be investigated, based on what they have learned about cells and bacteria.

Update the *Project Board*

5 min.

Students add newly learned information to the Project Board.

you think are important. Next to each of your ideas, write in which column of the *Project Board* you think your information belongs.

Also, now that you have learned more about cells and bacteria, write two new questions. What questions are you interested in answering?

Update the *Project Board*

Share the three things you think are important to add to the *Project Board* with the class. As a class, decide which information about cells and bacteria is important to add to the *Project Board*. If you add a fact to the *What are we learning?* column, be sure you have evidence to support it. Record this in the *What is our evidence?* column.

Then discuss the questions that you and your class think need to be answered about what makes you sick. Record these in the *What do we need to investigate?* column. Keep a record of what was added to the *Project Board* on your own *Project Board* page.

What's the Point?

In the last section, you learned that under favorable conditions, bacteria reproduce and multiply very quickly. Bacteria that cause disease multiply

Get Going

Have students share their three ideas with the class. Have the class decide which information about cells and bacteria is important to add to the *Project Board*. Make sure that for every fact added to the *What are we learning?* column, there is corresponding supporting evidence recorded in the What is our evidence? column. Have students share their questions and lead a class discussion leading to a consensus as to which questions need to be answered about what makes you sick. Add these questions to the *What do we need to investigate?* column. Remind students to keep a record of all of these additions on their own *Project Board* page.

Assessment Options

Targeted Concepts, Skills, and Nature of Science	How do I know if students got it?
Scientists often collaborate and share their findings. Sharing findings makes new information available and helps scientists refine their ideas and build on others' ideas. When another person's or group's idea is used, credit needs to be given.	**ASK:** Which ideas proposed by other groups did you find most useful in refining your own data analysis? **LISTEN:** Students should cite ideas such as focusing on variables that relate to the question.
Scientists must keep clear, accurate, and descriptive records of what they do so they can share their work with others, consider what they did, why they did it, and what they want to do next.	**ASK:** Why did your groups choose to organize your data in the way that you did? **LISTEN:** Students should make the connection between their method of organizing the data and more clearly showing relationships among variables that supported their conclusions.
Identifying factors that lead to variation is an important part of a scientific investigation.	**ASK:** What were the variables that Darrell tracked in his investigation? **LISTEN:** Students should be able to identify the date, food storage location, and the number of people that became ill.
Explanations are claims supported by evidence, accepted ideas, and facts.	**ASK:** What claims can you make about the cause of the gastrointestinal illness reported by customers of Uncle Joe's Restaurant and what is your supporting evidence? **LISTEN:** Students should be able to support their claims about the cause of the illness with evidence drawn from the data.

Targeted Concepts, Skills, and Nature of Science	How do I know if students got it?
Many diseases are caused by bacteria or viruses.	**ASK:** What evidence would support the claim that bacteria caused the gastrointestinal illness reported by customers of Uncle Joe's Restaurant? **LISTEN:** Students should identify knowledge from previous readings and recognize that the illness was associated with conditions that are favorable to the growth of bacteria – moisture and warm temperatures.
Students should be able to recognize and analyze patterns and trends.	**ASK:** Identify one trend that is shown by the data. **LISTEN:** Students should identify trends such as outside temperature increased, the number of sick people increased, fewer people became sick from food stored in the cooler than from food left out on the table.

Teacher Reflection Questions

- How could you tell that students understood the importance of supporting their conclusions with evidence? What opinions or overgeneralizations did your students include in their conclusions?
- Social interactions are powerful learning tools. Students who create positive bonds with peers build a strong self image and confidence necessary for success. What activities in this section fostered social interactions that created positive bonds between students?
- Managing group and class discussions can be challenging. When managing discussions, it is important to foster thoughtful criticism of ideas while maintaining respectful interactions. Were there instances in your class discussions when students were not interacting respectfully? How could you help students develop better social interaction skills?

SECTION 2.6 INTRODUCTION

2.6 Investigate

Where in Your Classroom Can You Find Bacteria?

◀ *2 class periods*

A class period is considered to be one 40 to 50 minute class.

Overview

In this section, students collaborate as a class to design an investigation to determine where bacteria can be found in the classroom. With you modeling best practices, the class develops a question to guide the investigation, predicts the outcome, identifies the independent and dependent variables as well as any controls, and develops a set of procedures. Students then run the investigation, collect the data, analyze it, and share the results with the class.

Targeted Concepts, Skills, and Nature of Science	Performance Expectations
Scientists often collaborate and share their findings. Sharing findings makes new information available and helps scientists refine their ideas and build on others' ideas. When another person's or group's idea is used, credit needs to be given.	Students should share their bacteria culture data with the class as they try to determine where bacteria can be found in the classroom.
Scientists must keep clear, accurate and descriptive records of what they do so they can share their work with others; consider what they did, why they did it, and what they want to do next.	Students should be able to write a clear and repeatable procedure for the investigation. Students should be able to record observations and data, quantify their observations of the bacteria cultures and organize it so that it can be easily shared with the class.
Scientific questions are directed toward objects and events that can be described, explained, or predicted by scientific investigations.	Students should be able to write a question to drive the bacteria investigation that meets the criteria for a good question.

Targeted Concepts, Skills, and Nature of Science	Performance Expectations
Identifying factors that lead to variation is an important part of a scientific investigation.	Students should be able to identify the independent and dependent variables in the bacteria culturing investigation and describe how other variables will be controlled.
Explanations are claims supported by evidence, accepted ideas, and facts.	Students should be able to analyze the bacteria culture data and propose explanations for the distribution of bacteria in the classroom that are supported by the data.
Students should be able to recognize and analyze patterns and trends.	Students should be able to analyze the class' bacteria culture data and find trends in the data. Students should be able to identify patterns associated with the locations of high bacteria concentrations.

Materials	
1 per group	Petri dish with nutrient agar
3-5 per group	Cotton swabs (enough for each group to have some- they need to be kept as sterile as possible)
1 per class	Warm area (or incubator) — dishes can be left in the open classroom or placed in another warm area. If incubator is not used, the growth of bacteria will be slower.
1 per class	Permanent markers or wax pencils to mark Petri dishes
1 per class	Adhesive tape to seal the lids of the Petri dishes
1 per group	Graph paper
1 per class	Bleach—essential, as all dishes need to be soaked in a bleach solution before disposal
1 per class	Garbage bags
1 per class	Class data table
1 per student	*My Experiment* page

Activity Setup and Preparation

Identify warm areas with sufficient space to store prepared Petri dishes for several days. Prepare agar dishes and run a trial investigation to determine the length of time needed to culture bacteria in your storage location.

Preparing Nutrient Agar

- Work in as sterile environment as possible and take care not to contaminate the Petri dishes or nutrient agar solution. Wash your work area and your hands.
- Follow directions on the package for preparing the powdered nutrient agar.
- Allow the nutrient agar solution to cool until it can be handled comfortably while wearing vinyl gloves. If the nutrient agar is poured while it is too hot, there will be excessive condensation on the cover of the Petri dish. Pour agar into the Petri dish until the agar just covers the bottom of the dish (approximately 20 ml). Place the Petri dish cover on the Petri dish immediately and allow the agar to solidify (fully gel). After the plate of agar has solidified, turn the dishes upside down and let them sit for 48 hours at room temperature to dry. Drying eliminates excess water in the agar and reduces the amount of condensation. Moisture on the cover will adversely affect results of later steps.
- Refrigerating prepared Petri dishes until they are used will inhibit bacterial growth.
- Wash your work area and your hands.

Using Petri dishes

- Allow refrigerated Petri dishes to warm to room temperature before using.
- Multiple samples can be cultured in one dish. Divide the dish into sectors by marking the outside of the bottom cover using a permanent marker or wax pencil. Mark the outside of the bottom agar dish, because the lid may rotate.
- Bacteria samples should be collected using a clean, untouched end of a cotton swab. Caution students not to touch the end of the swab that they will use to collect their sample.
- After selecting a site in the classroom to sample, wipe the area with a cotton swab.

LOOKING AHEAD

This investigation involves collecting samples and culturing them to determine if bacteria are present. Since the culturing will take more than one class period, you may want to start this activity and then move on to another activity. When the cultures have grown, you may return to this investigation and finish it. Following this investigation, students will perform a similar investigation independently, rather than as a class.

Students often have difficulty designing investigations. Scaffolding the process, as is done in this activity, helps students to develop the ability to do investigations independently. They will practice doing a similar investigation independently in Section 2.7.

- Lift the lid from the Petri dish just enough to admit the cotton swab and take care not to touch or breathe on the nutrient agar surface. If the lid is completely removed, it can become contaminated with bacteria from the environment. Gently brush/ touch the surface of the agar with the cotton swab and replace the lid. Record the place from which the sample was collected and the section of the Petri dish inoculated with that sample.
- Repeat this procedure for each of the classroom locations being investigating. Use a new cotton swab for each sample.
- After the Petri dishes have been inoculated, seal them. Petri dishes are usually sealed with two strips of adhesive tape from lid to base. Once sealed, DO NOT re-open until ready for disposal. Treat all bacteria as pathogens.

It is important that students do not collect bacteria samples from potentially hazardous places.

All students should wash their hands with a disinfectant detergent after sampling the bacteria.

VERY IMPORTANT—Make it clear to students that once Petri dishes are taped shut, they should NOT BE RE-OPENED. Only the teacher will re-open the Petri dishes, and then only during disposal.

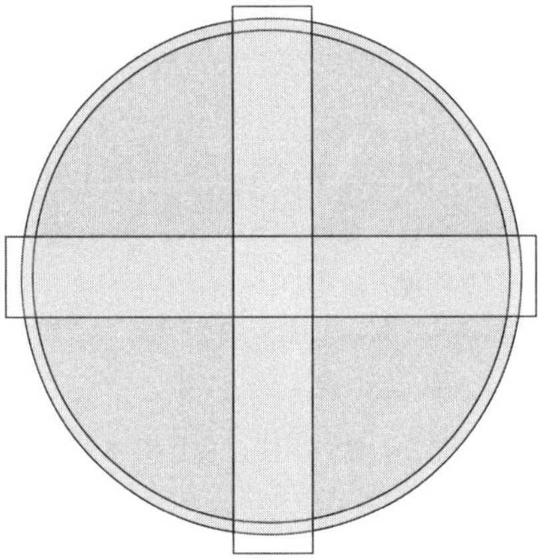

Seal Petri dishes with two strips of adhesive tape from lid to bottom, do not seal around edges.

- Petri dishes may be stacked upside down for storage in an incubator or a warm place until the following day. An incubator speeds growth, but is not critical to success. Without one, simply allow more time for bacterial growth.

Disposing of Petri dishes

- Petri dishes need to be decontaminated before disposal.
- The best way to decontaminate used Petri dishes is to autoclave them. Most autoclaves operate at 121.1°C (250°F) and 15 psi, which effectively kills bacteria.

- If an autoclave is not available, wear gloves and a face mask and open the Petri dishes in a large container and immediately submerge in full strength bleach. Leave Petri dishes in the bleach overnight. Bleach can be disposed down the sink with plenty of water. Rinse and re-close the dishes, wrap in plastic wrap and throw into trash or incinerator.

Homework Options

Reflection

- **Science Process:** Why were the Petri dishes labeled on the outside of the bottom dish? *(The lid may rotate so that markings become aligned with the wrong sample.)*
- **Science Process:** Why did you use a new cotton swab for every sample you collected? *(To prevent cross-contamination.)*
- **Science Content:** What conditions favor the growth of bacteria? *(An environment that is warm and moist, with abundant food.)*
- **Science Content:** Why is it important to wash your hands after sampling locations for bacteria? *(To prevent the spread of bacteria.)*

Preparation for 2.7

- **Science Process:** What part of designing this investigation did you find most challenging? *(Students typically cite formulating the question, identifying variable, writing conclusions.)* What would you do differently the next time you design an investigation? *(Answers will vary, but may include: look at exemplars of good questions, identify the variables before attempting to formulate the question, collaborate with others to make a list of factors that can change, connect conclusions to data and ask others to check their thinking when drawing conclusions.)*
- **Science Process:** What result of this investigation did you find most surprising? Why was it surprising to you? *(Answers will vary, but students should be able to articulate what made the result surprising to them.)*

NOTES

SECTION 2.6 IMPLEMENTATION

◀ *2 class periods**

2.6 Investigate

Where in Your Classroom Can You Find Bacteria?

You now know that harmful bacteria can be anywhere. There may be bacteria in your classroom. In this section, you will work as a class to develop an investigation that shows the presence of bacteria in your classroom. Then in the next section, you will follow a similar procedure to do an independent investigation to answer a question that interests you.

Microbiology is the study of microorganisms. Over many years, **microbiologists**, people who specialize in microbiology, have developed very specific techniques and procedures to study microbes. Before your class designs their investigation, read the sample procedure for growing bacteria. The procedure uses microbiology techniques adapted for a classroom.

microbiology: the study of microorganisms.

microbiologist: a scientist who specializes in microbiology.

Materials
- Petri dish with nutrient agar
- cotton swabs
- permanent or wax-pencil marker
- adhesive tape

Sample Procedure for Growing Bacteria

Scientists and technologists who study bacteria have certain materials and procedures they use. The materials listed here are similar to those they use. The list has materials that are suitable for your classroom. Here is a brief outline of a procedure that could be used for growing bacteria.

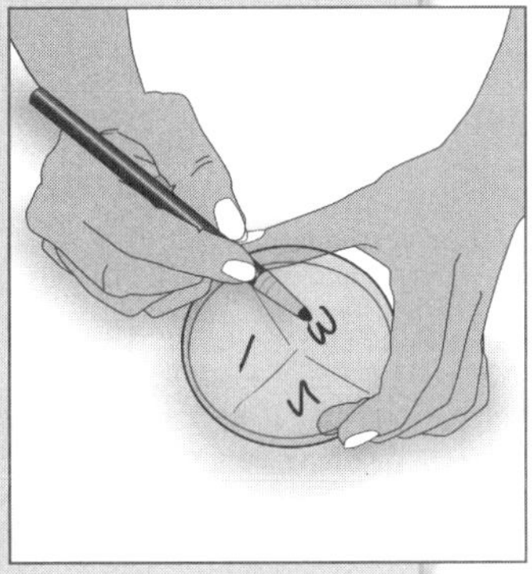

1. Use a wax pencil or permanent marker to divide the bottom half of a Petri dish containing nutrient agar into three sections. (Nutrient agar is a jelly-like substance that contains the nutrients bacteria need to grow.) By dividing the dish, you can use it for samples from more than one area. Carefully label each section so you know which area your bacteria were collected from.

GF 45 GOOD FRIENDS and GERMS

2.6 Investigate

Where in Your Classroom Can You Find Bacteria?

5 min.

Students are shown a sample procedure for growing bacteria. The class collaborates to design an investigation using this technique to determine where bacteria can be found in the classroom. Students collect samples from various places in the classroom, culture the samples, and identify places that harbor bacteria.

○ Engage

Ask students to summarize what they learned about bacteria in the last section (Darrell's experiment). Have students recall that some bacteria cause disease. Then ask students where in the classroom they think bacteria might live and why they think they might live there. Make a list of the students' ideas.

*A class period is considered to be one 40 to 50 minute class.

Get Going

Have students read the introduction. Make clear to students that this investigation will be done as a class, but that after they do this as a class, they will have the chance to run their own investigation.

Describe the process that they will engage in as a class: determine where they will test for bacteria, formulate a question about that place and bacteria to investigate, predict possible outcomes, identify variables, write the procedure they will follow, and then the entire class will do the same experiment. The class data (it will take about 2 days to incubate the bacteria) will be analyzed and conclusions drawn from the data.

Sample Procedure for Growing Bacteria

5 min.

Students are shown a sample procedure for growing bacteria.

growing ... procedure uses microb... niques adapted for a classroom.

• adhesive tape

Sample Procedure for Growing Bacteria

Scientists and technologists who study bacteria have certain materials and procedures they use. The materials listed here are similar to those they use. The list has materials that are suitable for your classroom. Here is a brief outline of a procedure that could be used for growing bacteria.

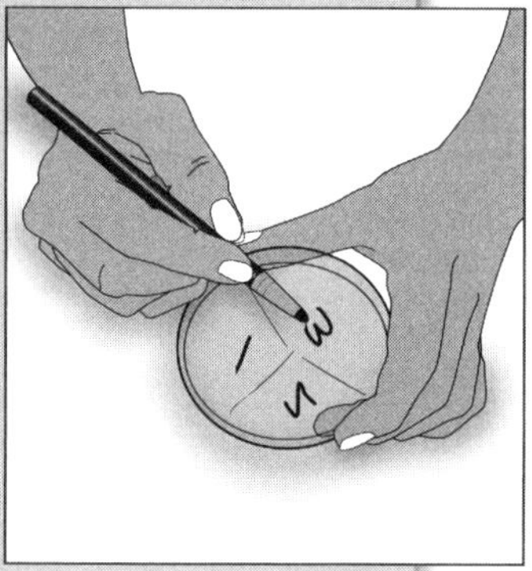

1. Use a wax pencil or permanent marker to divide the bottom half of a Petri dish containing nutrient agar into three sections. (Nutrient agar is a jelly-like substance that contains the nutrients bacteria need to grow.) By dividing the dish, you can use it for samples from more than one area. Carefully label each section so you know which area your bacteria were collected from.

GF 45 GOOD FRIENDS and GERMS

META NOTES

Students have difficulty designing and carrying out investigations independently. The instructional strategy used in this section was selected because it provides the teacher with the opportunity to model best practices while doing the investigation with the class. Students will have more success completing the independent investigation in the next section because in this section you have guided them through the process. Therefore, students need to be active participants in all aspects of the design process.

Guide

Describe the challenges associated with detecting the presence of bacteria. Have students recall the challenges they faced when observing cells under high power in the last activity. Point out that bacteria are smaller than cells, so small that they are difficult to see with a classroom microscope. Bacteria are measured in units of length called micrometers, or microns. One millimeter is equal to 1,000 microns. Some bacteria measure only one-half micron in length! In addition, the field of view of the microscope under high power covers a very small area. Trying to find a few bacteria on a microscope slide is like trying to find a grain of rice on a sandy beach.

Point out to students that there is, however, a way to detect bacteria based on how they grow. Have students recall that bacteria reproduce rapidly under favorable conditions and form colonies. Although one bacterium may

Do not open the Petri dishes after you have swabbed them.

You do not know what microbes are growing on the agar. They may be dangerous to your health.

All Petri dishes must remain tightly sealed. Your teacher will dispose of all the dishes when you have completed your investigation.

2. Use a cotton swab to collect a sample of bacteria from an area of interest. Wipe the area you want to test with the swab. Use a different cotton swab for each area you test. Be very careful that the end of the cotton swab does not touch you or anything else you are not testing.
3. Carefully lift the lid from the Petri dish. Lift up one side only enough to be able to gently swab the surface of the agar plate. Try not to breathe onto the surface.
4. Gently brush or touch the cotton swab to the surface of one of the labeled areas of the agar plate.
5. Repeat Steps 2 to 4 for the other areas you have decided to test.
6. Securely seal the lid to the Petri dish using adhesive tape.
7. Invert the Petri dish and place it in a warm place.
8. Check the Petri dish every 24 h (hours). Record your data.

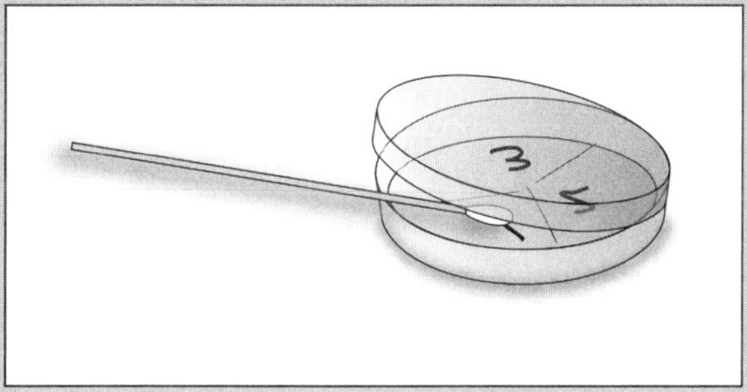

be hard to find, an entire colony can grow large enough to be seen without a microscope. Tell students that they will now read about a technique that can be used to grow colonies of bacteria.

Read through the sample procedure with the class. Emphasize safety precautions and the importance of limiting the chances of cross-contamination when collecting bacteria and when inoculating the Petri dishes.

Plan Your Investigation

15 min.

The class collaborates to design an investigation to determine where bacteria can be found in the classroom.

Plan Your Investigation

Read the questions in each part of the *My Experiment Page*. The purpose of the questions is to help you think about how to design an experiment. You used some of the questions on the *My Experiment Page* to guide previous investigations. You will use these questions to plan and guide the investigation for this section.

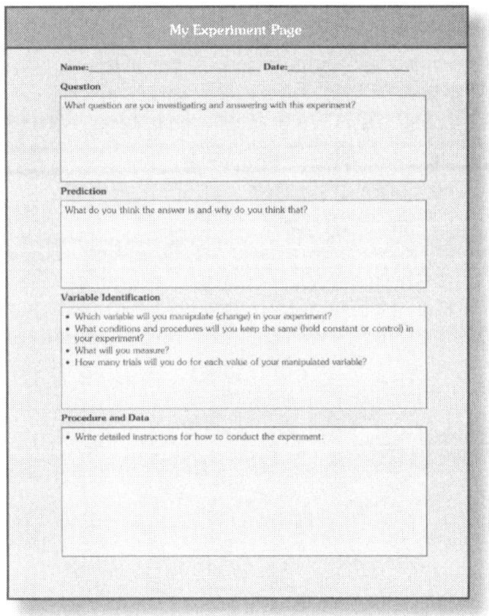

My Experiment Page

Name:__________ Date:__________

Question

What question are you investigating and answering with this experiment?

Prediction

What do you think the answer is and why do you think that?

Variable Identification

- Which variable will you manipulate (change) in your experiment?
- What conditions and procedures will you keep the same (hold constant or control) in your experiment?
- What will you measure?
- How many trials will you do for each value of your manipulated variable?

Procedure and Data

- Write detailed instructions for how to conduct the experiment.

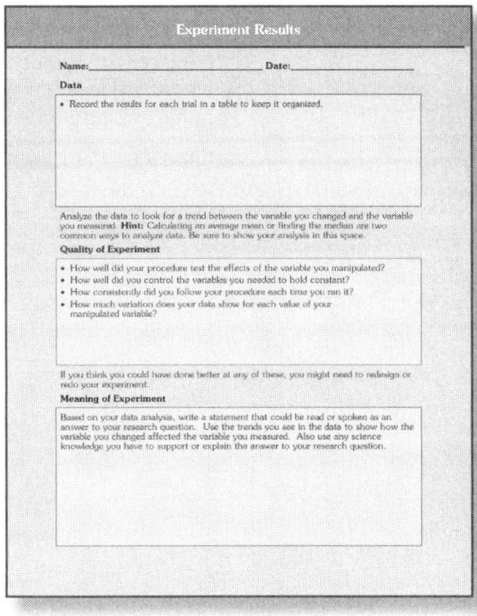

Experiment Results

Name:__________ Date:__________

Data

- Record the results for each trial in a table to keep it organized.

Analyze the data to look for a trend between the variable you changed and the variable you measured. **Hint:** Calculating an average mean or finding the median are two common ways to analyze data. Be sure to show your analysis in this space.

Quality of Experiment

- How well did your procedure test the effects of the variable you manipulated?
- How well did you control the variables you needed to hold constant?
- How consistently did you follow your procedure each time you ran it?
- How much variation does your data show for each value of your manipulated variable?

If you think you could have done better at any of these, you might need to redesign or redo your experiment.

Meaning of Experiment

Based on your data analysis, write a statement that could be read or spoken as an answer to your research question. Use the trends you see in the data to show how the variable you changed affected the variable you measured. Also use any science knowledge you have to support or explain the answer to your research question.

Question

- What question are you investigating and answering with this experiment?

With your class, talk about the different places you think there are a lot of bacteria. Then think of a question you could investigate about that place and bacteria. Record the question your class develops.

For example, some possible questions might involve classroom desks, pencils or pens, hands, the water fountains, or doorknobs. Your question might be: *Does my right hand or my left hand have more bacteria on it?*

○ Get Going

Distribute the *My Experiment* Page and review the experiment guidelines with the students. Emphasize identifying variables and writing conclusions as these are common areas of difficulty for middle-school students.

Draw students' attention to the class list created earlier identifying where in the classroom they think bacteria live. Conduct a class discussion with the goal of reaching a consensus on a location to be investigated for the presence of bacteria. Guide students to agree on a type of location rather than a specific location (for example: desktops rather than Ben's desk, or pens rather than Abby's pen).

Question

- What question are you investigating and answering with this experiment?

With your class, talk about the different places you think there are a lot of bacteria. Then think of a question you could investigate about that place and bacteria. Record the question your class develops.

For example, some possible questions might involve classroom desks, pencils or pens, hands, the water fountains, or doorknobs. Your question might be: *Does my right hand or my left hand have more bacteria on it?*

GF 47

GOOD FRIENDS and GERMS

⬡ Get Going

Have students read the *Question* section.

Ask students to suggest a good question to investigate about the location chosen by the class. Remind students of the criteria for a good question:

- We have to be interested in finding the answer.
- The question should require several resources to answer.
- The question should be related to the *Big Question.*
- Answering the question should require collecting and using data.
- The question should not have a yes/no answer.

Use a suggestion that does not meet all of these criteria as an example. A good example is a question like "Are there bacteria on my desk?" This question may meet some of the criteria, but it has a yes/no answer, therefore it does not meet all of the criteria. A better question might be "Are there more bacteria on my desk than on my chair?"

Lead a class discussion leading to agreement on a question to investigate. Then edit the question together with the class so that it is one that can be answered by collecting just two or three samples to compare. Guide the class to agree on the final wording of the question and have them record it on their *My Experiment* Page.

Prediction

10 min.

Prediction

- What do you think the answer is, and why do you think that?

You need to make a prediction and provide some reasons for your prediction. Look again at the question. What do you think the answer to the question is going to be? Record your answer, and then describe why you think that is going to be the answer. Be prepared to share and discuss your prediction with the class.

For example, one prediction might be "My left hand will have more bacteria on it because I am left-handed and use it more often."

Variable Identification

You will need to identify all the variables in your experiment.

- What variable will you change? This is the manipulated or independent variable. You should only change the value of one variable. It may be the area from where you collect bacteria.
- What variables will you measure? These are the responding or dependent variables. You might want to measure or observe the things that you think will be different depending on the value of your independent variable. You probably want to measure how much bacteria grew from each area you decided to test.
- How will you measure your responding or dependent variables? You need to decide how you will measure the amount of bacteria in each location.
- What conditions and procedures will you need to control in your experiment? That is, what conditions and procedures will you keep the same as you change the manipulated or independent variable?

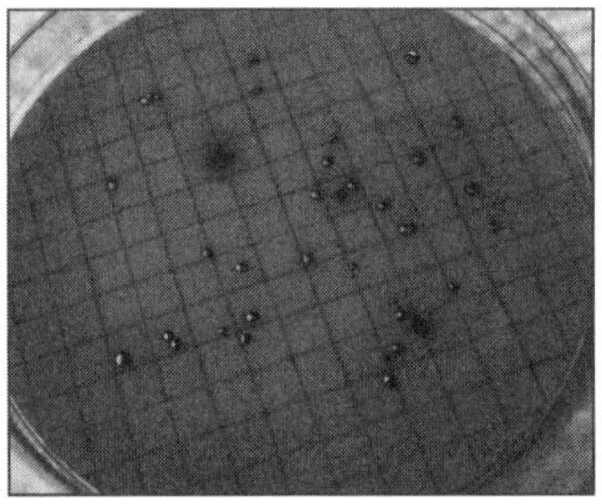

Bacteria growing in nutrient agar.

To answer the question, "Does my left or right hand have more bacteria?" your manipulated or independent variable will be the hand you collect from. The responding or dependent variable will be the amount of bacterial growth on the Petri dish. You will measure the responding or dependent variable by measuring the diameter of the bacterial growth. You will control the place on each hand from which you collect the bacteria.

⬡ Get Going

Have students read the *Prediction* section. Ask students to predict the answer to the question agreed upon by the class. Tell students to write down their prediction and also to write down the reason for their prediction. For example, if a student predicts that there will be more bacteria on their desk than on their chair, they should also write that they think this will be so because their hands touch their desk more frequently than they touch their chair.

Survey the class and make a list of different predictions. Have students copy the list of class predictions onto their *My Experiment* Page.

△ Guide

Guide students in re-writing their predictions in terms of what they think they will observe on the agar plates after they have incubated overnight.

...ediction wi...

For example, one predicti... be "My left hand w... bacteria on it because I am left-handed and use it more often."

Variable Identification

You will need to identify all the variables in your experiment.

- What variable will you change? This is the manipulated or independent variable. You should only change the value of one variable. It may be the area from where you collect bacteria.
- What variables will you measure? These are the responding or dependent variables. You might want to measure or observe the things that you think will be different depending on the value of your independent variable. You probably want to measure how much bacteria grew from each area you decided to test.
- How will you measure your responding or dependent variables? You need to decide how you will measure the amount of bacteria in each location.
- What conditions and procedures will you need to control in your experiment? That is, what conditions and procedures will you keep the same as you change the manipulated or independent variable?

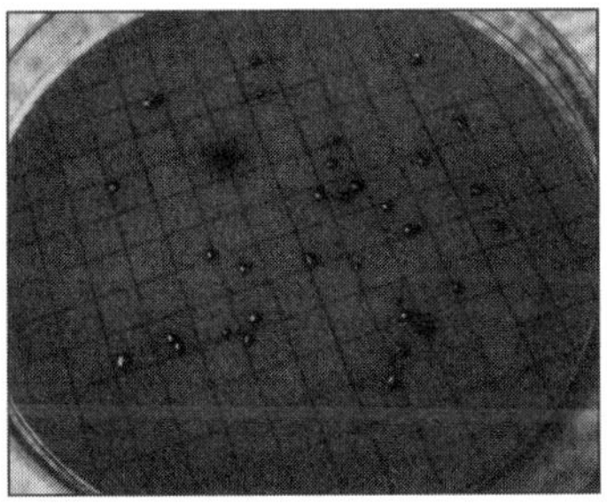
Bacteria growing in nutrient agar.

To answer the question, "Does my left or right hand have more bacteria?" your manipulated or independent variable will be the hand you collect from. The responding or dependent variable will be the amount of bacterial growth on the Petri dish. You will measure the responding or dependent variable by measuring the diameter of the bacterial growth. You will control the place on each hand from which you collect the bacteria.

GF 48

Project-Based Inquiry Science

Variable Identification

10 min.

⬡ Get Going

Have students read the *Variable Identification* section.

Ask students to identify the variables in this experiment and record them on their *My Experiment* Page.

△ Guide

Using the questions in the text, clarify the distinction between independent and dependent variables and how they will be measured or described. Discuss the conditions or procedures that need to be controlled, focusing on how this is to be accomplished.

Procedure and Data

Using the sample procedure as a guide, with your class, identify the steps of the investigation. Record detailed instructions for how to conduct the experiment. Someone else should be able to follow your instructions and duplicate your experiment. Include how to set up your experiment, how you will gather samples and record your data, how to measure the changes in your samples, and how to analyze your data.

Be careful when you are writing your procedure. Make sure you write it very accurately. Clearly record each step of the procedure. Make it clear enough that someone else could follow it. Pay attention to each part of the procedure and data list.

Run Your Experiment

Carry out your procedure. Be sure to follow the steps carefully. Record your results.

Analyze Your Data

Answer the following questions. As a class, discuss your answers.

1. Which area(s) that you tested had the most bacteria? What evidence do you have to support your claim?
2. Which area(s) that you tested had the least bacteria? What evidence do you have to support your claim?
3. How did the results of your investigation compare with your prediction? Suggest reasons for any differences. If your results surprised you, tell why.
4. Make a claim about the bacteria on the surfaces you tested. Your claim should answer the research question your class developed. Support your claim with evidence from your investigation.

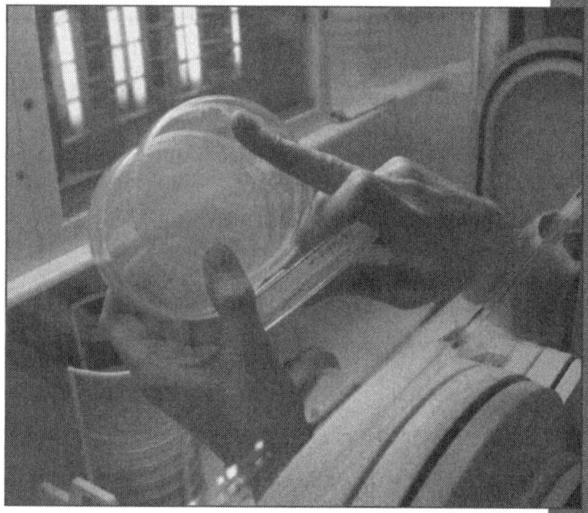

Scientists grow bacteria on agar in a Petri dish.

Procedure and Data

15 min.

Get Going

Have students read the *Procedure and Data* section. Remind students that they may use the sample procedure as a guide.

Ask students to independently write a set of procedures to follow in order to carry out the experiment. Have students share their procedure and lead a class discussion in which the class collaborates to develop a set of procedures that they will follow when carrying out this experiment. Encourage students to comment on each other's procedures, suggesting changes and noting what they think is done well. Edit the class procedure as you go along. When the class procedure is finalized, have students copy the procedure on their *My Experiment* Page.

⬡ Get Going

Demonstrate the steps of the class procedure, modeling proper technique for collecting bacteria samples.

- Remove a cotton swab from the package, being careful not to touch the end.
- Model incorrect handling of the cotton swab to the students by touching the end of the swab that you want to use with your hands. Ask students to explain why they should not handle the swab in this manner. *(Bacteria on your hand could contaminate the swab.)*
- Rub the area you want to test for bacteria with the clean, untouched end.
- Lift the lid of the agar plate just enough to allow the swab to enter. Ask students why you wouldn't want to remove it completely. *(Bacteria in the air could contaminate the agar plate.)*
- Gently but firmly rub the tip containing the sample against the surface of the agar.
- Record the location from which the sample was collected along with the sector of the agar plate to which it was applied.

Have students run the experiment by carrying out their procedure. After samples have been collected and applied to the agar plates, have students seal the plates with two strips of adhesive tape. Collect the agar plates, invert, and stack for convenient storage in a warm place. Agar plates should incubate for at least two days, before continuing to the *Analyze Your Data* section.

NOTE: This is an opportune time to conclude the first class session of this section.

Run Your Experiment

10 min.

Students collect samples to be cultured from various places in the classroom.

META NOTES

Unless the class carries out the procedure on a Friday, it is suggested that you continue on to *Section 2.8* while the bacteria cultures incubate for two days. You may want to have students observe the Petri dish and sketch its appearance after 24 hours and then resume this activity after 48 hours.

Analyze Your Data

20 min.

Students examine their bacteria cultures, record their observations, and analyze their data using the Analyze Your Data *questions.*

Record yo

Analyze Your Data

Answer the following questions. As a class, discuss your answers.

1. Which area(s) that you tested had the most bacteria? What evidence do you have to support your claim?
2. Which area(s) that you tested had the least bacteria? What evidence do you have to support your claim?
3. How did the results of your investigation compare with your prediction? Suggest reasons for any differences. If your results surprised you, tell why.
4. Make a claim about the bacteria on the surfaces you tested. Your claim should answer the research question your class developed. Support your claim with evidence from your investigation.

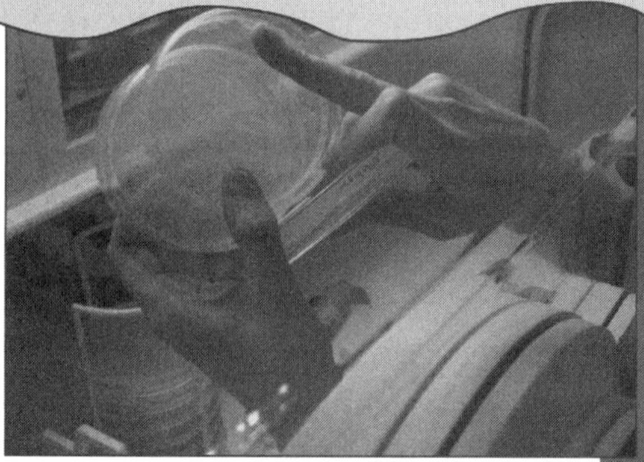

Scientists grow bacteria on agar in a Petri dish.

Get Going

Distribute the incubated agar plates and ask students to make observations. Ask students to suggest ways to quantify their observations. (count the number of individual colonies, measure the size of the colonies, put a graph paper transparency over the agar plate and count the number of boxes occupied by bacteria in each sector.)

Using their best suggestions, have students record their data and then share it with the class. Make a public list of the class's data.

Ask the students how they think the data should be organized in the data table. Ask them if they see any patterns. Ask them if they think a graph would be useful and how.

Have students read the *Analyze Your Data* section and record their answers to the questions.

Guide

Engage the class in a discussion of their answers to the questions. Encourage students to compare their results with their predictions. Ask them to identify any results that were surprising.

Discuss with students how to formulate a claim. First, have the students write factual statements they can make based on the data. These statements may assist in supporting their claim. The claim should answer the investigative question and it should be based only on the observations they made (i.e. the data). Invite students to share their claims and the evidence drawn from the data supporting that claim.

Reflect

Think about the following questions. Discuss your answers with the class.

- How reliable are your results? Support your answer with evidence.
- What are some of the sources of error that you may have introduced? For example, did all the colonies of bacteria that grew on the agar come from the area tested?
- In the next section, you will be planning and running your own investigation. What have you learned in this section that will help you design an investigation? What will you need to pay particular attention to in designing your investigation?

What's the Point?

In this investigation, you used microbiology techniques to answer your class's question about bacteria in your classroom. Microbiologists have spent many years developing procedures that produce reliable results. When they test an area for bacteria, they are very careful not to introduce bacteria from other sources onto the surface of the agar.

The investigation helped you think about where bacteria might be in your environment. The *Big Question* for this Unit is *How can you prevent your good friends from getting sick?* You probably found that there were bacteria in many parts of the classroom. Some of those bacteria could make you and your friends sick.

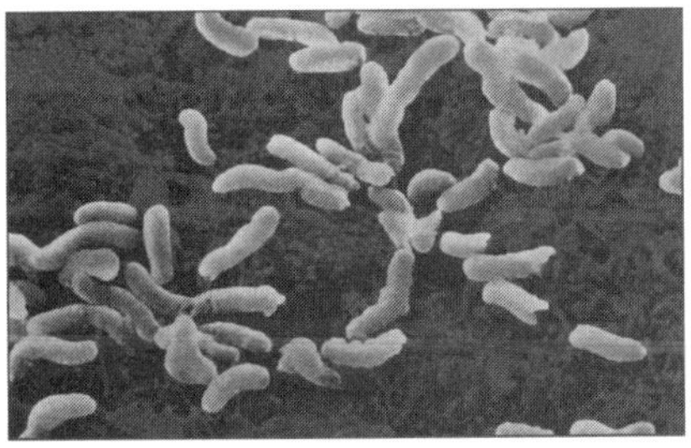

Heliobacter pyloris *are rod-shaped bacteria (artificially colored) that cause stomach ulcers in people.*

Your observations are similar to the observations that scientists make when they do investigations. Epidemiologists do investigations to track germs. Then they use the evidence from their investigations to make recommendations that could prevent people from getting sick.

Reflect

10 min.

Discuss possible sources of error and how they affect the validity of results using the Reflect *questions.*

⬡ Get Going

Have students read the questions and record their answers. Then, lead a class discussion focusing on possible sources of error and the validity of results.

Assessment Options

Targeted Concepts, Skills, and Nature of Science	How do I know if students got it?
Scientists often collaborate and share their findings. Sharing findings makes new information available and helps scientists refine their ideas and build on others' ideas. When another person's or group's idea is used, credit needs to be given.	**ASK:** Which ideas proposed by other groups did you find most useful in refining your own data analysis? **LISTEN:** Students should cite ideas such as focusing on variables that related to the question.
Scientific questions are directed toward objects and events that can be described, explained, or predicted by scientific investigations.	**ASK:** What makes a question a good one to investigate? **LISTEN:** Students should be able to identify the criteria for a good question listed in the student text.
Scientists must keep clear, accurate, and descriptive records of what they do so they can share their work with others, consider what they did, why they did it, and what they want to do next.	**ASK:** Why did your groups choose to organize your data in the way that you did? **LISTEN:** Students should make the connection between their method of organizing the data and more clearly showing relationships among variables that supported their conclusions.
Identifying factors that lead to variation is an important part of a scientific investigation.	**ASK:** What were the variables that Darrell tracked in his investigation? **LISTEN:** Students should be able to identify the date, food storage location, and the number of people that became ill.

Targeted Concepts, Skills, and Nature of Science	How do I know if students got it?
Explanations are claims supported by evidence, accepted ideas, and facts.	**ASK:** What claims can you make about the cause of the gastrointestinal illness reported by customers of Uncle Joe's Restaurant and what is your supporting evidence? **LISTEN:** Students should be able to support their claims about the cause of the illness with evidence drawn from the data.
Students should be able to recognize and analyze patterns and trends.	**ASK:** What did the locations in our classroom with the most bacteria have in common? **LISTEN:** Students should be able to identify patterns associated with the locations of high bacteria concentrations such as likelihood of contact with hands, warmer temperatures, or presence of moisture.

Teacher Reflection Questions

- In this activity, students practiced creating an investigation with your guidance. Did students understand the importance of writing clear, repeatable procedures?
- To which degree was the process of designing an investigation in this activity appropriate to the skill level of your students? Do your students need more guidance and practice or less?
- How could you evaluate whether your students are ready to move on to designing an investigation independently?
- What aspects of this investigation went well? Which aspects didn't go well? How could you improve this investigation?

NOTES

SECTION 2.7 INTRODUCTION

2.7 Investigate

Where Else Do Bacteria Live?

◀ *2 class periods*

A class period is considered to be one 40 to 50 minute class.

Overview

In this section, students work independently to develop a research question based on their findings in the last activity. Students then independently design an investigation, predict the outcome, identify the independent and dependent variables as well as any controls, and develop a set of procedures. Students share their procedure with the class in a *Plan Briefing* and make revisions based on feedback from their peers. Students then run the investigation, collect the data, analyze it, and share the results with the class during an *Investigation Expo*.

Targeted Concepts, Skills, and Nature of Science	Performance Expectations
Scientists often collaborate and share their findings. Sharing findings makes new information available and helps scientists refine their ideas and build on others' ideas. When another person's or group's idea is used, credit needs to be given.	Students should share their proposed procedure with the class and refine it based on feedback from their peers. Students should share their findings with the class and listen to the findings of others. They should be able to identify new information about bacteria that they have learned from the work of others and give credit to its source.
Scientists must keep clear, accurate, and descriptive records of what they do so they can share their work with others, consider what they did, why they did it, and what they want to do next.	Students should be able to write a clear and repeatable procedure for the investigation. Students should be able to record observations and data, quantify their observations of the bacteria cultures and organize them so that they can be easily shared with the class.
Scientific questions are directed toward objects and events that can be described, explained, or predicted by scientific investigations.	Students should be able to independently write a research question to direct their bacteria investigation.

Targeted Concepts, Skills, and Nature of Science	Performance Expectations
Identifying factors that lead to variation is an important part of a scientific investigation.	Students should be able to identify the independent and dependent variables in the bacteria investigation and describe how other variables will be controlled.
Explanations are claims supported by evidence, accepted ideas, and facts.	Students should be able to analyze the data collected during their investigation and make claims that are supported by the data.
Students should be able to recognize and analyze patterns and trends.	Students should be able to analyze their bacteria culture data and find trends in the data. Students should be able to identify patterns associated with the locations of bacteria in the classroom.

Materials	
1 per student	Petri dish with nutrient agar
3-5 per student	Cotton swabs (enough for each group to have some—they need to be kept as sterile as possible)
1 per class	Warm area (or incubator)—dishes can be left in the open classroom or placed in another warm area. If incubator is not used, the growth of bacteria will be slower.
1 per group	Permanent markers or wax pencil to mark Petri dishes
1 per group	Adhesive tape to seal the lids of the Petri dishes
1 per classroom	Fruit or other items suggested by students
1 per classroom	Cleaning agents: different soaps (name brand, generic, anti-bacterial)
1 per classroom	Cool area to grow bacteria (for temperature effects)
1 per class	Class data table
1 per student	*My Experiment* page

Activity Setup and Preparation

Refer to the directions for *Section 2.6.*

Homework Options

Reflection

- **Science Process:** Do research online to determine if any scientific research has been done relating to the question you chose to investigate. *(Answers will vary.)*
- **Science Content:** Find images of bacteria online or in magazines (look for ads for anti-bacterial soaps and disinfectants). Compare and contrast these images with the bacteria you observed growing in your Petri dish. *(The images are often photographs of stained and highly magnified bacteria, or an artist's conception of bacteria. The bacteria in the Petri dish are a colony made up of enormous numbers of individual bacteria. It is like looking at a person's hair from a block away; you can't see the individual hairs, but you can see the clump of hair covering the person's head.)*

Preparation for 2.8

- **Science Content:** You have just finished a series of investigations that has given you a lot of information about bacteria. Earlier, you read that viruses also cause disease. Based on what you know about bacteria, what do you think viruses are like? *(Student answers will vary, but should include characteristics similar to those of bacteria.)*

NOTES

NOTES

SECTION 2.7 IMPLEMENTATION

2.7 Investigate

Where Else Do Bacteria Live?

The investigation you just finished provided you with some evidence about where bacteria grow in your classroom. You may have been surprised by the results. What questions would you like to ask about bacteria based on what you have observed? You will now use what you observed about bacteria and designing an investigation to plan and run an investigation to answer your own research question.

Plan Your Experiment

There may have been something about the data from the last investigation that surprised you. You can turn this into a question that you can answer in this investigation. For example, you may have thought that a metal doorknob would have a lot of bacteria on it because so many people touch it. Instead, perhaps you found that there were less bacteria on the doorknob than on your desk. You could turn this into a question: "Are there more bacteria on a wooden surface than on a metal surface?"

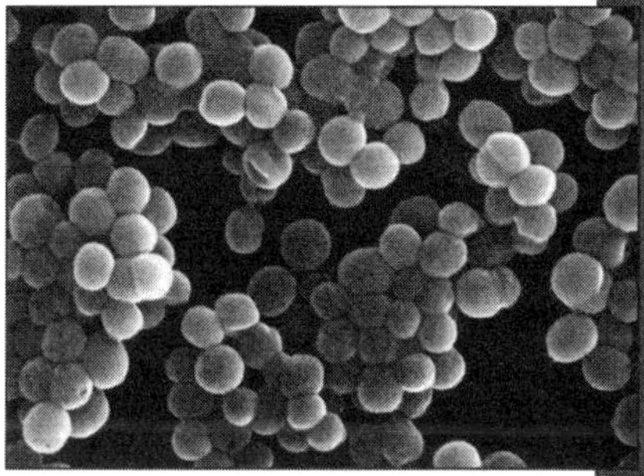

An area of skin as small as 6.5 cm² (1 square inch) may be home to more than half a million bacteria, such as Staphyloccus epidermis *that is shown here.*

Also, in Section 2.4, you generated new questions about bacteria. You can use some of these questions to help you develop a research question for this investigation.

With your group, discuss what research question you would like to investigate. Working within your group, develop a question, and record it on your *My Experiment* page.

Plan an investigation to answer your group's question. Use the *My Experiment Page* to help you plan your investigation. You may wish to review the part in Section 2.6 that explains how to use the questions and criteria on the *My Experiment Page*.

Create a numbered list of steps that you will use to gather bacteria, grow bacteria, and judge the amount of bacteria that has grown. Think about the precautions you must take and why they are important to follow. After you have recorded your procedure, you will present it to the class.

◄ *2 class periods**

2.7 Investigate

Where Else Do Bacteria Live?

5 min.

As a result of the previous investigation, students design their own investigations independently. They develop their research question and share it with their group. The group chooses a question to investigate, plans an investigation, and writes a procedure. Groups share their plans with the class and make revisions based on class discussion. Groups perform their investigations, analyze the results of the investigation, and communicate their results with the class in an Investigation Expo. *Students reflect upon what they have discovered as a result of their investigations.*

*A class period is considered to be one 40 to 50 minute class.

○ Engage

Briefly review the results of the investigation students have just completed. Ask students to identify any results they found surprising and specify what was surprising about those results.

Read the introduction with students, and review class safety rules.

Plan Your Experiment

10 min.

Students design an experiment to answer a research question agreed upon by the group.

△ Guide

Point out that in the last activity, you guided the class through the process of designing and carrying out an investigation. Inform students that they will now work independently to plan and carry out their own bacteria investigation.

Plan Your Experiment

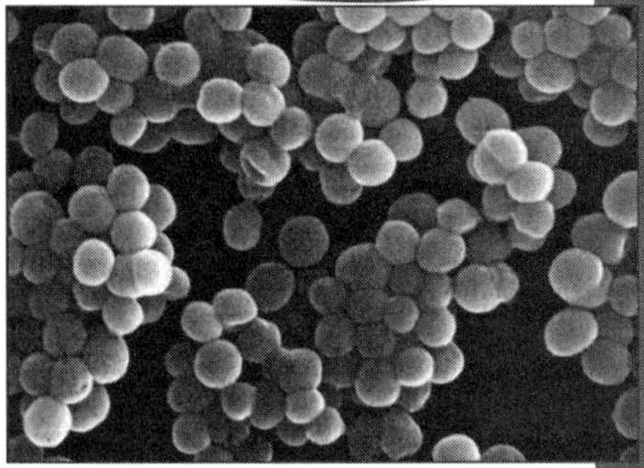

An area of skin as small as 6.5 cm^2 (1 square inch) may be home to more than half a million bacteria, such as Staphyloccus epidermis that is shown here.

There may have been something about the data from the last investigation that surprised you. You can turn this into a question that you can answer in this investigation. For example, you may have thought that a metal doorknob would have a lot of bacteria on it because so many people touch it. Instead, perhaps you found that there were less bacteria on the doorknob than on your desk. You could turn this into a question: "Are there more bacteria on a wooden surface than on a metal surface?"

Also, in Section 2.4, you generated new questions about bacteria. You can use some of these questions to help you develop a research question for this investigation.

With your group, discuss what research question you would like to investigate. Working within your group, develop a question, and record it on your *My Experiment* page.

Plan an investigation to answer your group's question. Use the *My Experiment Page* to help you plan your investigation. You may wish to review the part in Section 2.6 that explains how to use the questions and criteria on the *My Experiment Page*.

Create a numbered list of steps that you will use to gather bacteria, grow bacteria, and judge the amount of bacteria that has grown. Think about the precautions you must take and why they are important to follow. After you have recorded your procedure, you will present it to the class.

GF 51 GOOD FRIENDS and GERMS

⬡ Get Going

Distribute copies of the *My Experiment* Page and inform students that they should use it to guide them through the planning process.

Read the *Plan Your Experiment* section together with students. Point out connections between the tasks described in the reading and the sections of the *My Experiment* Page.

Have students work individually to formulate a research question. Encourage students to seek your guidance if they are experiencing difficulty in coming up with a research question.

△ Guide

Middle-school students often need help refining and refocusing broad, poorly defined questions. A useful strategy to help students generate experimental designs from general topics is to have students consider the following set of questions (the underlined word can be replaced with any general topic.)

- How do bacteria act? *(They grow.)*
- What materials do I have available for conducting experiments on bacteria? *(agar plates, classroom objects/locations, light, temperature)*
- How can I change the set of bacteria materials to affect the way bacteria act? *(collect bacteria samples from different objects/ locations, change temperature at which agar plates are incubated, incubate some agar plates in light and others in darkness, etc.)*
- How can I measure or describe the response of bacteria to the change? *(count the number of individual colonies, measure the diameter of the colonies, put a graph paper transparency over the agar plate and count the number of boxes occupied by colonies in each sector.)*

☐ Assess

As students are working, circulate and assess the questions they are formulating. Provide individual help to students that seem to be at an impasse.

⬡ Get Going

After students have formulated their research question, have students form small groups and share their question with the group. Then ask groups to discuss the questions proposed by the members of the group and to select one that the group will investigate. All group members should record this question on their *My Experiment* Page. Students should then complete the *Prediction* and *Variable Identification* sections of their *My Experiment* Page. Then, have groups write a procedure consisting of a set of numbered steps for carrying out the investigation.

☐ Assess

- Are students including a description of the reasoning behind their prediction?
- Have students distinguished between the independent and dependent variables in their investigation?
- Have students identified factors that need to be controlled?

Communicate Your Plan

15 min.

Students present their planned procedures to the class. Then, plans are critiqued and suggestions for improvement are made during a class discussion.

Revise Your Plan

5 min.

Students revise their planned procedures based on feedback received during a class discussion.

Communicate Your Plan

Plan Briefing

Share your investigation plan with the class. Also, listen to the plans of other groups carefully. Think about the question they are going to answer and how their plan is going to work. Have they met the criteria for good experiment design? If not, what could they do to improve their plan?

Revise Your Plan

With your group, revise your experiment plan based on the discussion you just had in class.

Your teacher must approve your procedure before you begin to run your experiment.

Run Your Experiment

With the approval of your teacher, carry out your investigation.

Analyze Your Data

After you have made your observations, discuss the results of your investigation with other members of your group. The questions on the next page should help you organize your discussion. Take notes on what others are saying during the discussion.

⬡ Get Going

Have students share their investigation plans with the class. Make clear to students that they are expected to respectfully critique their fellow students' plans based on the criteria for a good experimental design and to offer suggestions for improvement when they can.

Encourage groups to revise their plans based on feedback received during the class discussion. Then have groups present their revised plans to you. If needed, you should offer suggestions for further modification. Students should not begin their experiment without your approval.

Once the group's plan has received your approval, tell the group to carry out their experiment. After students have completed their procedure, the Petri dishes will have to incubate for two days. During this time, it is suggested that you complete *Section 2.8.*

Note: This is an opportune time to conclude the first class session of this section.

⬡ Get Going

Distribute the incubated Petri dishes and have groups make and record their observations. Have students read the *Analyze Your Data* section either individually or by taking turns reading aloud to their group.

Within their groups, have students discuss the results of their experiment focusing on the questions in the text. Inform students that they are expected to keep notes on what is said during the discussion.

NOTES

Run Your Experiment

10 min.

Students carry out the experiment they have designed using their revised and approved procedures.

Analyze Your Data

15 min.

Students present their planned procedures to the class. Then plans are critiqued and suggestions for improvement are made during a class discussion.

Communicate Your Results

20 min.

Students present the results of their group's experiment to the class.

- What variable were you investigating? What were you investigating about that variable? How did you vary it to determine its effects?
- List all of the variables you tried to hold constant in your experiment.
- Do you think that the data set you collected was useful in determining the effect your variable had on the growth of bacteria? Describe why or why not.
- What is the answer to your research question? How confident are you of your answer?

Communicate Your Results

Investigation Expo

Recall that an *Investigation Expo* is designed to help you present results of an investigation. Scientists present results of investigations to other scientists. This lets the other scientists build on what was learned. Each group in your class investigated a different question. Others in your class will want to know your results to help them answer the *Big Question: How can you prevent your good friends from getting sick?*

There are several things your classmates will want to know about your investigation. These include the following:

- the question you were trying to answer in your investigation
- your prediction
- the procedure and what makes it a fair test
- your results and how confident you are about them
- your interpretation of the results and how confident you are of them

To prepare for an *Investigation Expo*, make a poster that includes all of the five items listed above. Present them in a way that will make it easy for someone to make sense of your poster. Others should be able to identify what you have done and what you found out.

There are two parts to an *Investigation Expo*: presentations and discussions. As you look at posters and listen to other groups present their work, ask questions you need answered to understand the results and to satisfy yourself that the results and conclusions others have drawn are trustworthy. Be sure that you trust the results that other groups report.

△ Guide and Assess

Have students read the first two paragraphs and emphasize for students the underlying purpose of an *Investigation Expo.* Remind students that scientists regularly engage in this type of activity and highlight the benefits of sharing findings to both the individual scientist and to the scientific community as a whole.

Assess whether students comprehend their task in an investigation expo by asking:

- "What are the two main parts of an *Investigation Expo*?" *(presentations and discussions)*
- "What are the five items to be included in each poster and presentation?" *(question, prediction, procedure, results, and your interpretation of the results)*

⬡ Get Going

Have students prepare their posters and presentations as described in the student text.

△ Guide

Before beginning the presentations, ask the class for some examples of questions that they might ask during the presentation. Point out to students that they do not have to be able to answer every question; questions often come up during scientific presentations that may require further investigation before they can be answered. However, they should be able to answer questions about what they did, what they observed, what they think the results mean and why.

⬡ Get Going

Have groups make their presentations to the class.

During their presentations, help students with their questioning by modeling appropriate questions. Also, model appropriate discussion techniques and help redirect the discussion if needed. Avoid dominating the discussion; keep the discussion student-centered, not teacher-centered.

NOTES

Reflect

5 min.

Students consider what they have learned about bacteria and experimental design as a result of their investigation.

Reflect

Answer the following questions. Be prepared to share your answers with the class.

1. What did you learn about where bacteria live?
2. Why do you think they live in those places?
3. What trends did you see in the data of other groups about where bacteria live?
4. Think about the investigation you planned. Was it a fair test? If you do not think you ran a fair test as you had planned, describe how you would change your procedure if you had a chance to run the experiment again.

What's the Point?

The investigations in this section and the previous section helped you understand more about bacteria. You have learned about where they might be present. The data and conclusions you had for your investigations may have been difficult to make sense of. This is true in many science investigations.

Sometimes it is helpful to do an investigation more than once so that there is more data to analyze. Sometimes it is helpful to share your data with others so they can offer their ideas. This is why you do *Investigation Expos*. In these presentations, you hear about other people's results and combine them with your own to better understand the claims. This gives you the opportunity to see how many parts of the puzzle come together.

⬡ Get Going

After all of the groups have presented, have students read the *Reflect* questions and write down their answers. Lead a class discussion of students' responses to the *Reflect* questions. Elicit students' ideas about what they think they have learned about bacteria and about experimental design.

Assessment Options

Targeted Concepts, Skills, and Nature of Science	How do I know if students got it?
Scientists often collaborate and share their findings. Sharing findings makes new information available and helps scientists refine their ideas and build on others' ideas. When another person's or group's idea is used, credit needs to be given.	**ASK:** What changes did you make in your procedure based on feedback you received during the class discussion? **LISTEN:** Students should identify specific changes they made to their procedure.
Scientists must keep clear, accurate, and descriptive records of what they do so they can share their work with others, consider what they did, why they did it, and what they want to do next.	**ASK:** Which observations of your bacteria colonies were you not able to express as a measurement? How did you record those observations? **LISTEN:** Students should identify characteristics that are difficult to quantify such as color, texture, shape, and how colonies interact at borders with other colonies. Students should recognize that such observations should be recorded as detailed notes.
Scientific questions are directed toward objects and events that can be described, explained, or predicted by scientific investigations.	**ASK:** Which of the criteria for a good research question direct it toward an investigation rather than library research? **LISTEN:** Students should cite the requirement that is involved in collecting and recording data.
Identifying factors that lead to variation is an important part of a scientific investigation.	**ASK:** Why should all of the variables except the independent variable in an investigation be controlled? **LISTEN:** Students should connect changing more than one variable at a time with uncertainty about which change caused the response in the dependent variable.

Targeted Concepts, Skills, and Nature of Science	How do I know if students got it?
Explanations are claims supported by evidence, accepted ideas, and facts.	**ASK:** State one claim that is supported by the data collected during your investigation. **LISTEN:** Students should be able to make claims that are supported by the data they collected.
Students should be able to recognize and analyze patterns and trends.	**ASK:** What trends in the data did you predict before you carried out your investigation? How did your prediction compare with the actual data collected? **LISTEN:** Students should be able to state their initial prediction and cite data that either supports or refutes their prediction.

Teacher Reflection Questions

- In this section of *Learning Set 2*, students worked independently to design an investigation. Which aspects of the task did your students find most challenging? What support did you provide for students struggling with those aspects of the task?
- Class discussions played a key role in two major parts of this activity – the *Plan Briefing* and the *Investigation Expo*. How did you use these occasions to model methods of interacting respectfully during a group discussion and critiquing ideas rather than people when offering feedback?
- One technique commonly used to help students develop social interaction skills is to provide them with sample gambits (remarks made to open or redirect a conversation). For example, "Do I understand you correctly to say…?" or "I'm not sure that is right, I thought it was…." or "What else makes you believe…?" What gambits could you suggest to your students for use when critiquing the work of others in the class? In what other types of activities would your students benefit from having samples of gambits they could use when interacting?

SECTION 2.8 INTRODUCTION

2.8 Read

Some Germs Are Viruses

◀ *1 class period*

A class period is considered to be one 40 to 50 minute class.

Overview

Students are presented with information about another disease-causing agent, viruses. Viruses and bacteria are compared and contrasted, and students examine information about a variety of diseases caused by viruses, including: how they are spread, the effect of the disease on the body, and whether or not a cure has been found. Students then consider whether viruses are alive and the way in which viruses reproduce is described.

Targeted Concepts, Skills, and Nature of Science	Performance Expectations
Scientists often collaborate and share their findings. Sharing findings makes new information available and helps scientists refine their ideas and build on others' ideas. When another person's or group's idea is used, credit needs to be given.	Students propose arguments for and against viruses being considered alive.
The cell is the basic unit of structure and function in all living things. Cells are usually microscopic in size.	Students compare and contrast the structures and functions of a cell with those of a virus.
Disease is a breakdown in the structures or functions of an organism. Some diseases are the result of failures of a body system. Other diseases are the result of damage by infection from other organisms (germs).	Students describe some of the effects of viruses on the body.
Many diseases are caused by bacteria or viruses.	Students describe how viruses interact with cells and how viruses reproduce. Students should be able to identify diseases caused by viruses.

Materials	
1 per classroom	Optional – projections of images in student text

Homework Options

Reflection

- **Science Content:** How are viruses like and unlike living things? *(Like living things, viruses have genetic material and reproduce. Unlike living things, viruses are not made of cells; reproduce only within the cells of living hosts; once replicated within the host cell viruses do not continue to grow; viruses rely entirely on the host cell's ability to produce the energy needed to manufacture the virus.)*
- **Science Process:** In the previous investigation, you grew bacteria on agar in a Petri dish. Why would you not be able to grow viruses using this technique? *(Viruses require a living host cell to reproduce, and nutrient agar is not made of living cells.)*

Preparation for 2.9

- **Science Process:** What "shots" are required in order to be able to attend school? *(Common immunizations required by school districts include: DPT - diphtheria, pertussis, and tetanus [whooping cough]; MMR - measles, mumps, rubella [German measles]; varicella [chicken pox, hepatitis B and polio].)*Why do you think school districts require students to have these shots in order to attend school? *(To prevent children from getting the disease and spreading it to others.)*

SECTION 2.8 IMPLEMENTATION

2.8 Read

Some Germs Are Viruses

Any organism or particle that can get inside of you and make you sick is an **infectious agent**. You learned that some, but not all, bacteria are infectious agents. A virus is another example of an infectious agent. Like bacteria, they are microbes. They cannot be seen by the unaided human eye. However, in many ways viruses are different from bacteria. Viruses are much smaller than bacteria. They cause different illnesses. They work differently. The illnesses caused by viruses are treated differently. Viruses, although very small, cause a lot of changes in your body. They can also affect other animals and plants.

infectious agent: something that can get inside your body, multiply, and cause disease.

host: an organism (animal or plant) that harbors (provides food and a place to stay) for another organism, such as a virus.

What Are Viruses?

Viruses are particles that can cause many different types of diseases in plants and animals. Unlike bacteria, viruses cannot survive on their own. They need a **host** to survive. Viruses can only reproduce inside cells. Each virus has a way to reproduce once it is inside an organism's cell.

Viruses cause many different diseases. Some, like the cold virus, make you uncomfortable. Others, like the polio virus, can stop your body from working as it should.

There are many different types of viruses. Some viruses are able to infect only certain types of plants and animals. Some viruses infect only plants, others infect only animals, and still others infect only bacteria. For example, the bacteriophage infects bacteria. The feline immunodeficiency virus infects only cats.

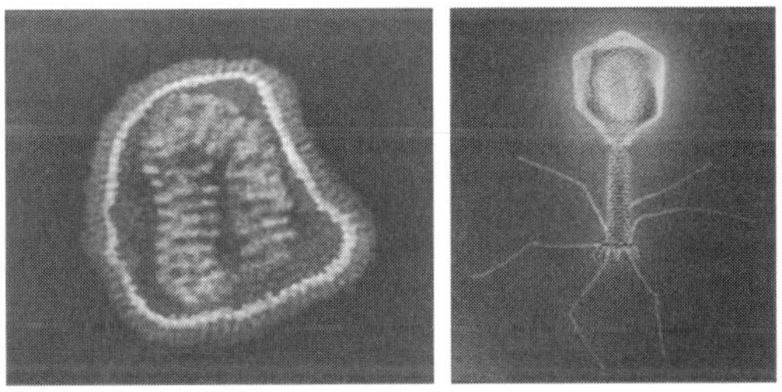

Viruses have many different shapes.

◀ *1 class period**

2.8 Read

Some Germs Are Viruses

5 min.

Students read about viruses – another microbe that causes communicable diseases. Students are given an overview of the structure of a virus, how viruses function and reproduce, and how they can be transmitted.

META NOTES

Activating prior knowledge prepares students for what they are about to learn. It also provides you with information about where students are in their understanding of viruses and alerts you to any misconceptions that may need to be addressed.

○ Engage

Before students begin reading, activate prior knowledge by asking a series of questions designed to elicit their ideas about viruses.

*A class period is considered to be one 40 to 50 minute class.

TEACHER TALK

"From your experiences, what can you tell the class about viruses?

What do you think viruses look like? and why do you think they look like that?

How do you think viruses are like or unlike bacteria?"

Make a class list of students' ideas about viruses.

⬡ Get Going

Read the introduction together with the class. Emphasize the ways in which viruses differ from bacteria.

What Are Viruses?

10 min.

Students read about a virus' need for a host in order to survive, diseases caused by viruses, the difficulties associated with treating viral diseases and the spread of viruses via droplet transmission.

...ther animals and p...

What Are Viruses?

Viruses are particles that can cause many different types of diseases in plants and animals. Unlike bacteria, viruses cannot survive on their own. They need a **host** to survive. Viruses can only reproduce inside cells. Each virus has a way to reproduce once it is inside an organism's cell.

Viruses cause many different diseases. Some, like the cold virus, make you uncomfortable. Others, like the polio virus, can stop your body from working as it should.

There are many different types of viruses. Some viruses are able to infect only certain types of plants and animals. Some viruses infect only plants, others infect only animals, and still others infect only bacteria. For example, the bacteriophage infects bacteria. The feline immunodeficiency virus infects only cats.

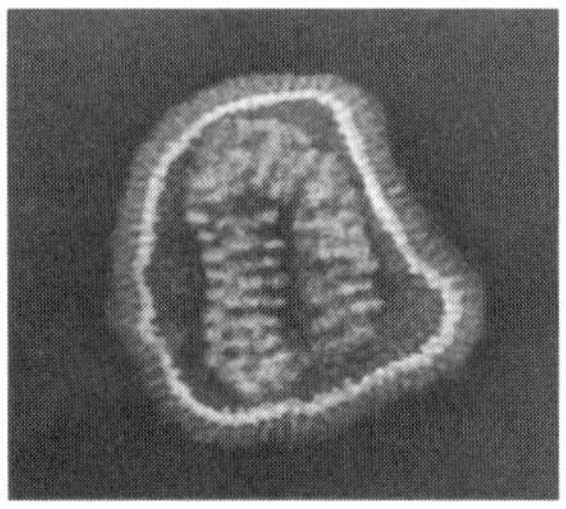

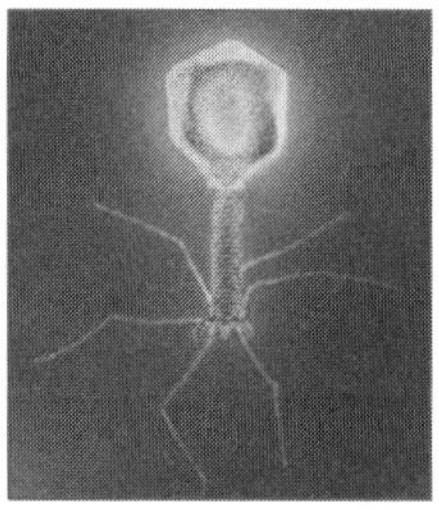

Viruses have many different shapes.

...cause dise...

host: an organism (animal or plant) that harbors (provides food and a place to stay) for another organism, such as a virus.

GF 55 GOOD FRIENDS and GERMS

⬡ Get Going

As students read this section, consider asking them to make a graphic organizer showing the main idea of each paragraph.

Have students read the table showing some common diseases caused by viruses, and have them use the information to update their *Communicable Disease Information Table.*

PBIS Learning Set 2 • **What Kinds of Things Make You Sick?**

droplet transmission: a way that an infectious disease can be transmitted. Droplets containing an infectious agent (bacteria or viruses) are released into the air when an infected person sneezes, coughs, talks, or exhales. They then come into contact with another person's eye, nose, or mouth.

Because of the number of viruses and the ways they can change over time, it is difficult to treat illnesses caused by viruses (viral illnesses).

Treatment for illnesses caused by viruses depends on the illness. Some diseases, like the common cold and the flu, are hard to treat and impossible to cure. You might take some "cold medicine." The medicine will make you *feel* better. It may relieve your symptoms. However, it does not make you better. For that, you have to wait for the virus to die out. For a cold, this usually takes about seven days. A lot of illnesses caused by viruses are like that. Viruses are very hard to remove from your body. For most viral illnesses, you just have to wait until the virus dies out. The table gives you information about some common viral illnesses.

Viruses are easily moved from one person to another the same way bacteria are spread. Many viruses infect other people by **droplet transmission**. Other viruses are spread through contact with open sores.

Disease	Caused by and spread by	Effect on body	Cure or not?	Picture
influenza (also called "the flu")	caused by: influenza virusspread by: person to person in droplets of coughs and sneezes, droplet spread	discomfort, high fever, can lead to other diseases	no cure, treat the symptoms to make more comfortable; yearly immunization for prevention	
common cold	caused by: rhinovirus, adenovirus spread by: droplet spread	discomfort but not as much as the flu, can cause other diseases	no cure	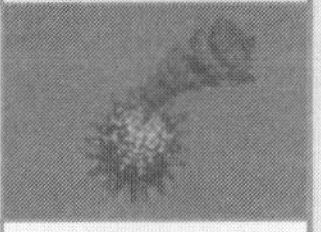
chickenpox	caused by: varicella virus spread by: droplet spread	scratchy itchy spots on body, fever, headache	no cure, immunization for prevention; becoming less common because of recent immunization requirements	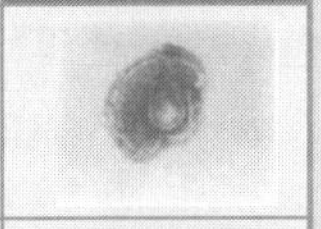
smallpox	caused by: variola virus spread by: saliva droplet spread or direct contact with sores	small itchy spots, high fever, death	no cure, immunization for prevention; almost eradicated from the planet	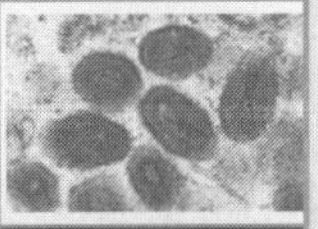
polio	caused by: poliomyelitis virus spread by: feces, hand to mouth	minor to severe muscle weakness and paralysis	no cure, immunization for prevention, almost eradicated from the planet	

GF 56

Project-Based Inquiry Science

△ Guide

Lead a class discussion of the main ideas that students have identified. Emphasize that viruses are particles that parasitize living host cells; they cannot survive and reproduce on their own. Elicit that there are many kinds of viruses and that some are only able to infect certain types of plants or animals. Draw students' attention to actions that can result in droplet transmission, and actions they can take to minimize droplet transmission.

Are Viruses Alive?

10 min.

Students consider some of the ways in which viruses are like and unlike living things. The mechanism by which viruses reproduce is described in detail.

2.8 Read

Are Viruses Alive?

What makes something alive? When you were reading about bacteria, you read that bacteria reproduce, use energy, and grow. Viruses can also reproduce, use energy, and grow. However, unlike bacteria, viruses can only do this inside another cell.

To reproduce, a virus attaches itself to the cell membrane of a host cell. It then either injects (inserts) its genetic material into the host cell or enters the host cell. The genetic material contains all the information required to make new viruses. Inside the cell, the genetic material causes the host cell to produce the necessary components to make new viruses. The components made by the host cell then assemble to make new viruses. The new viruses are released from the host cell. The exact way that the new viruses are made and released varies from virus to virus. However, they all follow the sequence described. A virus must use the structures and metabolism of a host cell to reproduce.

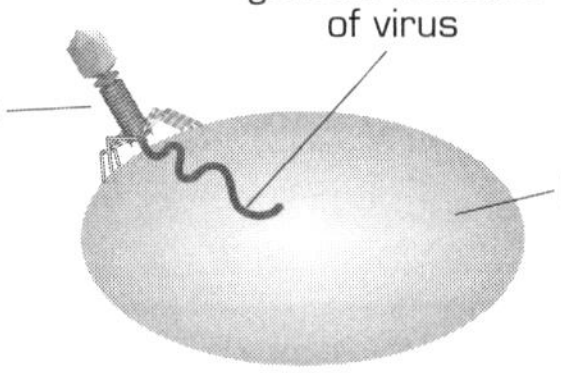

Scientists have learned a lot about viruses in the last 50 years. This has happened because new tools have made it possible to see the viruses. Very powerful microscopes, called electron microscopes, have made it possible for scientists to see viruses and identify how they work. Microscopes are so powerful now that scientists can see viruses as they move. This has made it possible to develop new ways to prevent viral illnesses.

Stop and Think

1. Explain one way that viruses can spread from person to person.
2. In what ways are viruses different from bacteria?
3. Describe the sequence of events that take place when a virus reproduces.
4. In your opinion, are viruses alive? Give reasons for your opinion.

What's the Point?

Viruses are one type of microbe that cause illness. There are many different types of viruses. Some viruses infect certain animals and others may infect certain plants. There is no cure for many viral illnesses. However, by learning about how viruses spread, you can prevent other people from getting sick.

GF 57

GOOD FRIENDS and GERMS

Get Going

Ask students how they could tell if something is alive. Make a class list of students' ideas. Have students read this section and then discuss the question "Are viruses alive?" with their group. Encourage students to refer to the class list of the characteristics of living things during their discussion. Have groups make a list of evidence that support the idea that viruses are alive and evidence that support the idea that viruses are not alive.

△ Guide

Have students read the *Stop and Think* questions and record their answers. When students have finished answering the questions, hold a brief class discussion focusing on whether or not viruses are alive. Create a class list stating what students think and the evidence that supports their claim.

Assessment Options

Targeted Concepts, Skills, and Nature of Science	How do I know if students got it?
Scientists often collaborate and share their findings. Sharing findings makes new information available and helps scientists refine their ideas and build on others' ideas. When another person's or group's idea is used, credit needs to be given.	**ASK:** When the class discussed the question "Are Viruses Alive?", which ideas proposed by other groups did you find most persuasive? **LISTEN:** Students should be able to cite arguments for and against viruses being considered alive. Students may argue for the ability to reproduce, use energy and grow. Others may argue against, citing that viruses are not made of cells, cannot reproduce, use energy, or grow outside of a living host cell.
The cell is the basic unit of structure and function in all living things. Cells are usually microscopic in size.	**ASK:** How does the structure of a virus differ from the structure of a cell? **LISTEN:** Students should be able to compare and contrast cell structures and functions with those of a virus. Responses should include: like cells, viruses contain genetic information; unlike cells, viruses do not have cytoplasm or organelles that carry out life functions.

Stop and Think

10 min.

Hold a class discussion of the characteristics of viruses and whether or not students think viruses are alive.

NOTES

Targeted Concepts, Skills, and Nature of Science	How do I know if students got it?
Disease is a breakdown in the structures or functions of an organism. Some diseases are the result of failures of a body system. Other diseases are the result of damage by infection from other organisms (germs).	**ASK:** How do viruses affect the body? **LISTEN:** Students should be able to describe some of the effects of viruses on the body listed in the table on page 56 of the student text.
Many diseases are caused by bacteria or viruses.	**ASK:** What are some of the illnesses caused by viruses? **LISTEN:** Students should be able to identify viral diseases such as influenza, the common cold, chickenpox, smallpox and polio. **ASK:** Describe how viruses interact with cells and reproduce. **LISTEN:** Students should describe how viruses attach to the cell membrane of a host cell and inject their genetic material, which then directs the cell to produce necessary components and assemble them to make a new virus.

Teacher Reflection Questions

- When the class discussed the question "Are viruses alive?", what conceptions about "life" did you encounter? How could you help students refine their understanding of these concepts?
- In what ways were students productively engaged during this section of the *Learning Set?* How could you tell?
- How are you encouraging students to listen and respond to the remarks of their peers during discussions in large and small groups?
- To what extent did your questioning foster critical and creative thinking?

SECTION 2.9 INTRODUCTION

2.9 Case Study

Investigating Smallpox

1 class period

A class period is considered to be one 40 to 50 minute class.

Overview

Students examine a case study of a disease caused by a virus, smallpox. The case study describes both the characteristics of the disease and the historic role smallpox played in the development of immunization by vaccination.

Targeted Concepts, Skills, and Nature of Science	Performance Expectations
Scientists often collaborate and share their findings. Sharing findings makes new information available and helps scientists refine their ideas and build on others' ideas. When another person's or group's idea is used, credit needs to be given.	Students will describe how scientists have built on the ideas of Edward Jenner to create vaccines for many diseases.
Explanations are claims supported by evidence, accepted ideas, and facts.	Students will provide evidence from Edward Jenner's experiment that supports the claim that having had cowpox provides immunity to smallpox.
Many diseases are caused by bacteria or viruses.	Students identify a variety of diseases caused by viruses, including smallpox, polio, rubella, diphtheria, and pertussis.
Immunization is a medical treatment that helps protect you from disease.	Students should be able to define immunization and describe the role of Edward Jenner in the historical development of immunization by vaccination.
Vaccination is the process by which a person is given a vaccine – a substance that protects a person from a disease.	Students should be able to distinguish between a vaccine and a vaccination.

Materials	
1 per classroom	Optional – projections of images in student text

Homework Options

Reflection

- **Science Process:** Ask your parents, or call your doctor and find out what immunizations you have been given. For which of the diseases mentioned in the case study have you been vaccinated? For which have you not been vaccinated? *(Students have typically received DPT, MMR, polio, and varicella immunizations. They have not received smallpox immunization.)*
- **Science Content:** Why do you think school districts do not require that children be immunized against smallpox? *(Smallpox has been eradicated; according to the CDC, the last known case occurred in Somalia in 1977.)*
- **Science Process:** In 1806, President Thomas Jefferson wrote a letter to Edward Jenner in which he said "future generations will know by history only that this loathsome disease has existed." Why do you think it took until 1977 for Jefferson's prediction to come true? *(Trust in vaccination came slowly and in order to eradicate smallpox, vast numbers of people worldwide had to be vaccinated.)*

Preparation for 2.10

- **Science Content:** How did case studies play a role in Edward Jenner's work on immunization? *(Jenner analyzed the information in the case studies of cowpox and concluded that there was evidence suggesting that having cowpox prevented a person from getting smallpox. Jenner then designed and carried out an experiment to test this idea.)*
- **Science Process:** Why would Jenner's experiment be unacceptable to the scientific community today? *(It is considered unethical to experiment on humans with life-threatening infectious disease agents.)*

SECTION 2.9 IMPLEMENTATION

2.9 Case Study

Investigating Smallpox

Bacteria and viruses can make you sick. The illnesses they cause can be very serious. Over the last 200 years, the number of bacterial and viral illnesses that can change your life has dropped dramatically. For example, 200 years ago most people got smallpox, almost as many as now get the flu. Unfortunately, many of those who were infected died of smallpox. Smallpox has been mostly **eradicated,** and you do not hear about people getting it anymore. Read about smallpox below. Use what you learn to fill in the *Communicable-Disease Information Table* with information about smallpox.

eradicated: wiped out.

Smallpox

Symptoms and Spread

Smallpox is a serious, contagious, and often fatal infectious disease. Smallpox is caused by the variola (vair-ee-OH-luh) virus. Humans are the only natural hosts of the variola virus. The symptoms include a rash and high fever. Smallpox is often recognized by the type of rash it causes.

The rash is in the form of small blisters. The blisters are filled with fluid and crusted over. This sounds like chickenpox, but the blisters look and feel different. The smallpox rash usually appears on exposed parts of the body: the face, arms, palms, lower legs, and soles of the feet. Other symptoms of smallpox include fever, headache, backache, and fatigue (tiredness).

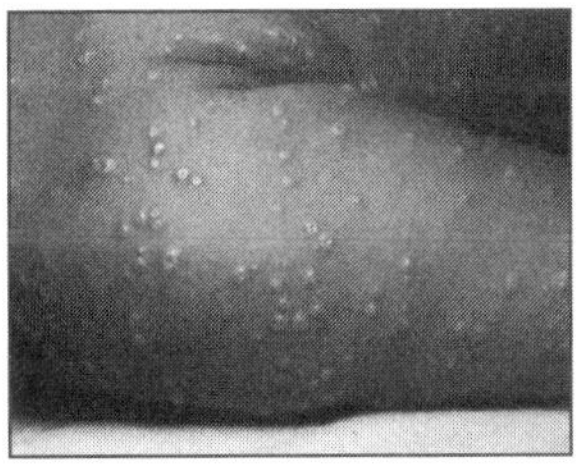

The "pox" means "spotted." It refers to bumps that appear on an infected person, as shown in the picture.

Generally, face-to-face contact is required to spread smallpox from one person to another. Smallpox also can be spread through direct contact with infected bodily fluids. If someone touches an object such as bedding or clothing that has smallpox on it, he or she may get infected. Sometimes, smallpox has been spread by viruses carried in the air. Smallpox is not known to be transmitted by insects or animals.

GF 58

Project-Based Inquiry Science

◄ *1 class period**

2.9 Case Study

Investigating Smallpox

5 min.

Students examine a case study of smallpox, the contributions of Edward Jenner, and his role in the development of immunization by vaccination.

○ Engage

Ask students to summarize what they have learned about viruses. Elicit the following:

- Viruses are very small particles.
- Viruses are infectious agents that cause disease.
- Viruses can only survive and reproduce inside a living host cell.

*A class period is considered to be one 40 to 50 minute class.

- It is difficult to treat illnesses caused by viruses because there are so many types of viruses and viruses can change over time.

Ask students how they think scientists might have studied diseases caused by viruses before it was possible for them to actually see viruses.

⬡ Get Going

Before reading, ask students if anyone they know has ever had smallpox and what they think they know about smallpox. Read the introduction together with the students. Consider reading it aloud to the class. Ask students why they have never seen a case of smallpox. Then have students take out their *Communicable Disease Information Table* and tell the students to have it ready so that as they read, they can add information about smallpox.

Smallpox: Symptoms and Spread

15 min.

information about smallp

Smallpox

Symptoms and Spread

Smallpox is a serious, contagious, and often fatal infectious disease. Smallpox is caused by the variola (vair-ee-OH-luh) virus. Humans are the only natural hosts of the variola virus. The symptoms include a rash and high fever. Smallpox is often recognized by the type of rash it causes.

The rash is in the form of small blisters. The blisters are filled with fluid and crusted over. This sounds like chickenpox, but the blisters look and feel different. The smallpox rash usually appears on exposed parts of the body: the face, arms, palms, lower legs, and soles of the feet. Other symptoms of smallpox include fever, headache, backache, and fatigue (tiredness).

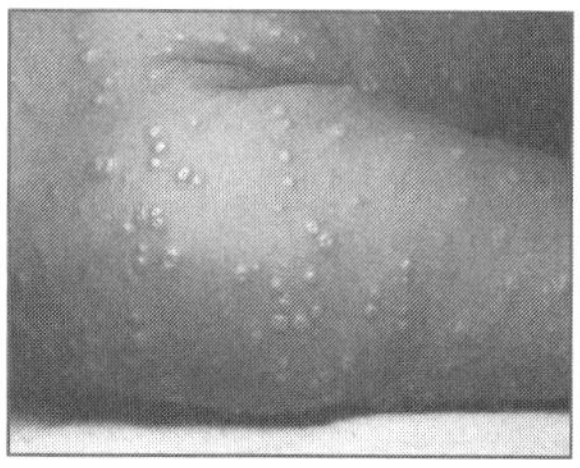

The "pox" means "spotted." It refers to bumps that appear on an infected person, as shown in the picture.

Generally, face-to-face contact is required to spread smallpox from one person to another. Smallpox also can be spread through direct contact with infected bodily fluids. If someone touches an object such as bedding or clothing that has smallpox on it, he or she may get infected. Sometimes, smallpox has been spread by viruses carried in the air. Smallpox is not known to be transmitted by insects or animals.

GF 58

Project-Based Inquiry Science

⬡ Get Going

Have students read the section. Remind students that as they are reading, they should add information about smallpox to their *Communicable Disease Information Table.*

Consider having students read aloud and pausing during the reading when relevant information is given so that students can add it to the table.

△ Guide

The reading selection generally follows the same sequence as the columns in the *Communicable Disease Information Table.* Guide students to add the following information to the table:

Name of disease	Cause (bacterial or viral)	How does it spread?	Which body systems does it affect?	What are the symptoms?	How is it treated? Is there a vaccine?	Other notes
smallpox	Variola virus	Direct contact with infected person or infected body fluids; contact with infected objects such as bedding or clothing	Skin, muscular	rash in the form of small blisters, high fever, headache, backache, feeling tired	rest and fluids to prevent dehydration during fever	Vaccination can provide immunization

NOTES

Immunization

15 min.

A person with smallpox becomes contagious when they get a fever. A person becomes more contagious when they get the rash. At this stage, the infected person is usually very sick and not able to move around in the community. The infected person is contagious until the last smallpox scab falls off.

Immunization

Have you ever heard of smallpox? Has anyone in your family had smallpox? You probably answered that no one in your family has had smallpox. This is because scientists have discovered how to prevent smallpox through **vaccination**. Edward Jenner was one of the first scientists to discover how to protect people from diseases through vaccination. He listened to stories about who got sick and who did not get sick and made sense of the reports he heard. He figured out how **immunization** might work based on these reports. These reports are called **case studies**.

The year was 1796. There were no cars, no telephones, and no electricity in houses. Edward Jenner heard that milkmaids, girls who milked cows, did not get smallpox. However, milkmaids did get a disease similar to smallpox from milking the cows. It was called cowpox. Unlike smallpox that killed many victims, the milkmaids only had a few blisters, felt a little tired, and had some aches.

vaccination: the process by which a substance that protects a person from a disease is given.

immunization: a medical treatment that helps protect you from disease.

case study: an observation of a person or group to use as a model.

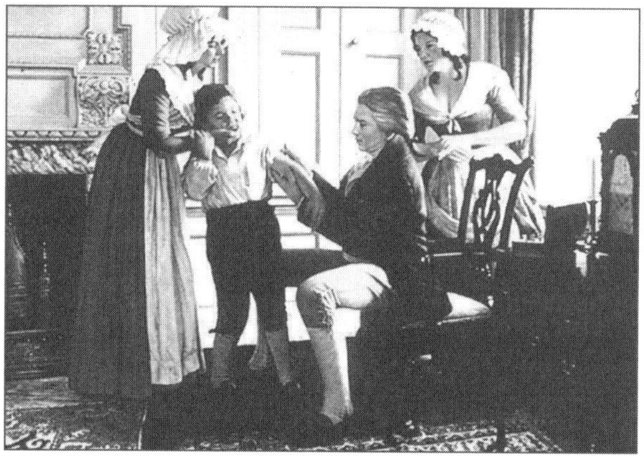

This picture shows Jenner transferring fluid from a cowpox blister into a cut on a boy's arm.

Get Going

Before reading, ask students what they think they know about vaccination. Make a list of students' ideas. Consider having students read this section in pairs or in groups. After students have completed reading this section, ask groups to add information about the diseases described on page GF 60 to their *Communicable-Disease Information Table.* Let students know that even though there is not enough information to complete all of the columns on the table now, they should add these diseases because more information will be given later in the Unit.

He thought that maybe cowpox protected them. He decided to try an experiment. With the permission of the boy's father, Jenner gave an eight-year-old boy the cowpox virus that the milkmaids got. The boy had a few aches but did not get very sick. Later, Jenner exposed the boy to smallpox. The boy did not get sick. It seemed that the cowpox virus could protect the boy from smallpox. Several months later, Jenner again exposed the boy to smallpox disease. Again the boy stayed healthy. Jenner had found a way to prevent smallpox. The cowpox virus was acting as a **vaccine**.

vaccine: a substance that protects a person from a disease; from the Latin *vacca* for cow.

Trust in vaccinations came very slowly. But, little by little, people came to believe that the cowpox vaccine could prevent smallpox. Up until the 1970s, everyone was vaccinated against smallpox. This vaccination practice stopped because smallpox has been eradicated from the general population.

Before vaccines, parents in the United States could expect a whole variety of diseases to affect their children:

- *Polio*, a virus, would paralyze 10,000 children.
- *Rubella (German measles)*, a virus, would cause birth defects and mental retardation in as many as 20,000 newborns.
- *Measles*, a virus, would infect about 4 million children, killing 3,000.
- *Diphtheria*, a virus, would be one of the most common causes of death in school-aged children.
- *Haemophilus influenzae type b (Hib)*, a bacterial infection, would cause meningitis in 15,000 children, leaving many with permanent brain damage.
- *Pertussis (whooping cough)*, caused by a virus, would kill thousands of infants.

Just two generations ago, when your grandparents were children, they feared polio, rubella, and measles. Your parents also worried about measles, mumps, and rubella. Since Edward Jenner, vaccinations have been made for many different diseases. Vaccines have become commonplace in the last hundred years. In many cases, vaccines have eliminated diseases that killed or severely disabled people. Most children are given vaccinations as part of their health care in the first two years of their lives.

△ Guide

After students have updated the *Communicable-Disease Information Table*, discuss Edward Jenner's investigation with the class. Ask students to identify the question Edward Jenner was investigating, the procedure he followed, and the observations made during the investigation. Ask students to state a claim that would be supported by the data collected during Jenner's investigation. Make it clear to students that Jenner did not know how or why immunity worked, only that it did. Also, ask students why they think people were slow to trust vaccinations. Guide students to understand that it took many years for scientists to determine the mechanism by which vaccination makes a person immune to a disease. Let students know that in *Learning Set 3* they will investigate exactly how the body becomes immune to a disease.

Ask students why Jenner's experiment would probably not be allowed today. Discuss the ethical issues surrounding experimentation on humans.

☐ Assess

During the discussion, listen for the following:

- Jenner's question: "If a person gets cowpox, will that prevent them from getting smallpox?"
- The steps in the investigation carried out by Jenner.

  1. Infect the boy with cowpox.
  2. Observe that the boy did in fact get cowpox.
  3. Expose the boy to smallpox.
  4. Several months later, expose the boy to smallpox again.

- Results:

  Claim: If you get cowpox, you will be immune to smallpox.

  Evidence: After getting cowpox, the boy did not get smallpox when exposed to it.

- Ethical issues include, but are not limited to:

  The boy might have gotten smallpox and died.

  Does a parent have the right to permit their child to participate in a potentially life-threatening experiment?

How Important Are Vaccinations?

Immunization has saved millions of lives in the last 200 years. In 1980, smallpox was declared eradicated. Polio has not been a problem for almost 50 years. These diseases can almost not be found. This means that people live longer, healthier, and better lives. Today, scientists continue to invent vaccinations for deadly illnesses. Some of those illnesses are bacterial (caused by bacteria) and some are viral (caused by viruses). Recently, scientists have developed a vaccine against a virus that can cause cancer in women.

What's the Point?

Scientists have developed many ways to combat some viral and bacterial diseases. Using vaccinations, they have eradicated some disabling and deadly diseases. Smallpox is one of those. Polio, chicken pox, measles, mumps, and rubella have also almost been eradicated. Vaccinations play an important role in maintaining people's health. Vaccinations are usually given to children early in their lives.

Scientists use new technology to see viruses and bacteria and to figure out how those microbes work. Then they are able to invent new ways of stopping them from working. These new scientists "stand on the shoulders of giants" like Edward Jenner. These scientists build on the work of Jenner, who used observations and case studies to figure out that immunization was possible.

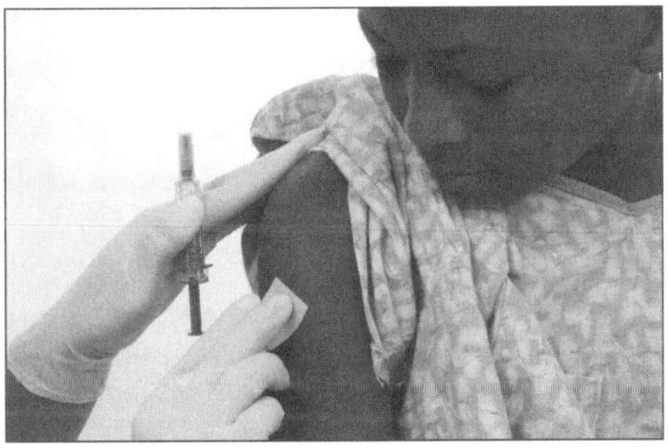

How Important Are Vaccinations?

5 min.

Once students have learned of Jenner's experimentation they will continue to see how it has affected today's world.

△ Guide

After students have read this section, hold a class discussion focusing on the following questions:

1. Is it still important to get your vaccinations?
2. What would happen if you didn't get your vaccinations?

META NOTES

This section of the Learning Set connects the development of the first vaccine with the importance of vaccines today. Vaccines are required for school entry and before going to other countries because they are important in preventing disease and the spread of disease. Discussion of these last two questions sets the stage for the next activity.

Assessment Options

Targeted Concepts, Skills, and Nature of Science	How do I know if students got it?
Scientists often collaborate and share their findings. Sharing findings makes new information available and helps scientists refine their ideas and build on others' ideas. When another person's or group's idea is used, credit needs to be given.	**ASK:** From 1798 to 1801, Jenner published four papers describing his cowpox/smallpox experiments and also made presentations to the medical community in London. How did these actions help advance science and medicine? **LISTEN:** Students should recognize that sharing his findings made new information available and helped other scientists build on his ideas to create vaccines for many other diseases.
Explanations are claims supported by evidence, accepted ideas, and facts.	**ASK**: What evidence did Edward Jenner obtain from his experiment that supported his claim that getting cowpox provides immunity to smallpox? **LISTEN:** Students should cite the two instances in which, after getting cowpox, the boy was exposed to smallpox, but did not get the disease.
Many diseases are caused by bacteria or viruses.	**ASK:** In addition to smallpox, what are some other viral diseases for which vaccines have been developed? **LISTEN:** Students identify a variety of diseases caused by viruses for which there are vaccines, including polio, rubella, measles, diphtheria, meningitis and pertussis.
Immunization is a medical treatment that helps protect you from disease.	**ASK:** What does it mean when we say a person has been immunized? **LISTEN:** Students should define immunization as a medical treatment that helps protect you from a disease.

Targeted Concepts, Skills, and Nature of Science	How do I know if students got it?
Vaccination is the process by which a person is given a vaccine – a substance that protects a person from disease.	**ASK:** What did the vaccine used by Edward Jenner contain? How did Jenner vaccinate the boy? **LISTEN:** Students should identify fluid from a cowpox blister as the vaccine and transferring the fluid to a cut on the boy's arm as the process of vaccination.

Teacher Reflection Questions

- In what ways does this case study set the stage for the next activity of this *Learning Set?*
- Are my assessment techniques appropriate for the amount and type of information needed by my students to be successful?
- Am I providing sufficient opportunity and time for the children to work independently, in pairs and in small groups?
- How am I encouraging students' questions and curiosity?

NOTES

NOTES

SECTION 2.10 INTRODUCTION

2.10 Case Study

Disease Overviews

◀ $1\frac{1}{2}$ *class periods*

A class period is considered to be one 40 to 50 minute class.

Overview

Students work in small groups to examine case studies of infectious diseases. Each group examines a case study for a single disease and adds information about that disease to their *Communicable-Disease Information Table*. Groups then make class presentations in which they share what they have learned about that disease. Students add information about the diseases presented by other groups to their own *Communicable-Disease Information Table*.

Targeted Concepts, Skills, and Nature of Science	Performance Expectations
Scientists often collaborate and share their findings. Sharing findings makes new information available and helps scientists refine their ideas and build on others' ideas. When another person's or group's idea is used, credit needs to be given.	Students share information from their group's case study with the class and use information presented by other groups to update their own *Communicable-Disease Information Tables*.
Disease is a breakdown in the structures or functions of an organism. Some diseases are the result of failures of a body system. Other diseases are the result of damage by infection from other organisms (germs).	Students identify the effects that specific diseases have on body systems during class presentations of the case studies they have examined.
Many diseases are caused by bacteria or viruses.	Based on case study presentations to the class, students will identify whether a disease is caused by a virus or a bacteria.
Vaccination is the process by which a person is given a vaccine – a substance that protects a person from a disease.	During class case study presentations, students will identify chickenpox as a disease that can be prevented by vaccination.
An antibiotic is a drug that kills or prevents the growth of bacteria.	Students will identify antibiotics as a treatment for bacterial diseases.

Materials	
1 per classroom	Optional – projections of images in student text

Homework Options

Reflection

- **Science Process:** Research an infectious disease that has not been mentioned in this Unit. Add the disease to your *Communicable-Disease Information Table* and any information that you found which answers the questions on the table. Write a one-page summary of your findings to share with the class.
- **Science Content:** Why are antibiotics not an effective treatment for a disease caused by a virus? *(antibiotics kill bacteria, not viruses)*

Preparation for Back to the Big Question

- **Science Process:** What new information about disease prevention did you obtain by listening to presentations made by other students? *(Sample answers include: additional diseases that can be treated with antibiotics, personal hygiene practices that prevent disease spread, there can be viral and bacterial forms of the same disease.)*

SECTION 2.9 IMPLEMENTATION

2.10 Case Studies

Disease Overviews

Edward Jenner used case studies to learn more about the viral illness smallpox. On the following pages, you will read case studies about five people with different diseases. You will learn a lot about the diseases and the people who have them.

Reading some of the case studies might be difficult because they contain a lot of medical terms. For example, in the first reading, about chickenpox, the word varicella-zoster is used. This is the medical term for chickenpox. One way to read this type of word is to pronounce it the best you can. Get the main idea of the word, and then keep reading. As long as you get the main ideas of the case study, you do not need to pronounce or remember every word exactly.

Procedure

The case studies that follow give you information about five different diseases.

1. Read the case study (or case studies) your teacher assigns you.
2. As you are reading the case study, look for the information that helps you answer the questions in the Communicable-*Disease Information Table*. You will need to know
   - the name of the disease,
   - whether it is caused by a bacteria or a virus,
   - how it spreads,
   - how it affects your body systems,
   - the symptoms, and
   - how it is treated.

In the column about how it is treated, tell whether the disease has a vaccine. A sample entry into the table is shown. Notice the questions you should try to answer in each column.

◀ *1 class period**

2.10 Case Study

Disease Overviews

5 min.

Students consider how case studies can be used to learn about diseases.

Engage

Ask students to identify diseases that they would be interested in studying and what they think causes those diseases. Record students' ideas and discuss them. You may want to assign students to investigate diseases that will not be examined in the case studies as part of a homework assignment.

*A class period is considered to be one 40 to 50 minute class.

△ Guide

Have students recall the nature of a case study from the previous activity. Emphasize the need to not get sidetracked by medical terminology as they examine their case study.

every word exactly.

Procedure

The case studies that follow give you information about five different diseases.

1. Read the case study (or case studies) your teacher assigns you.
2. As you are reading the case study, look for the information that helps you answer the questions in the Communicable-*Disease Information Table*. You will need to know
   - the name of the disease,
   - whether it is caused by a bacteria or a virus,
   - how it spreads,
   - how it affects your body systems,
   - the symptoms, and
   - how it is treated.

In the column about how it is treated, tell whether the disease has a vaccine. A sample entry into the table is shown. Notice the questions you should try to answer in each column.

Procedure

20 min.

Groups are assigned a disease case study and they are informed of the expectations of the task. Then students work together in their groups to examine the case study and record information about it on their Communicable-Disease Information Table.

Have students read the *Procedure* they are to follow when examining the case studies, and have them discuss the expectations of the task. Draw students' attention to the sample *Communicable-Disease Information Table,* and highlight both the questions for each column and the elaboration given below each question.

Either allow groups to select the case study that they would like to examine (ensuring that each of the five case studies will be examined by a group), or assign each group one of the five case studies to examine.

□ Assess

Assess whether students understand what is expected of them when they examine their case study.

- Ask students to identify the information about the disease that they should be looking for when examining their case study. Listen for the six bulleted items listed in the student text.
- Ask students how they will record the information they find. Student responses should mention recording the information in the appropriate column on their *Communicable-Disease Information Table.*

2.10 **Case Studies**

Be a Scientist

Case Study: Chickenpox

India is two years old. She woke up this January morning with a low fever, and she feels terrible. She also has spots all over her. The spots are red and quite large and are mostly on her belly, face, chest, and back. They itch a lot. India is finding it hard not to scratch the spots. The last time her mother looked at India's belly and chest, there were more.

India likes to play with her three-year-old cousin. Last week her cousin could not play with her because he had the chickenpox. Now India's mom is sure she also has chickenpox.

India has two older brothers. India's mother is worried about them now.

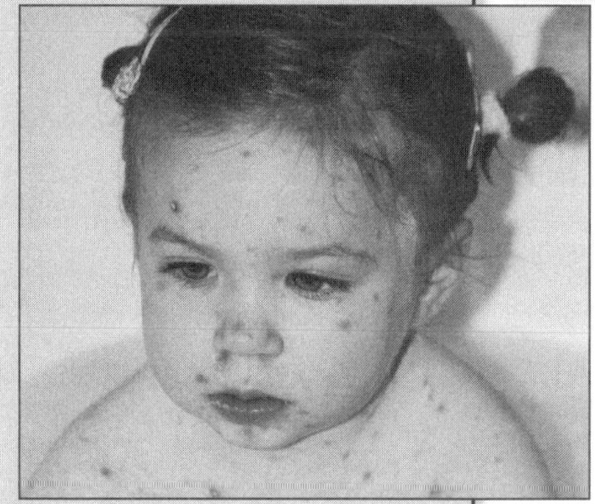

Communicable Disease Information Table 2.10.1

Name: ______________________ Date: ______________

Name of disease	Cause (bacterial or viral)	How does it spread?	Which body systems does it affect?	What are the symptoms?	How is it treated? Is there a vaccine?	Other notes
Write the scientific name and the common name. People usually say that they have "the flu," which is a shorter way to say that they have "influenza."	Many diseases are caused by either a virus or a bacterium.	Some diseases are spread through blood, body fluids (like saliva), air, contaminated food, and animal bites.	Some diseases affect several systems. List the main one(s).	The symptoms are what help you to determine that you have a particular disease.	"Treat the symptoms" means doing something so the symptoms will not be as bad. Sometimes, you can treat a disease, sometimes you can only treat the symptoms.	Use this column to write anything you think is interesting or important to remember about a particular disease.
common cold	viral	Body fluids in the air or on surfaces—blowing your nose, touching your hands to other people or objects, sneezing or coughing.	respiratory	cough, body aches, congestion	It can't be cured. You can only relieve the symptoms to feel better, using medicine to reduce the fever and to stop the runny nose, and cough.	

Chickenpox was once a common childhood disease. The number of cases of chickenpox in the United States has been reduced since a vaccine was introduced in 1995.

GF 63

GOOD FRIENDS and GERMS

TEACHER TALK

“As you examine your case study, what specific information about the disease should you be looking for? What are you expected do with the information about a disease that you find in your case study?”

Learn about Chickenpox

Chickenpox is a common disease caused by a virus called the varicella-zoster virus (VZV). Varicella-zoster virus spreads in the air through coughs or sneezes or through contact with fluid from inside the chickenpox blisters.

Chickenpox, which occurs most often in late winter and early spring, is a communicable disease. It can be spread very easily. Chickenpox is very contagious. If your brother gets chickenpox and you have not had it yet, there is an 80%–90% chance that you will get it too.

Although it is more common in persons under the age of 15, anyone can get chickenpox. A person has only one episode of chickenpox in his or her lifetime.

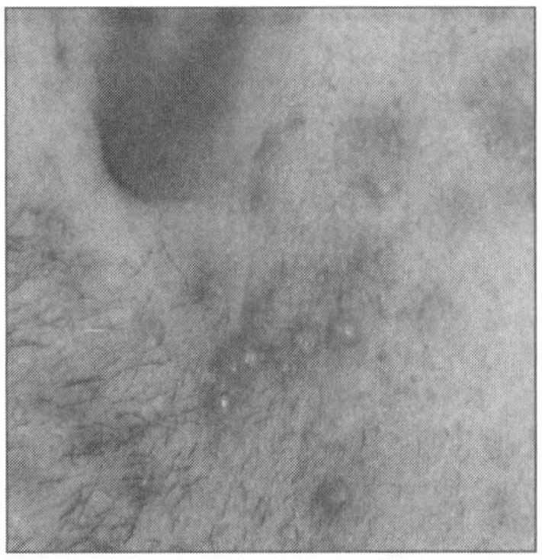

One of the symptoms of chickenpox is an itchy rash.

If you have a younger brother or sister, they may never get chickenpox. For the last few years, the chickenpox vaccine has been required of all children before they go to school. The vaccine gives you some of the chickenpox virus, but not enough to get you sick. Your body thinks you have had chickenpox and builds up defenses. You cannot get chickenpox again.

One of the symptoms of chickenpox is a red, itchy rash on the skin. The chickenpox rash usually appears first on the abdomen or back and face. Then it spreads to almost everywhere else on the body, including the scalp, mouth, nose, and ears.

Some people can have only a small rash, while others have a serious rash that spreads all over their body. The rash may begin as many small, red bumps that look like pimples or insect bites. These bumps are about two to four millimeters wide. They develop into thin-walled blisters filled with clear fluid. The fluid in the blisters becomes cloudy. The blister wall breaks, leaving open sores. The sores finally crust over to become dry, brown scabs.

Some infected people have a fever, abdominal pain, or a vague sick feeling a day or two before the rash appears. These symptoms may last for a few days. The fever stays in the range of 37.7°C to 38.8°C (100°F to 102°F). Younger children often have milder symptoms and fewer blisters than older children or adults.

⬡ Get Going

Have students work in small groups to examine their assigned case study.

Be a Scientist

Case Study: Lyme Disease

Last week, Shane took a walk in the woods with his dog. Shane lives in Maine where there are lots of fields and forests. Shane loves to take frequent walks to look at the wildlife and get some fresh air. This morning Shane woke up with a sore spot on his leg. It looks like a **bull's-eye**. He also has a fever and headache and his muscles hurt. Shane is going to stay home from school today.

Shane's dad called the doctor, who said to look very carefully at the center of the bull's-eye. Shane's dad used a magnifying glass to look at his leg. In the center of the bull's-eye, Shane's dad found a small insect. He carefully pulled it out, making sure that he got the whole insect.

The doctor prescribed some antibiotics for Shane, because Shane's illness is caused by bacteria. The medicine will help kill the bacteria that caused Shane's disease. It may take a long time for Shane to feel better.

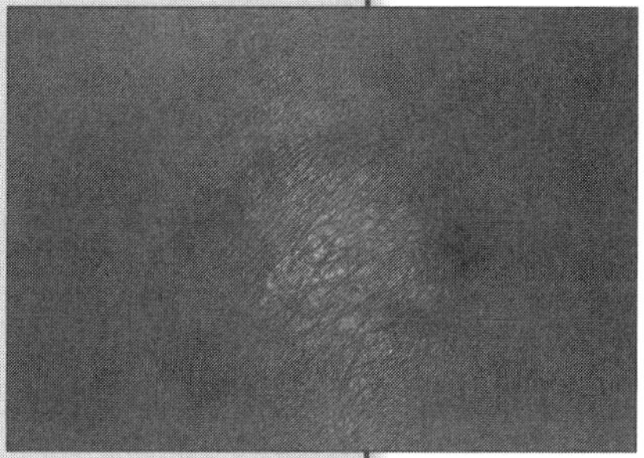

The first symptom of Lyme disease is a bull's-eye rash.

bull's-eye: a circular spot, usually black, at the center of a target. The bull's-eye is surrounded by concentric circles.

Learn about Lyme Disease

Lyme disease is caused by bacteria that are found in some ticks. The ticks live in wooded areas and fields. This is one reason why it is so important to check yourself for ticks after being outside, even in your backyard or neighborhood park.

Within one to two weeks of being bitten by the tick, infected people may have a "bull's-eye" rash. They may also have a fever, a headache, and muscle or joint pain. Some people have Lyme disease and do not have any early symptoms. Other people have a fever and other "flu-like" symptoms without a rash.

Not all ticks carry the bacteria that cause Lyme disease. A tick can become infected with the bacteria after it bites and feeds on an infected animal, often a deer. The bacteria live inside the tick and are passed into the animal that is bitten. Sometimes that bitten animal is a person.

☐ Assess

Monitor groups' progress. Check to see that all group members are recording information on their *Communicable-Disease Information Table* and that the information is in the appropriate column.

If Lyme disease is left untreated, after several days or weeks, bacteria may spread throughout the body of an infected person. Then there may be many more symptoms. An infected person may get rashes on other parts of the body. Symptoms also may include pain that seems to move from joint to joint. There may be signs of inflammation of the heart or nerves.

If the disease is still not treated, infected people can get additional symptoms. These can include swelling and pain in major joints, or mental changes. Some symptoms might not show up until months after getting infected. The treatment usually involves using different types of antibiotics.

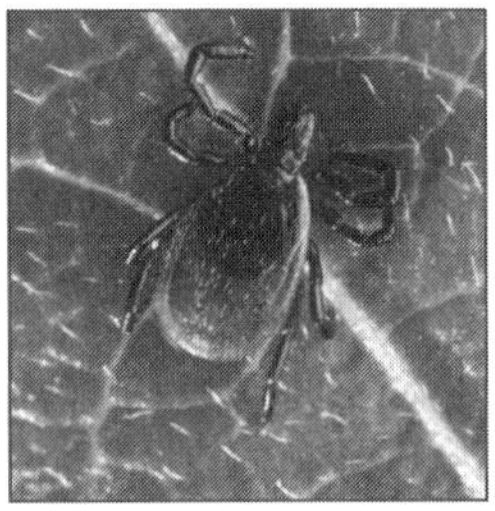

Adult female deer ticks like the one in this picture may carry the bacteria for Lyme disease.

While most of the cases of Lyme disease in the United States are in the Northeastern part of the United States, this disease is found all over the world and is a concern across the United States. Montana is the only state that has never reported a case of Lyme disease. During spring and summer, you will often hear news stories about Lyme disease. It is important to be careful if you live in an area that has deer ticks, because they may carry the bacteria that cause Lyme disease.

Be a Scientist

Case Study: *E. coli* Infections

It was August and September, 2006, and calls began to come into doctors' offices and hospitals around the country. People were complaining of diarrhea, fever, cramps, and vomiting. Most people felt very sick for several hours and missed school or work. Many people were hospitalized. Over the next few days, there were more calls from people all over the country. Officials counted calls from 25 states. Eventually, by the first of October, 199 people had gotten sick. Three people had died.

The calls from sick people began arriving about August 19. Investigators from the Food and Drug Administration (FDA) spoke with many of the sick people. Thinking that the illnesses were being caused by something in food, the FDA used a survey to identify what foods the people had eaten. The survey had hundreds of foods on it. People told the FDA what they had eaten in the last several days. When all the data were analyzed, spinach came out as a food most people had eaten three to four days before they got sick. The FDA began to look for contaminated spinach. They sent out the press release shown on the next page.

FDA officials also warned that *E. coli (Escherichia coli)* could be spread among people who did not maintain good hygiene. *E. coli* is found in the stools of people who are infected. This is especially a problem for people who take care of children who are not toilet trained. The FDA reminded people to wash their hands well after using the bathroom and after taking care of small children.

FDA Warning on Serious Foodborne E.coli O157:H7 Outbreak

One Death and Multiple Hospitalizations in Several States

The U.S. Food and Drug Administration (FDA) is issuing an alert to consumers about an outbreak of E. coli O157:H7 in multiple states that may be associated with the consumption of produce. To date, preliminary epidemiological evidence suggests that bagged fresh spinach may be a possible cause of this outbreak.

Based on the current information, FDA advises that consumers not eat bagged fresh spinach at this time. Individuals who believe they may have experienced symptoms of illness after consuming bagged spinach are urged to contact their healthcare provider.

"Given the severity of this illness and the seriousness of the outbreak, FDA believes that a warning to consumers is needed. We are working closely with the U.S. Centers for Disease Control and Prevention (CDC) and state and local agencies to determine the cause and scope of the problem," said Dr. Robert Brackett, Director of FDA's Center for Food Safety and Applied Nutrition (CFSAN).

E. coli O157:H7 causes diarrhea, often with bloody stools. Although most healthy adults can recover completely within a week, some people can develop a form of kidney failure called Hemolytic Uremic Syndrome (HUS). HUS is most likely to occur in young children and the elderly. The condition can lead to serious kidney damage and even death. To date, 50 cases of illness have been reported to the Centers for Disease Control and Prevention, including eight cases of HUS and one death.

At this time, the investigation is ongoing and states that have reported illnesses to date include: Connecticut, Idaho, Indiana, Michigan, New Mexico, Oregon, Utah, and Wisconsin.

FDA will keep consumers informed of the investigation as more information becomes available.

	Number if oersins with *E.Coil* infection
Arizona (AZ)	8
California (CA)	2
Colorado (CO)	1
Connecicut (CT)	3
Idaho (ID)	7
Illinois (IL)	2
Indiana (IN)	10
Kentucky (KY)	8
Maryland (MD)	3
Maine (MD)	3
Michigan (MI)	4
Minnesota (MN)	2
Nebraska (NE)	11
New Mexico (NM)	5
Nevada (NV)	2
New York (NY)	11
Ohio (OH)	25
Pennsylvania (PA)	10
Tennesse (TN)	1
Utah (UT)	19
Viraginia (VA)	2
Washington (WA)	3
West Viarginia (WV)	1
Wisconsin (WI)	45
Wyoming (WY)	1

Learn about *E. coli*

Escherichia coli is the formal scientific name for *E. coli*. It is pronounced as esh-er-ick-ee-eh cole-eye, but it is usually referred to as *E. coli*. Symptoms of *E. coli* include watery or bloody diarrhea, fever, abdominal cramps, nausea, and vomiting. Illness may be mild or severe. Very young children and very old or sick people are more likely to have severe symptoms, sometimes including kidney failure and death.

E. coli is a common cause of foodborne illness. An estimated 73,000 cases of *E. coli* infection and approximately 60 deaths occur in the United States each year. Most illness has been associated with eating undercooked, contaminated ground beef. Person-to-person contact in families and childcare centers is also an important mode of transmission. Infection can also occur after drinking milk that has not been **pasteurized**. A person can also be infected after swimming in or drinking sewage-contaminated water.

There are several things around your house you can do to prevent *E. coli* infection. For example, it is important not to use the same cutting board for raw meat and other foods. If you do, you can transfer bacteria that were on your meat to your vegetables. You might not get sick from the meat, because you cook it, but you can get sick from the vegetables, because you may not cook them. You also should use a different plate to carry uncooked meat out to the barbeque grill and to carry the cooked meat back inside.

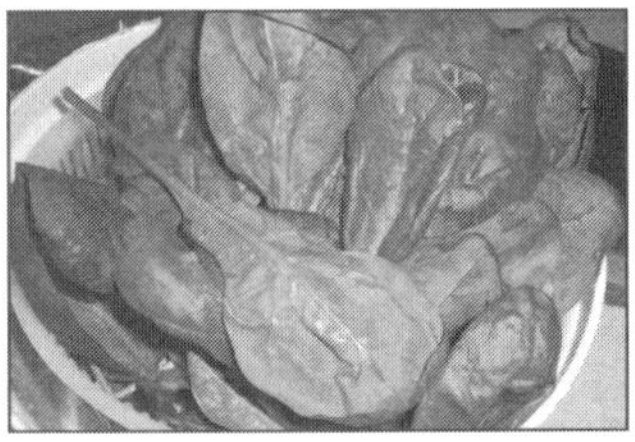

Uncooked spinach can sometimes harbor E. coli.

Be a Scientist

Case Study: Mononucleosis

A fifteen-year-old high-school student named Ally has been complaining about feeling ill for about three days. Ally's mom has decided it is time to take Ally to the doctor. The doctor examines Ally and asks her to describe her symptoms. Ally says she feels very tired and her neck is sore and stiff. The doctor feels her neck in the back and on the sides. She asks Ally to bend her neck forward. Then the doctor asks Ally to lie down, and she presses on her belly. She notices that Ally cannot bend her neck as far as she is supposed to and that the glands in the sides of her neck are lumpy and swollen. When the doctor looks at Ally's throat, she notices that it is very swollen and red. There is no drainage from her nose (the drainage might indicate a cold). When the doctor pressed Ally's belly, she could feel that Ally's spleen was slightly swollen. Ally did not have any skin rash and did not have a fever. No one Ally knows has been sick lately.

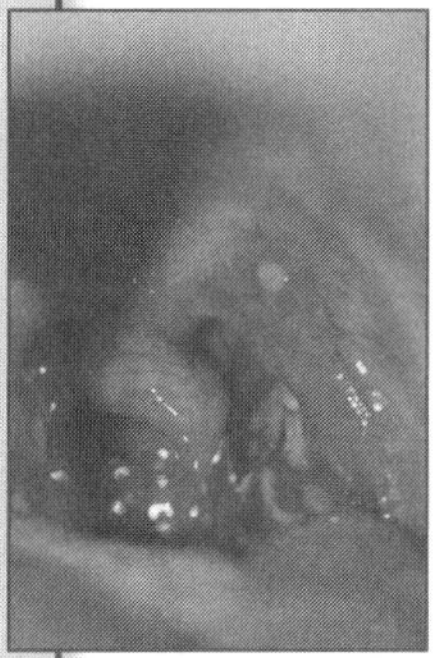

One symptom of mono is white patches on the back of the throat.

Learn about Mononucleosis

Infectious mononucleosis (mah-noh-new-klee-oh-sis) is often called mono. It is sometimes known as the kissing disease. It can spread through kissing, but it can also spread through coughing, sneezing, or sharing a glass or cup. The Epstein-Barr virus causes mononucleosis. However, similar signs and symptoms are sometimes caused by cytomegalovirus (CTV). Full-blown mono is most common in adolescents and young adults. Signs of mono can include fever, sore throat, headaches, white patches on the back of your throat, swollen glands in your neck, feeling tired, headaches, sore muscles, and not feeling hungry. A sick person does not need to have all of these symptoms to have mono.

pasteurized: to heat food to a temperature that is high enough to kill most harmful bacteria.

sputum: matter that is coughed up and mixed with saliva.

The virus that causes mononucleosis is transmitted from one person to another through the saliva of an infected person. The virus can continue to be excreted in saliva and **sputum** (matter that you cough up) for as long as six months after the infected person has gotten better. This is one reason why it is important not to share things, like straws or lip gloss, at school with your friends. A person does not need to feel sick to make other people sick with this disease.

Mono makes you feel tired and achy, but it usually is not very serious. The virus remains in your body for life. Most people have been exposed to the Epstein-Barr virus by the time they are 35 years old. Because they have had the virus, they are immune and cannot get mononucleosis again.

Be a Scientist

Case Study: Pneumonia

An elderly man reports to his nurse that he has had a bad cold for about ten days. He has been coughing and has had a high fever with shaking and chills. He tells the nurse that he does not understand why the cold will not go away. He has been coughing a lot, and the coughing produces sputum that is rust-colored. This morning when he woke up, he had a terrible shooting pain in the right side of his chest. He is also having difficulty breathing.

The nurse collects some of the sputum, gives the man oxygen through a tube in his nose, and admits him to the hospital. The man begins a treatment of antibiotics. The nurse sends the sputum to the laboratory to have it tested for bacteria. When all the data comes back from the lab, the doctor tells the nurse that the man has bacterial pneumonia. He probably caught it by breathing in the bacteria coughed into the air by someone else.

Learn about Pneumonia

pus: a white, or slightly yellow or green substance that your body develops in response to an infection; it is made up of dead skin, white blood cells (that fight infection), and some bacteria.

Pneumonia is a serious infection or inflammation of the lungs. Air sacs in the lungs fill with **pus** and other liquid so that oxygen has trouble getting into the blood. If there is too little oxygen in the blood, a person's body cells cannot work properly. Because of this, and because infection can spread throughout the body, pneumonia can cause death.

Until 1936, pneumonia was the number one cause of death in the United States. Since then, antibiotics are used so that fewer people die from pneumonia. There are 30 different causes of pneumonia. The most common causes are bacteria and viruses. Some of the similarities and differences are described in the following sections.

Bacterial Pneumonia

Pneumonia bacteria are present in some healthy throats. When body defenses are weakened in some way, the bacteria can multiply and cause serious damage. Body defenses are weakened when a person is very young or old, when the person is not in good health, when the person is in a weakened condition like just after having surgery, or when the person has impaired immunity. When a person's defenses are weakened (sometimes people say their resistance is lowered), bacteria work their way into the lungs and inflame the air sacs.

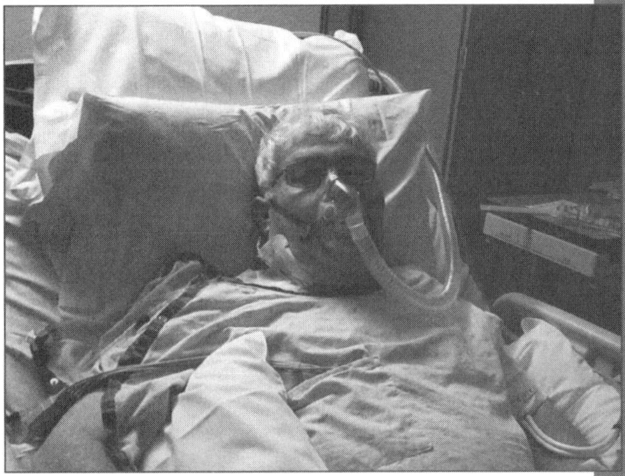

Sometimes the symptoms of bacterial pneumonia happen gradually. Sometimes they happen suddenly. A person may experience shaking chills, chattering teeth, severe chest pain, and a cough that produces rust-colored or greenish sputum. Body temperature may rise as high as 40.5°C (105°F). The person sweats profusely, and breathing and pulse rate increase rapidly. Lips and nail beds may have a bluish color due to lack of oxygen in the blood. In the most serious cases, the person may be confused or delirious. Streptococcus pneumoniae is the most common cause of bacterial pneumonia. It is one form of pneumonia for which a vaccine is available.

The symptoms of pneumonia can range from mild to serious, depending on the age of the person and the type of pneumonia.

Viral Pneumonia

Half of all pneumonias are believed to be caused by viruses. More and more viruses are being identified as the cause of respiratory infection. Some produce pneumonia, especially in children. Most of these pneumonias are not serious and last a short time. The virus invades the lungs and multiplies. There are almost no physical signs of lung tissue becoming filled with fluid. Many people who get viral pneumonia are those who have breathing conditions or who are pregnant.

The initial symptoms of viral pneumonia are fever, a dry cough, headache, muscle pain, and weakness. Within 12 to 36 hours, there is increasing breathlessness. The cough becomes worse and produces a small amount of mucus. There is a high fever, and a person's lips may look bluish. In extreme cases, the patient has a desperate need for air and extreme breathlessness. Sometimes viral pneumonia is complicated by an invasion of bacteria, and the patient has two types of pneumonia.

Communicate

30 min.

Guide students through the reading about sharing their case study and facilitate groups' class presentations.

Communicate

Share Your Case *Study*

Each of the case studies in this section presented a different example of a communicable disease. Since each group read only one case study, it is important that the information is shared with the class. Use the information you recorded on your *Communicable-Disease Information Table* to make presentations to your class about what you learned from the case study you read. Your classmates will be relying on you to learn about different examples of communicable diseases. It is important for you to include the following information:

- the name of the disease
- whether it is caused by a bacteria or a virus
- how it spreads
- how it affects your body systems
- the symptoms
- how it is treated

As a presenter, it is important that your presentation include all of the above points. As you listen to your classmates' presentations, it is just as important that you hear and understand this information. If you do not understand something, or if you think presenters left out something important, ask questions. Be careful to ask your questions respectfully, and do not interrupt your classmates' presentations.

What's the Point?

Learning about different diseases helps you understand how you can get sick. It is interesting to see how easily some diseases are spread from one person to another.

The chickenpox virus can infect others when an infected person coughs or sneezes. The bacterium that causes Lyme disease can spread to others by the bite of a deer tick. *E. coli* is a bacterium usually found in contaminated food and is responsible for causing gastrointestinal infections. Mononucleosis, a viral disease, can be spread through the saliva of an infected person for up to six months after the person is feeling better. Bacterial pneumonia, which can be treated with antibiotics, usually occurs in people who are already sick with something else. Viral pneumonia is usually less severe than bacterial pneumonia.

△ Guide

Point out why it is important for each group to share their information with the class. Briefly review the information about their disease that they will be expected to share with their classmates. Check whether groups need any equipment or materials for their presentation.

Emphasize the importance of not interrupting during a presentation and of interacting respectfully during group discussions. Encourage students to ask a question when they think something important has been left out or if they do not understand something.

⬡ Get Going

Have groups make their class presentations. As groups present, add the information they share to the class *Communicable-Disease Information Table.*

☐ Assess

Monitor groups' presentations for comprehensiveness and encourage presenters to cite supporting evidence from the case study they read. A sample *Communicable-Disease Information Table* is shown below:

Name of disease	Cause (bacterial or viral)	How does it spread?	Which body systems does it affect?	What are the symptoms?	How is it treated? Is there a vaccine?	Other notes
Chickenpox	Varicella-zoster virus	in the air through coughs and sneezes, contact with fluid from blisters	Skin, respiratory	red itchy rash usually appearing on abdomen first or face and neck. low fever, abdominal pain	Rest There is a vaccine for this disease.	
Lyme disease	Bacteria	by bite from an infected tick	Circulatory	bull's eye looking spot at bite on skin, headache, fever, muscle pain, inflammation of heart and nerves	antibiotic	
E. Coli infections	Escherichia coli Bacteria	between people who did not maintain good hygiene, through food, contaminated surfaces.	digestive	diarrhea, fever, cramps, vomiting. In some cases can cause kidney failure	NOTE: case study doesn't list treatment but you can tell students antibiotics and rest - or assign it as homework	Can be fatal.
Mononucleosis	Epstein-Barr virus	through saliva, coughing, sneezing, sharing food or drink, kissing	digestive	fever, sore throat, headaches, white patches on the back of the throat, swollen glands in neck, feeling tired, sore muscles, not hungry	rest	Can still be spread for up to 6 months after infected person has recovered from symptoms.

Assessment Options

Targeted Concepts, Skills, and Nature of Science	How do I know if students got it?
Scientists often collaborate and share their findings. Sharing findings makes new information available and helps scientists refine their ideas and build on others' ideas. When another person's or group's idea is used, credit needs to be given.	**ASK:** How did you personally benefit from groups' sharing information from their case studies? **LISTEN:** Students should recognize that they obtained new information that could be used to update their own *Communicable-Disease Information Tables*.
Disease is a breakdown in the structures or functions of an organism. Some diseases are the result of failures of a body system. Other diseases are the result of damage by infection from other organisms (germs).	**ASK:** What information shared during class presentations of the case studies helps you to determine which body systems are affected by a disease? **LISTEN:** Students should identify the symptoms of the disease and connect symptoms with specific body parts or systems. For example, poxes are damaged skin cells, a cough or runny nose indicates problems with the respiratory system.
Many diseases are caused by bacteria or viruses.	**ASK:** Based on the case studies presented to the class, what new bacterial diseases did you add to your *Communicable-Disease Information Table?* What new viral diseases did you add? **LISTEN:** Students should identify Lyme disease, *E. coli*, and some forms of pneumonia as bacterial diseases; and chickenpox, mononucleosis and some forms of pneumonia as viral diseases.
Vaccination is the process by which a person is given a vaccine – a substance that protects a person from a disease.	**ASK:** For which of the diseases you heard about during the class presentations is a vaccine available? **LISTEN:** Students will identify chickenpox as a disease that can be prevented by vaccination.

Targeted Concepts, Skills, and Nature of Science	How do I know if students got it?
An antibiotic is a drug that kills or prevents the growth of bacteria.	**ASK:** Why would antibiotics not be an effective treatment for mononucleosis? **LISTEN:** Students will identify antibiotics as a treatment for bacterial diseases and recognize that they are not effective against viruses, such as the one that causes mononucleosis.

Teacher Reflection Questions

- Which facets of the *Big Question* were addressed by having students examine case studies?
- How effective were the case studies in focusing students on the different kinds of infectious agents? On disease spread?
- What were the strengths and weaknesses of the class presentations made by your students? How could the management of the class presentations have been improved?

NOTES

NOTES

Learning Set 2

Back to the Big Question

◂ *2 class period*

A class period is considered to be one 40 to 50 minute class.

Overview

Students read a summary of the key ideas of *Learning Set 2* and reflect on what they have learned. Students update the *Project Board* making sure to include evidence to support information they added to the *What are we learning?* column. Students then consider how what they have learned will help them to answer the *Big Question* and focus on updating the last column, What does it mean for the challenge or question?

Targeted Concepts, Skills, and Nature of Science	Performance Expectations
Scientists often collaborate and share their findings. Sharing findings makes new information available and helps scientists refine their ideas and build on others' ideas. When another person's or group's idea is used, credit needs to be given.	Students share what they claim to have learned and the evidence that supports the claim as they fill out the class *Project Board*.
Explanations are claims supported by evidence, accepted ideas, and facts.	As students identify ideas that should be added to the *What are we learning?* column of the *Project Board*, they provide corresponding supporting evidence in the *What is our evidence?* column.

Materials	
1 per class	*Project Board*

Homework Options

Reflection

- **Science Process:** Which of the diseases examined during this *Learning Set* were able to spread by droplet transmission due to coughing or sneezing? *(cold, flu, chickenpox, smallpox, mononucleosis, pneumonia)* What simple practices could help prevent the spread of these diseases? *(covering your mouth and nose with a handkerchief when coughing or sneezing, frequent hand washing)*
- **Science Process:** What are the benefits of keeping a record of what you learn about a disease on the *Communicable-Disease Information Table? (It summarizes information that will be needed to meet the two challenges of this* Learning Set *and makes it more easily accessible. It also organizes the information so that relationships can be more easily seen.)*

Preparation for Learning Set 3

- **Science Content:** Give two examples of a symptom that is associated with more than one disease and identify the diseases that have that symptom in common. *(Examples may include, but are not limited to: FEVER – chickenpox, cold, flu, Lyme disease; RASH – smallpox, chickenpox, Lyme disease; SORE THROAT – cold, flu, mononucleosis.)* Why do you think different diseases have similar symptoms? *(Because they affect the same body system(s).)*

NOTES

BACK TO THE BIG QUESTION IMPLEMENTATION

Learning Set 2

Back to the Big Question

How can you prevent your good friends from getting sick?

The *Big Question* for this Unit is *How can you prevent your good friends from getting sick?* To help you answer this question, you looked at cells. Your body is made up of cells. To understand what happens to your body when you get sick, you need to know about what makes up a cell and how a cell functions.

You know that bacteria and viruses can make you sick and cause disease in your body. You also know that some bacteria are needed by your body to keep you healthy. Bacteria and viruses are both microbes (microorganisms). You cannot see them without a microscope. Bacteria are single-celled organisms. You need to know how a cell functions to understand how bacteria function.

Viruses are even smaller than bacteria. You need to use a special type of microscope to see them. There is no cure for diseases caused by viruses. Preventing the viral disease from spreading is the best way to stay healthy.

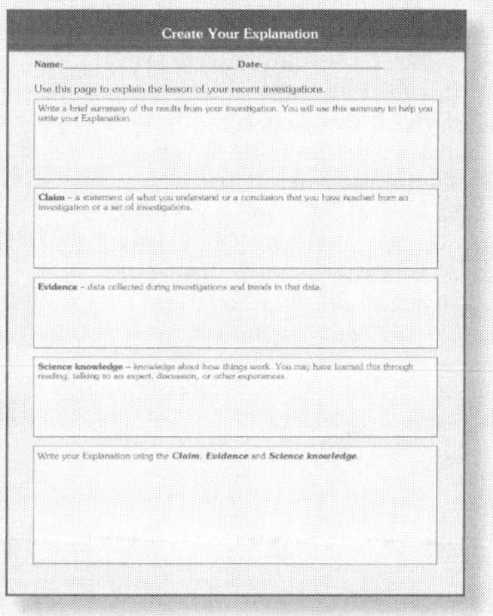

Explain

You know so much more now about how you can get sick from communicable diseases. It is time to revise your explanations of how you can get sick from communicable diseases. Begin by revisiting the claims you made earlier. Work on making them more specific. You might want to add specifics to your claim about how you can get sick from a communicable disease by touching everyday objects or by eating everyday foods. What you've read about specific communicable diseases can help you make your claims more specific.

◀ *1 class period**

Learning Set 2

Back to the Big Question

5 min.

Students reflect upon what they have learned, construct explanations for how they get sick using Create Your Explanation *pages and share their explanations with the class. Students update the recommendations they made in* Section 1.3. *Students identify three things that should be added to the* Project Board.

○ Engage

Begin by asking students why it was important to record information in the *Communicable-Disease Information Table* throughout the *Learning Set.* (Because it is easy to forget details or lose track of information when you are involved in a series of investigations.) Connect this with the importance of updating the *Project Board* throughout the Unit and maintaining focus on the *Big Question.*

*A class period is considered to be one 40 to 50 minute class.

△ Guide

Have students read the section summarizing the main ideas of *Learning Set 2.* You might consider having students take turns reading to the class, and discussing the content as you go along. Guide students through the reading, asking them to connect main ideas identified in the reading with specific activities in the *Learning Set.*

Example:

Main idea: Your body is made up of cells.

Activity: Observe cells with the microscope.

You kn… and viruses can … nd cause … your body. … know that some bacteria … eeded by your body … you healthy. Bacteria and viruses are both microbes (microorganisms). You cannot see them without a microscope. Bacteria are single-celled organisms. You need to know how a cell functions to understand how bacteria function.

Viruses are even smaller than bacteria. You need to use a special type of microscope to see them. There is no cure for diseases caused by viruses. Preventing the viral disease from spreading is the best way to stay healthy.

Explain

You know so much more now about how you can get sick from communicable diseases. It is time to revise your explanations of how you can get sick from communicable diseases. Begin by revisiting the claims you made earlier. Work on making them more specific. You might want to add specifics to your claim about how you can get sick from a communicable disease by touching everyday objects or by eating everyday foods. What you've read about specific communicable diseases can help you make your claims more specific.

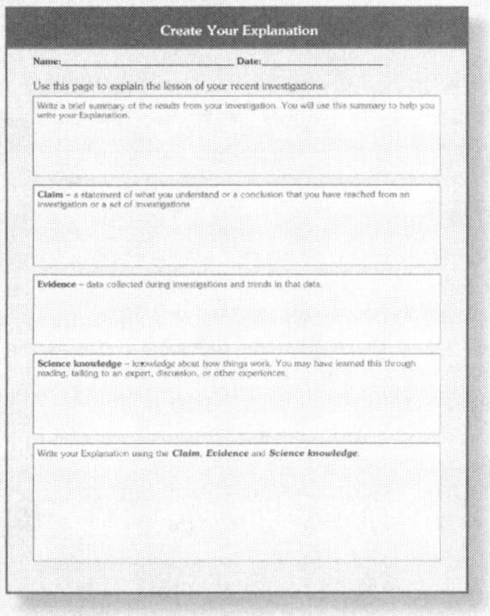
Create Your Explanation

Name:________ Date:________

Use this page to explain the lesson of your recent investigations.

Write a brief summary of the results from your investigation. You will use this summary to help you write your Explanation.

Claim – a statement of what you understand or a conclusion that you have reached from an investigation or a set of investigations.

Evidence – data collected during investigations and trends in that data.

Science knowledge – knowledge about how things work. You may have learned this through reading, talking to an expert, discussion, or other experiences.

Write your Explanation using the ***Claim***, ***Evidence*** and ***Science knowledge***.

GF 73 GOOD FRIENDS and GERMS

Explain

15 min.

Groups construct explanations for how you get sick using the Create Your Explanation page.

△ Guide

Once groups have considered what they now know about how you get sick from communicable diseases, have them make a list of claims they made earlier. Make clear that their first task is to make their claims more specific.

Use the examples given in the student text as the basis for a brief discussion of how claims can be more specific. For example, students may have made the claim that some bacteria can make you sick. This claim could be made more specific by identifying a specific bacteria and the specific disease it causes.

Record your new more specific claim on a new *Create Your Explanation* page. Then record evidence and science knowledge that support your claim, and write a statement that connects your claim and your evidence and science knowledge. Make sure your statement discusses the role bacteria and viruses play in getting you sick and how they are passed between people. Remember that a good explanation can convince someone else of your claim. If your statement doesn't seem convincing, revise your claim so your evidence and the science knowledge you know support it. Be prepared to share your claims and explanations with the class.

Communicate

Share Your Explanation

Share your explanation with the class. Make sure to be clear about your claim, the evidence and science knowledge that support it, and your explanation statement. As you listen to the explanations of others, make sure you understand them completely and agree with them. If you do not understand something, or if someone has not been clear, ask them to add more detail to what they are saying. If you think someone is wrong, respectfully disagree, being sure to give evidence to support your disagreement.

Revise Your Explanation

After everyone has presented, as a class, develop an explanation about how you get sick from communicable diseases that everybody agrees with.

Recommend

Using your explanations as a guide, revisit your recommendations, and develop more specific recommendations about staying well. Your recommendations should answer the same two questions you answered in Section 1.3.

- How can I keep from getting sick with a communicable disease?
- How can I prevent my friends from getting sick with a communicable disease?

Remember that you now know much more about which everyday objects have a lot of germs on them and the locations where you are especially likely to come into contact with germs. For each recommendation you made previously, decide if you still think it is a good recommendation. If you think it is but you can state it better, restate your recommendation. Then add any

A claim could also be made more specific by describing how a communicable disease could be transmitted from one person to another by contact with everyday objects or eating everyday food. For example, students may have made the claim that bacteria are present in their classroom. This claim could be made more specific by identifying specific places in the classroom or by describing specific ways in which bacteria in the classroom could enter their bodies.

Have groups write each new more specific claim on a separate *Create Your Explanation* page and record evidence and science knowledge obtained from their investigations and readings that support that claim.

Once the evidence and science knowledge supporting the claim has been recorded, review the parts of a good explanation with the class: the claim; the evidence; the science knowledge; and a logical statement tying the claim to the evidence. Make clear that their next task is to write an explanation of how you get sick from communicable diseases. The explanation should focus on the role bacteria and viruses play in making you sick and how they are passed from person to person. Point out that their explanation statement should be clear and persuasive – anyone reading it should be convinced of the claim being made.

☐ Assess

Check each group's explanations and ask them what claims they are making and what evidence and science knowledge they are using to support the claim. Some possible explanations are:

- You can get a gastrointestinal illness by eating unrefrigerated food because warm temperatures are favorable to the growth of bacteria and bacteria that grow on the food can get into your body when you eat it.
- You can get a cold from someone if they do not cover their mouth when they cough, because droplets containing rhinovirus or adenovirus are released when they cough. If the virus in those droplets come in to contact with your eyes, nose or mouth, it can attach itself to one of your cells, inject its genetic material and reproduce.

Students' evidence should be based on their investigations and science knowledge obtained from readings or what you discussed with them during the Unit.

NOTES

> explanations with the class.
>
> **Communicate**
>
> *Share Your Explanation*
>
> Share your explanation with the class. Make sure to be clear about your claim, the evidence and science knowledge that support it, and your explanation statement. As you listen to the explanations of others, make sure you understand them completely and agree with them. If you do not understand something, or if someone has not been clear, ask them to add more detail to what they are saying. If you think someone is wrong, respectfully disagree, being sure to give evidence to support your disagreement.
>
> **Revise Your Explanation**

Communicate

15 min.

Groups present their explanations to the class and the class evaluates the evidence and reasoning behind each explanation.

⬡ Get Going

When groups have finished writing their explanations, have each group present their explanations to the class. Students should take notes on the explanations presented by other groups and write down any questions they have concerning the explanation.

△ Guide

As groups present their explanations, encourage students to ask questions about things they do not understand and point out logical inconsistencies or missing evidence they perceive in the explanation.

Guide the class in a brief summary discussion of their explanations offered by the groups. Ask students what evidence and reasoning they found most convincing and which explanations they can all agree upon.

> understand them completely and agree with them. If you do not understand something, or if someone has not been clear, ask them to add more detail to what they are saying. If you think someone is wrong, respectfully disagree, being sure to give evidence to support your disagreement.
>
> **Revise Your Explanation**
>
> After everyone has presented, as a class, develop an explanation about how you get sick from communicable diseases that everybody agrees with.
>
> **Recommend**
>
> Using your explanations as a guide, revisit your recommendations, and develop more specific recommendations about staying well. Your recommendations should answer the same two questions you answered in

Revise Your Explanation

5 min.

Groups present their revised explanations to the class and the class collaborates to create explanations they can all agree upon.

△ Guide

Make a list of the class explanations about how you get sick and edit them as the class revises each statement. Guide the class to develop an explanation about how you get sick from communicable diseases that everyone can agree upon.

NOTE: This would be an opportune time to conclude the first class session.

Recommend

10 min.

Students evaluate their earlier recommendations and, if necessary, restate them to be more specific or more complete. Students then create new recommendations based on what they now know and complete a Create Your Explanation *for each that provides evidence and science knowledge to support each recommendation.*

> ...sure to give evidence to support your disagreement.
>
> ### Revise Your Explanation
>
> After everyone has presented, as a class, develop an explanation about how you get sick from communicable diseases that everybody agrees with.
>
> ### Recommend
>
> Using your explanations as a guide, revisit your recommendations, and develop more specific recommendations about staying well. Your recommendations should answer the same two questions you answered in Section 1.3.
>
> - How can I keep from getting sick with a communicable disease?
> - How can I prevent my friends from getting sick with a communicable disease?
>
> Remember that you now know much more about which everyday objects have a lot of germs on them and the locations where you are especially likely to come into contact with germs. For each recommendation you made previously, decide if you still think it is a good recommendation. If you think it is but you can state it better, restate your recommendation. Then add any

△ Guide

Remind students of the recommendations they made in *Section 1.3* in response to the two questions. Have students evaluate each of their earlier recommendations and decide if they still think it is a good recommendation. Encourage students to consider ways in which they could re-state the recommendations to make them more specific or more complete.

⬡ Get Going

Have students make any necessary edits to their earlier recommendations to make them more specific and more complete. Once students have finished editing their earlier recommendations, ask them to develop any new recommendations they may have based on what they now know.

For each new recommendation, have students complete a new *Create Your Explanation* page using their recommendation as the claim and adding any evidence and science knowledge that supports it. Then have them develop an explanation statement connecting their recommendation to the evidence and science knowledge. Make clear to students that they should be prepared to present their recommendations to the class, and should be able to support them with evidence and science knowledge.

□ Assess

Check to see that students have evaluated and made any necessary edits to their earlier recommendations. Ask students to describe the reasoning behind any changes they did or did not make. Check whether students' new recommendations and explanations are connected to evidence and science knowledge that supports it.

Back to the Big Question

new evidence or science knowledge that supports it. Revise your explanation statement to make it more complete.

You might also want to make new recommendations. Use a new *Create your Explanation* page for each. State your claim as a recommendation and add evidence and science knowledge that support it. Then develop a statement that connects your recommendation to your evidence and science knowledge. Be prepared to share your recommendations with the class and to support them with evidence and science knowledge.

Communicate

Share Your Recommendation

Share your recommendations with the class. When you present, be clear about the evidence, science knowledge, and explanation that support each of your recommendations. As you listen, be sure that you understand each recommendation your classmates make and that you are convinced the recommendation is good. If the evidence or science knowledge does not support it well, respectfully disagree with the presenter and report the evidence or science knowledge that tells you the recommendation is not convincing.

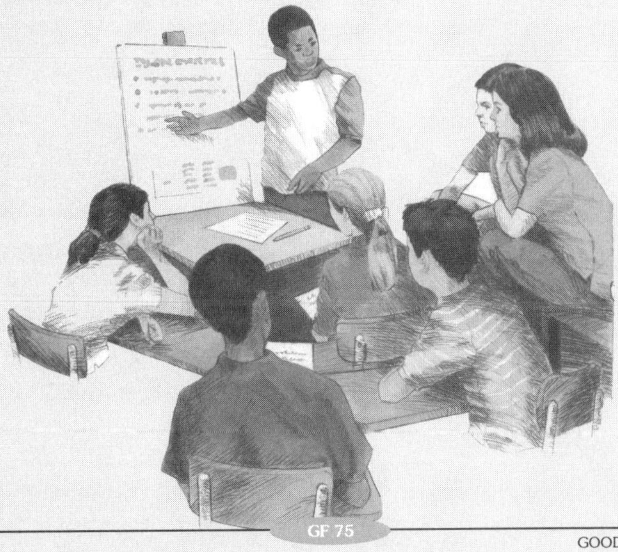

Communicate

15 min.

Students share their recommendations with the class and the class collaborates to create recommendations that everyone can agree upon.

○ Get Going

When students have finished writing their new recommendations and explanations, have them share their explanations with the class. Remind students that when they share their recommendations, they should clearly state the evidence and science knowledge that supports it.

△ Guide

As recommendations are being presented, encourage students to ask questions about things they do not understand and to point out places where they think something is missing from explanations. Guide the class to create recommendations that everyone in the class can agree upon.

NOTES

Conference

With your group, identify three things you have read about in this *Learning Set* that you think should be added to the *Project Board* and the evidence that supports each.

Update the *Project Board*

The *What are we learning?* column on the *Project Board* helps you pull together what you have learned. Share the things you think are important to add to the *Project Board* and your evidence for each. As a class, decide which of these things belong in the *What are we learning?* column. Remember to include evidence in the *What is our evidence?* column. You must fit the pieces together to help you address the challenge. Your *Big Question* was *How can you prevent your good friends from getting sick?* The last column, *What does it mean for the challenge or question?*, is the place to record how learning about bacteria and viruses can help you answer the *Big Question*. Remember to keep a record of what was added to the class *Project Board* on your own *Project Board* page.

How can you prevent good friends from getting sick?

What do we think we know?	What do we need to investigate?	What are we learning?	What is our evidence?	What does it mean for the challenge or question?

Conference

5 min.

Students collaborate in groups to identify three things they have read about in this Learning Set *that should be added to the* Project Board.

⬡ Get Going

Working collaboratively in their groups, ask students to identify three things they have read about in this *Learning Set* that they think should be added to the *Project Board*.

For each thing that the group thinks should be added to the *Project Board*, ask the group to write down all supporting evidence.

Update the *Project Board*

15 min.

Students update the Project Board *with any supporting evidence that they have learned in this* Learning Set *and consider how this new information helps them answer the smaller question for this* Learning Set *and the* Big Question *for the Unit.*

you think should be ad... *Project Board* ... nce that supports each.

Update the *Project Board*

The *What are we learning?* column on the *Project Board* helps you pull together what you have learned. Share the things you think are important to add to the *Project Board* and your evidence for each. As a class, decide which of these things belong in the *What are we learning?* column. Remember to include evidence in the *What is our evidence?* column. You must fit the pieces together to help you address the challenge. Your *Big Question* was *How can you prevent your good friends from getting sick?* The last column, *What does it mean for the challenge or question?*, is the place to record how learning about bacteria and viruses can help you answer the *Big Question*. Remember to keep a record of what was added to the class *Project Board* on your own *Project Board* page.

How can you prevent good friends from getting sick?				
What do we think we know?	What do we need to investigate?	What are we learning?	What is our evidence?	What does it mean for the challenge or question?

△ Guide

Lead a class discussion focusing on what groups think should be added to the *What Are We Learning?* and *What Is Our Evidence?* columns of the Project Board now that they have completed *Learning Set 2*. During the discussion, guide students to identify what they have learned about *"What kinds of things make you sick?"* and where that information should be recorded on the *Project Board.*

Guide students by asking:

- What have we learned about cells?
- Where should we put the information about cells?

Facilitate a class discussion of the kinds of things that make us sick and guide students to specifically associate evidence with their claims.

Example:

Claim: Some bacteria cause diseases.

Evidence: Information in case studies.

Claim: Bacteria are present in many places in our classroom.

Evidence: Bacteria culturing activity.

Hold a discussion focusing on how students think what they have learned will help them answer the *Big Question, How can you prevent your good friends from getting sick?* Record students' ideas to the *What does it mean for the challenge or question?* column.

Teacher Reflection Questions

- How did you guide the reading? What would you improve upon the next time you guide students through a reading?
- How did you get students to contribute to the discussion of what should be added to the *Project Board?* What are some other techniques that you could try?
- What questions or techniques did you find most successful in helping students connect what they have been learning to the B*ig Question?*

NOTES

NOTES

LEARNING SET 3 INTRODUCTION

Learning Set 3

What Happens to You When You Get Sick?

◀ *17 class periods*

A class period is considered to be one 40 to 50 minute class.

Students investigate and read about the structural and functional levels of organization in living things from cells to organ systems. Students then build and run models of organ systems and connect the models to what they are learning by reading about organ systems. As students read about diseases that affect various organ systems, they repeatedly return to the Communicable-Disease Information Table *created in* Learning Set 2 *to add new understandings about diseases and how they affect body systems.*

Overview

Learning Set 3 opens with students reflecting on what they have investigated in previous Learning Sets and connecting it to what will be investigated in the upcoming *Learning Set. Section 3.1* introduces the concept of a system and elicits students' prior knowledge of organ systems. Students then develop questions to investigate that will help them better understand body systems. Students update the *Project Board* to include what they think they know about body systems and what needs to be investigated.

Section 3.2 guides students through the levels of organization of multicellular organisms from cells to organ systems and identifies some of the major organ systems of the human body.

In *Section 3.3*, students investigate their breathing rate. Students measure breathing rate while resting and after exercise, share their results with the class, and analyze class data to identify patterns and trends. In *Section 3.4*, students connect their experiences breathing to the respiratory system and investigate its structure and function. Students read about the respiratory system, build a working model of the system, and use the model to simulate breathing. Students then connect the parts of the model to the parts of the respiratory system and how the various parts work together in the process of breathing. Then, students read how diseases affect the functioning of the respiratory system. Students add information about respiratory-system diseases to the *Communicable-Disease Information Table* and update the *Project Board.*

In Section 3.5, students investigate their heart rate. Students measure heart rate while at rest and after activity, they share their results with the class, and analyze the results for patterns and trends. Here again, students' concrete experiences with heartbeat, serve as a foundation for investigating the circulatory system in the next section. In *Section 3.6*, students investigate how the circulatory system works. After reading about the connection between the circulatory system and the respiratory system, students assume the role of blood cells and engage in a simulation of circulation in the body. Students then connect the model used in the simulation with the human circulatory system. Additional readings extend students' knowledge of the circulatory system and make connections between disease and circulation. Students add information about circulatory-system diseases to the *Communicable-Disease Information Table* and update the *Project Board.*

In *Section 3.7*, students build a scale model of the digestive system as a story is read aloud to them describing the journey of food through the various parts of the system. Students then move beyond the model and consider connections between the digestive system and the circulatory system. Further readings elaborate on the effects of disease on the digestive system, the role the digestive plays as a route of entry for germs, and the preventing of the spread of digestive-system diseases. Students then read about the nervous system. Students add information about digestive-system and nervous system diseases to the *Communicable-Disease Information Table* and update the *Project Board.*

In *Section 3.8*, students read about the immune system. The various methods employed by the body to prevent germs from entering and to combat germs once they have entered are described. Groups then make present one part of the immune system using a poster they made, showing that part of the immune system, its function, and how it works together with other parts of the immune system. The class then collaborates to construct a more complete understanding of how the parts of the immune system work together to keep you healthy. Here again, students add information to the *Communicable-Disease Information Table* and update the *Project Board.* The *Learning Set* concludes with students engaging in an activity in which they work in small groups to make drawings of body systems showing organs in their correct size and position in relation to the body. Students then overlay the transparencies; collaborate within their group to identify ways in which the body systems are interconnected and how they interact. Groups then present their findings to the class in a Communicate Your Ideas: *Solution Showcase.*

Targeted Concepts, Skills, and Nature of Science	Section
Scientists often collaborate and share their findings. Sharing findings makes new information available and helps scientists refine their ideas and build on others' ideas. When another person's or group's idea is used, credit needs to be given.	All sections
Scientists must keep clear, accurate, and descriptive records of what they do so they can share their work with others, consider what they did, why they did it, and what they want to do next.	All sections
Models are a representation of something in the world. Simulations use a model to imitate, or act out, real-life situations.	3.4, 3.6,
Systems are made up of parts that move or work together as a whole.	3.1
The cell is the basic unit of structure and function in all living things. Cells are usually microscopic in size.	3.2
Cells vary in size, shape and function and are organized to work together more effectively in multicellular organisms. Levels of organization include cells, tissues, organs, and organ systems	3.2
An organ system is a group of body organs, each of which performs a specific function, that work together to carry on one or more life functions.	3.2, 3.4, 3.6, 3.7, 3.8
Tissues, organs, and organ systems help to provide all cells with nutrients, oxygen, and waste removal.	3.4, 3.6,
Foods contain substances called nutrients that serve as fuel and building material for all organisms.	3.7
Foods contain a variety of nutrients, which are vital to the survival of an organism, including: carbohydrates, fats, vitamins, proteins, minerals, and water.	3.7
Energy in foods is measured in Calories.	3.7
The digestive system is made up of organs that break down food mechanically and chemically into molecules that can be absorbed and transported to cells.	3.7
The excretory system functions in the disposal of dissolved waste molecules, the elimination of liquid and gaseous wastes, and the removal of excess heat energy.	BBQ, MTL

Targeted Concepts, Skills, and Nature of Science	Section
The respiratory system supplies oxygen to the blood and removes carbon dioxide from the blood.	3.4
The circulatory system moves substances to and from cells in response to changing demands.	3.5, 3.6
The nervous and endocrine systems interact to control and coordinate the body's responses to changes in the environment, and to regulate growth, development, and reproduction. Hormones are chemicals produced by the endocrine system; hormones regulate many body functions.	BBQ, MTL
Locomotion, necessary to escape danger, obtain food and shelter, and reproduce, is accomplished by the interaction of the skeletal and muscular systems, and coordinated by the nervous system.	BBQ, MTL
The male and female reproductive systems are responsible for producing sex cells necessary for the production of offspring.	BBQ, MTL
Disease breaks down the structures or functions of an organism. Some diseases are the result of failures of a body system. Other diseases are the result of damage by infection from other organisms.	3.4, 3.6, 3.7, 3.8
Specialized cells protect the body from infectious disease. The chemicals they produce identify and destroy microbes that enter the body.	3.8
Identify structure and function relationships in organisms	3.4, 3.6, 3.7, 3.8
Identify cause-and-effect relationships.	3.3
Identify pulse points and pulse rates.	3.5

NOTES

Understanding for Teachers

Cells vary in size and shape. Cells are specialized to perform specific functions. The shape of a cell is related to its function. A group of cells similar in structure and function which work together is called a tissue. Tissues are named for their function and location. The table below shows some typical cells of human tissues.

Cell	Function
Blood	Carries oxygen to body cells and carbon dioxide to lungs.
• Red	Engulfs and destroys germs.
• White	
Connective	Makes up body framework and aids in body movement.
• Bone	Bone - cells embedded in collagen and the mineral calcium phosphate which makes bone hard.
• Cartilage	Cartilage - cells embedded in a firm, yet flexible matrix of collagen fibers and a rubbery, gelatinous substance called chondrin.
• Dense fibrous	Dense fibrous - composed of a thick network collagen fibers that form strong, ropelike structures such as ligaments and tendons.
• Loose fibrous	Loose fibrous - widely spaced fibers of collagen that holds organs in place and attaches epithelial tissue to other underlying tissues.
Reproductive	Combine to form new living things.
• Egg	Carries genetic information from female.
• Sperm	Carries genetic information from male.
Epithelial	Covers surface of body, inside surface of nose, throat, trachea and digestive system. Protects the body and produces secretions.
Muscle	Produces movement through contraction and relaxation.
• Skeletal	Skeletal - Elongated, striated fibers under voluntary control.
• Smooth	Smooth – unstriated, spindle-shaped fibers not under voluntary control.
• Cardiac	Cardiac – branched, striated fibers found in heart tissues; not under voluntary control
Nerve	Carries impulses which cause muscles to act and detect information about our surroundings. Some nerve cells are several feet long.

NOTES

LEARNING SET 3 IMPLEMENTATION

Learning Set 3

What Happens to You When You Get Sick?

When you are sick, your body does not feel normal. Your throat might hurt, your nose might be stuffy, and it might be hard to breathe. Different illnesses make you feel bad in different ways. Recall that these indications that you are sick are called *symptoms*. Doctors and nurses use symptoms as clues to determine the disease you might have. Symptoms are connected to body systems.

You first modeled how a disease could spread. Then you learned that bacteria and viruses are microbes that can make you sick and cause disease. In this *Learning Set*, you will investigate some of the body systems that are affected by communicable diseases. You will also look at the immune system, which helps you stay healthy. You will see how microbes can move within your body and how the symptoms you feel are caused. This will help you answer the question, *What happens to you when you get sick?*

The amazing thing about your body is how all the parts work together. You can walk and talk and move because of the ways the parts of your body work. Air moves into your lungs. The oxygen in the air then moves into your blood. The blood moves through your body. It brings the oxygen to each cell where it is needed. The way your body works also allows you to spread germs and to transmit diseases to others.

In this *Learning Set*, you will be learning more about each of your body systems. As you build models of the body systems, you will think about how the parts of the systems work together to keep you healthy. You will also learn about the role each system plays in spreading disease. Learning how your body works will help you stay healthier and help you avoid making others sick.

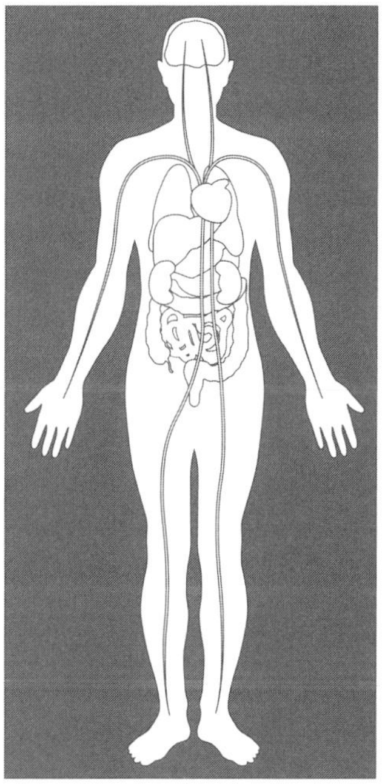

The human body is made up of many systems, including the respiratory, circulatory, and digestive systems.

◄ $\frac{1}{2}$ *class period**

Learning Set 3

What Happens to You When You Get Sick?

20 min.

Students consider the smaller question for Learning Set 3, *What Happens to You When You Get Sick?*

△ Guide

Briefly review the key ideas of the previous two *Learning Sets* and connect them to the question the class will investigate in this *Learning Set.*

TEACHER TALK

“We began this Unit by investigating how disease spreads and then we investigated the microbes that cause disease. Now we are going to focus on what happens to your body when you get sick.”

*A class period is considered to be one 40 to 50 minute class.

Ask students how they can tell that someone is sick. Make a class list of observations that indicate that a person is sick.

TEACHER TALK

"Imagine that your friend walks into class. You take one look at your friend and say "Are you OK? You look really sick." Or, imagine that a friend calls you up on the telephone and after listening for a minute or so you say, "What's wrong? You sound really sick."

What does 'sick' look like? What does 'sick' sound like? What causes a person to look and sound a certain way when they are sick?"

Hold a brief class discussion creating a T-chart with observations on one side and the body part(s) they affect on the other. Guide students to recognize that what they are observing are changes in the body due to an illness. As students fill out the chart guide them to recognize that the body parts affected by a particular illness are part of the same body system. For example, students might suggest that a person looks pale or flushed when they are sick, and could be guided to connect this with blood flow and to connect blood flow with the heart and blood vessels. Emphasize that the body systems work together.

Point out to students that understanding how their body works will help them understand what happens to them when they get sick.

Remind students that the *Big Question* of the Unit is *How can you prevent your good friends from getting sick?* Ask students to think about this question as they read the introduction to *Learning Set 3*.

Read through the introduction with the students. Consider having students read aloud. Emphasize that the smaller question they will be focusing on in this *Learning Set* is *What Happens to You When You Get Sick?* and that as they pursue answers to this question, they will investigate the organ systems that make up their body and how the functioning of those systems are affected by disease.

SECTION 3.1 INTRODUCTION

3.1 Understand the Question

Thinking about What Happens to You When You Get Sick

◀ *1 class period*

A class period is considered to be one 40 to 50 minute class.

Overview

This section introduces the concept of a system and elicits students' prior knowledge of organs and body systems. Students write down what they think they know about the organs that comprise several body systems. Then, students formulate questions that might help them answer the guiding question of *Learning Set 3*. Students share their ideas and questions with their group. Working collaboratively within their groups, students evaluate whether their questions meet the criteria for a good question. Each group then selects their two most interesting questions to share with the class and the *Project Board* is updated.

Targeted Concepts, Skills, and Nature of Science	Performance Expectations
Scientists often collaborate and share their findings. Sharing findings makes new information available and helps scientists refine their ideas and build on others' ideas. When another person's or group's idea is used, credit needs to be given.	Students share their individual questions with their group and their group's questions with the class as they consider *What do we think we know?* and *What do we need to investigate?* when they fill out the class *Project Board*.
Scientists must keep clear, accurate, and descriptive records of what they do so they can share their work with others, consider what they did, why they did it, and what they want to do next.	Students participate in a class discussion updating their *Project Board*, which is an organizing record of their ideas and what they are learning.

Materials	
1 per class	*Project Board*

Activity Setup and Preparation

This activity begins with students working individually to formulate questions and ends with a class discussion leading to updating the *Project Board*. Decide how you would like students to keep a record of individual and group work. Make sure the class *Project Board* is posted where everyone can see it.

Homework Options

Reflection

- **Science Content:** When the members of your group shared their ideas about organs and body systems, which organs were they able to name, but did not know how it worked or to which organ system it belonged? *(Answers will vary, but may include organs such as the pancreas or gall bladder.)*
- **Science Process:** Which of the criteria for a good question helps you refine a question so that it is one that can be answered by carrying out an investigation? *(The criterion that states that question requires collecting and using data.)*

Preparation for 3.2

- **Science Content:** What types of cells do you think make up your hand? *(Typical responses include: skin, bone, blood, nerve, and muscle.)*

SECTION 3.1 IMPLEMENTATION

3.1 Understand the Question

Thinking about What Happens to You When You Get Sick

The question for this *Learning Set* is *What happens to you when you get sick?* To help you answer this question, you will be learning about body systems. It is a good idea to think about what you know about systems, and what you know about the systems in your body. It is also important to think about what you are unsure about and what you would like to investigate.

Get Started

A system is like a team of players. Each player on the team has a job to do. One player on the team may be very important and talented. However, without the other players, he or she cannot get the job done.

Think about some systems that you already know. For example, you may know about the transit system in your community (buses, subway, trains). You may know about sound systems, emission systems on cars, alarm systems, or irrigation systems for a farm. Each system has parts. Each of the parts has a job, or function. In a system, the parts work together to do a job. If you think of a bicycle as a system, the pedals, chain, seat, handlebars, and you work together to make a bicycle move.

What do you already know about your body systems? You probably know the names of some organs and some body systems. Record what you know about the names of organs in each of these body systems:

- the respiratory system,
- the circulatory system, and
- the digestive system.

3.1 Understand the Question

Thinking about What Happens to You When You Get Sick

5 mins.

Students consider the smaller question for Learning Set 3: What Happens to You When You Get Sick?

△ Guide

Read through the introduction with the students. Emphasize that the focus of this Learning Set will be the question *What happens to you when you get sick?* Point out that to answer this question they are going to investigate organ systems – how they work and how they are affected by the germs that cause communicable disease.

TEACHER TALK

"In the last two *Learning Sets*, you investigated communicable diseases, how they spread, and the types of germs that cause communicable diseases. You also considered some of the ways in which body systems are affected by communicable diseases. In this *Learning Set* you are going to focus on the question What happens to you when you get sick? In order to answer this question you are going to focus more closely on how body systems work. Understanding how body systems work and will help you better understand what is happening to you when you get sick."

Get Started

15 min.

Students consider the concept of a system and share prior knowledge of organs and body systems. Students then formulate questions about what happens to you when you get sick that can be answered through scientific inquiry. Working in small groups, students evaluate one another's questions to be sure that they meet the criteria for a good question. Each group comes up with two questions that will help guide their investigation into what happens to you when you get sick.

△ Guide and Assess

Have students read the first part of the *Get Started* section dealing with systems. During the reading, pause from time to time and assess whether students are grasping the concept of a system by asking questions such as:

- What are the two key characteristics of a system? *(system has parts, parts work together as a whole)*
- What is an example of a system not mentioned in the reading? Why do you think this is a system? *(i.e. a fishing rod – it has parts such as a rod, a reel, fishing line, etc. and those parts work together as a whole)*

Have students carefully examine the photograph of the bicycle and ask them to think about it as a system. Ask students to identify the different parts and describe what each part does. Then ask students to describe how the different parts work together as a whole to make the bicycle move. Help students to recognize that even non-moving parts, such as the seat or the frame, contribute to the overall functioning of the system. Emphasize that the concept of a "system" is an important theme that runs through all of science and engineering, and it is an important theme in communicable disease because diseases often target specific body systems. Also point out that the symptoms of a disease help us to identify the affected body system and provide important clues as to what the disease might be. Once a disease has been identified, one can determine how it spreads and how it can be prevented.

(Here are some examples of organs to get you started: heart, intestines, skin, brain, lungs, stomach, veins, esophagus, and trachea. Don't worry if you don't know all of these. During the next few days, you will be learning more about each of these systems and organs.)

On your own, think of several questions that might help you answer the question for this *Learning Set*. Develop two questions that might help you understand body systems better.

In your group, share your ideas about what you think you know about body systems. Listen carefully to all the ideas presented. Then share your two questions with your small group. Carefully consider each question and decide if it meets the criteria for a good question. With your group, refine the questions that do not meet the criteria. Choose the two most interesting questions to share now with the class. Give your teacher the rest of the questions so they might be used later.

Update the *Project Board*

Share your group's ideas about body systems with the class. You can record these ideas in the *What do we think we know?* column. Next, share your group's two questions with your class. Your teacher will add your questions to the *Project Board*. In this *Learning Set*, you will work to answer some of these questions.

How can you prevent good friends from getting sick?

What do we think we know?	What do we need to investigate?	What are we learning?	What is our evidence?	What does it mean for the challenge or question?

⬡ Get Going

Have students read the next paragraph and make a list of organs that they think are part of the three organ systems listed in the student text: the respiratory system, the circulatory system, and the digestive system. The point here is to elicit prior knowledge that will enable you to better prepare students for further learning and guide students' investigation of body systems.

Read the remainder of the *Get Started* section together with the class and discuss the criteria for the questions to be written. Check that students understand each of the criteria. Then, have students work individually to formulate their questions. Before having students share their questions, describe what the group is expected to accomplish during the discussion. Make clear that they are expected to evaluate whether questions meet all of the criteria for a good question. Point out that you expect to hear groups discussing questions such as:

- How does this question relate to the *Big Question* of the Unit? How does it relate to smaller question of this Learning Set?
- In what way does answering this question involve collecting and using data?
- What resources are required to answer this question?
- How could this question be changed so that it would meet all of the criteria?

Then, have students share their questions with the group. As questions are shared, the group should evaluate whether each question meets the stated criteria.

☐ Assess

As groups are working, listen to the substance of their discussions and evaluate whether they are focusing on the criteria for a good question.

When you visit a group, ask questions such as:

- "What are the five criteria you are using to evaluate questions?"
- "What is one question suggested so far that you think did not meet one of the criteria for a good question?"
- "Which of the criteria was not met by this question?"
- "How could this question be changed so that it would meet all of the criteria for a good question?"
- "Who is keeping track of the group's questions?"
- "How are you keeping track of good ideas brought up during the discussion?"

the questi... not meet the criteria. C... wo most interestin... questions to share now with the class. Give your teacher the rest of the questions so they might be used later.

Update the *Project Board*

Share your group's ideas about body systems with the class. You can record these ideas in the *What do we think we know?* column. Next, share your group's two questions with your class. Your teacher will add your questions to the *Project Board*. In this *Learning Set*, you will work to answer some of these questions.

How can you prevent good friends from getting sick?

What do we think we know?	What do we need to investigate?	What are we learning?	What is our evidence?	What does it mean for the challenge or question?

Update the *Project Board*

15 min.

Students review the Project Board and add questions generated to the Project Board.

△ Guide and Assess

Lead a discussion in which groups share their two most interesting questions with the class. The goal of this discussion is to engage the class in evaluating whether the questions meet the stated criteria. It also provides the basis for connecting the readings and investigations in *Learning Set 3* to questions raised by the students. Challenge students to support the questions they propose by describing how they meet the criteria.

Assess students' ability to generate questions that meet the criteria and are suitable for scientific investigation.

Assessment Options

Targeted Concepts, Skills, and Nature of Science	How do I know if students got it?
Scientists often collaborate and share their findings. Sharing findings makes new information available and helps scientists refine their ideas and build on others' ideas. When another person's or group's idea is used, credit needs to be given.	**ASK:** What responses by other students to your questions were most helpful in revising your own questions? **LISTEN:** Students should cite examples of valid comments relating to meeting the criteria for questions.
Scientists must keep clear, accurate, and descriptive records of what they do so they can share their work with others consider what they did, why they did it, and what they want to do next.	**ASK:** In what ways will the questions you added to the *Project Board* help you to search for answers to the *Big Question?* **LISTEN:** Students should cite examples of how answering the questions added to the Project Board will help them to obtain information that will enable them to address the two challenges associated with the *Big Question*.
Systems are made up of parts that move or work together as a whole.	**ASK:** Would a single cell be considered a system? **LISTEN:** Students recognize that cells are comprised of parts called organelles that work together as a whole. Guide students to view a cell as both a complex system in and of itself and as a part of a larger system.

Teacher Reflection Questions

- How did you ensure that students clearly understood what they were expected to accomplish during small group discussions?
- Which example of a system did you find most effective for helping students grasp the concept of a system? What made that example so effective?
- What ideas about organs or organ systems that came up in class discussions indicate a lack of understanding that needs to be addressed by further instruction?
- In the next section, students will read about the levels of organization in multicellular organisms – cells, tissue, organs and organ systems. Preliminary research indicates that it may be easier for students to understand that the cell is the basic unit of structure (which they can observe) than that the cell is the basic unit of function (which has to be inferred from experiments). How could you help students develop a clearer understanding of the cell as a basic unit of function?

NOTES

NOTES

SECTION 3.2 INTRODUCTION

3.2 Read

From Cells to Systems

◀ *1 class period*

A class period is considered to be one 40 to 50 minute class.

Overview

In this section, students read about the organization of cells in multicellular organisms. The levels of organization from cells to tissue, organs and organ systems are described and their interrelatedness is discussed.

Targeted Concepts, Skills, and Nature of Science	Performance Expectations
Scientists often collaborate and share their findings. Sharing findings makes new information available and helps scientists refine their ideas and build on others' ideas. When another person's or group's idea is used, credit needs to be given.	Students share their ideas about the structure and function of tissues, organs and organ systems.
Scientists must keep clear, accurate, and descriptive records of what they do so they can share their work with others, consider what they did, why they did it, and what they want to do next.	Students make notes of important ideas as they read.
The cell is the basic unit of structure and function in all living things. Cells are usually microscopic in size.	Students give examples of the functions of different types of cells and connect the cell function to a body function.
Cells vary in size, shape, and function and are organized to work together more effectively in multicellular organisms. Levels of organization include cells, tissues, organs, and organ systems	Students describe each level of organization in multicellular organisms in the *Stop and Think* activity.
An organ system is a group of body organs, each of which performs a specific function, that work together to carry on one or more life functions.	Students identify the organs comprising several major organ systems in the human body.

Homework Options

Reflection

- **Science Content:** What is the difference between a cell and a tissue? *(Cells are individual units; a tissue is a group of cells that are similar in structure and function which work together.)*
- **Science Content:** What types of tissues must work together for you to move your finger? What role does each tissue play in the act of moving your finger? *(Nerve: signal from brain/stimulate muscle; blood: supply food and oxygen; muscle: changes length to create force that pulls on bone; connective: connect muscle to bone and hold bones together at joints; bone: provides a rigid surface against which force can be exerted by muscle; epithelial: covers and protects surface)*
- **Science Content:** How does the shape of nerve cells aid in performing their function? *(The long, narrow wire-like shape of a nerve cell's axons enables the same amount of cell material to cover a greater linear distance than other shapes. This helps nerve cells to carry impulses to widely separated parts of the body.)*

Preparation for 3.3

- **Science Content:** Survey your friends and family asking: "When you sleep, do you usually breathe through your mouth or your nose?" *(Results typically indicate that most people breathe through their nose when at rest.) Why do you think some people are mouth breathers and some are nose breathers? (Mouth breathing typically results from physical constriction of the nasal passages or nasal congestion due to mucus buildup.)*

SECTION 3.2 IMPLEMENTATION

3.2 Read

From Cells to Systems

You learned that the building blocks of all living things are cells. Unlike the bacteria you read about that can be single-celled organisms, you are made up of trillions of cells. The types of cells in your body are varied in size, shape, and how they function. (Function means how they work or the job they do.) Each different type of cell is like a "specialist." Cells in your body are specialized to do different things. However, they must work together to keep your body functioning properly. In this section, you will read about how the cells in your body are organized.

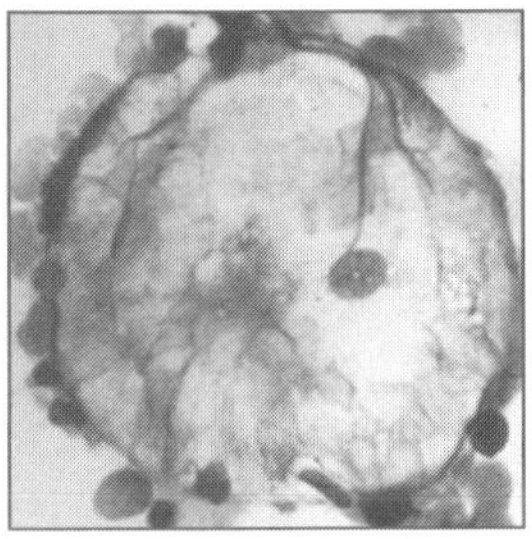

A marrow fat cell.

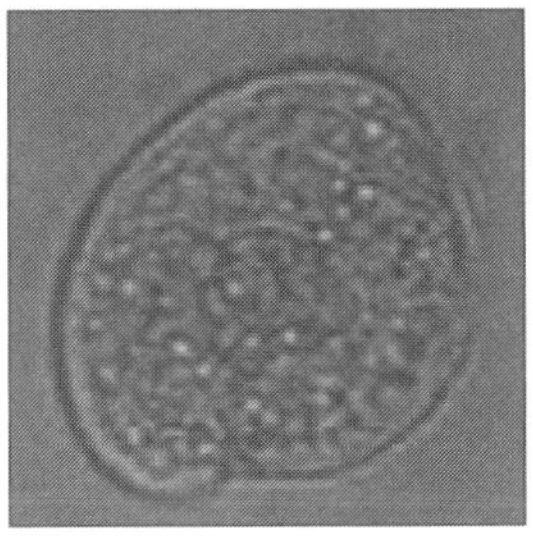

An epithelial cell.

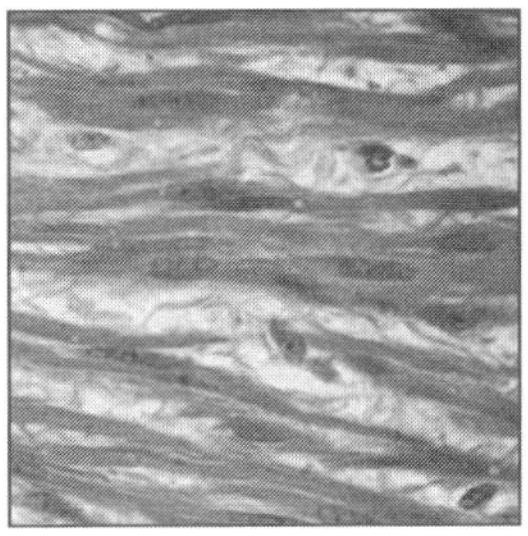

A type of muscle tissue made up of muscle cells.

tissues: groups of cells that are similar in structure and function.

contract: shorten.

organs: structures that have a specific function and are made up of different tissues.

organ systems: groups of organs that have related functions.

How Are Cells Organized?

The cells in your body have many shapes and functions. Cells that are similar in structure and function are called **tissues**. Muscle is an example of one type of tissue. It is made of muscle cells. Muscle tissue is able to **contract**. That is what makes parts of your body able to move.

Tissues are organized into larger structures called **organs**. Organs are groups of different tissues that have a particular job. For example, your heart is an organ made up of muscle, nerve, fat, and other tissues. The nerve tissue gets the muscle tissue in the heart to contract. That makes blood move through the body.

Organs are grouped together in **organ systems**. The organs in a system have related functions. The circulatory system includes the heart, veins, and arteries. They work together to get the job done.

◀ $\frac{1}{2}$ *class period**

3.2 Read

From Cells to Systems

10 min.

Students read about the levels of organization in multicellular organisms from cells to organ systems.

△ Guide

Read this section with the class or have students read aloud. Review with students the observations of cells they made in *Learning Set 2.* Ask students why bacterial, animal, and plant cells could be considered systems. Guide students to recognize that cells have parts, each with a specific function, that work together as a whole. Then, have students recall that the cells they observed had different shapes and sizes. Guide students to make connections between the shape of a cell and the function it performs. The following optional activity may help students make these connections in a more concrete way.

*A class period is considered to be one 40 to 50 minute class.

- Show students a ball of clay close to the size of a golf ball and hold it up against a 4"x 6" index card. Point out that the ball of clay does not cover the entire card. Then, ask students how they could cover an entire index card with the ball of clay. Students typically suggest flattening the clay to form a sheet large enough to cover the index card. Point out that just by changing the shape of the clay, they were able to accomplish the task with the same amount of clay.
- Next, draw two points on the chalkboard about a half meter apart. Roll the sheet of clay back into a ball, hold it up between the two points, and ask students how they could connect the points with the ball of clay. Students typically suggest rolling the clay into a long, thin wire-like strand. Point out that again, just by changing the shape of the clay, they were able to accomplish the task with the same amount of clay. Guide students to recognize the connection between structure and function. Guide students to extend this idea to the cells that make up our bodies – the shapes and sizes of cells are related to their function in the body.
- Have students recall the flat, pancake-like shape of the skin (epithelial) cells they observed. Ask how this shape helps skin cells perform the function of covering the body surface. Have students recall the thin, tentacle-like axons that extending from the nerve cells they observed. Ask how this shape helps nerve cells carry messages from one part of the body to another.

Emphasize to students that our bodies are composed of many types of cells, which have many different shapes and sizes that help them to perform different functions.

NOTES

to keep your body functioning properly. In this section, y about how the cells in your body are organized.

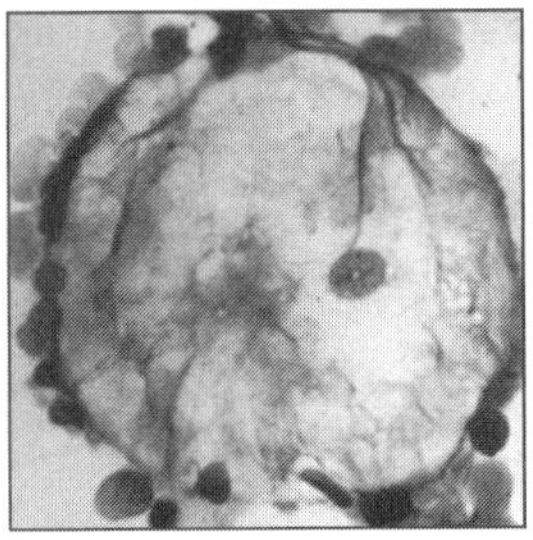

A marrow fat cell.

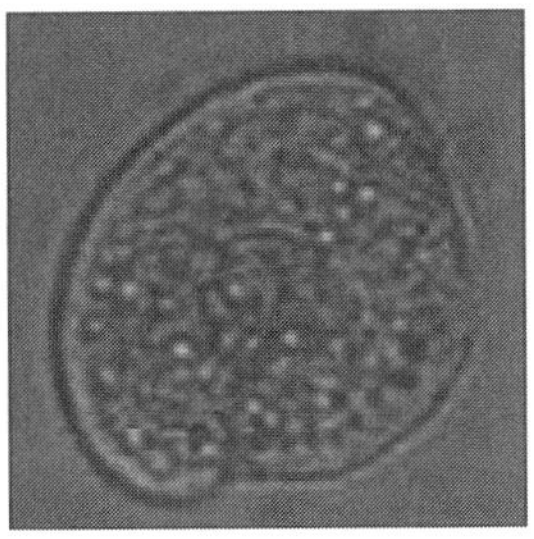

An epithelial cell.

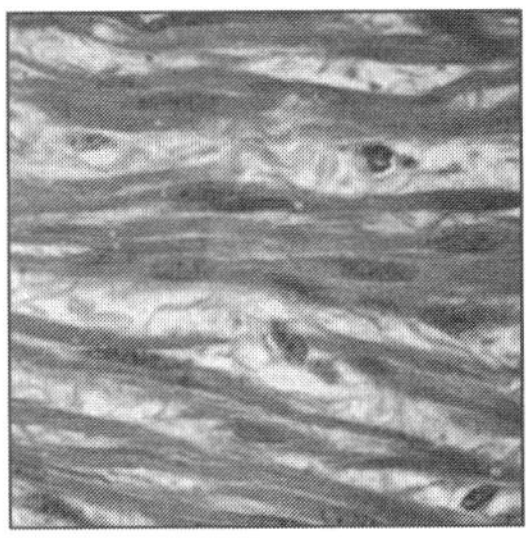

A type of muscle tissue made up of muscle cells.

tissues: groups of cells that are similar in structure and function.

contract: shorten.

organs: structures that have a specific function and are made up of different tissues.

organ systems: groups of organs that have related functions.

How Are Cells Organized?

The cells in your body have many shapes and functions. Cells that are similar in structure and function are called **tissues**. Muscle is an example of one type of tissue. It is made of muscle cells. Muscle tissue is able to **contract**. That is what makes parts of your body able to move.

Tissues are organized into larger structures called **organs**. Organs are groups of different tissues that have a particular job. For example, your heart is an organ made up of muscle, nerve, fat, and other tissues. The nerve tissue gets the muscle tissue in the heart to contract. That makes blood move through the body.

Organs are grouped together in **organ systems**. The organs in a system have related functions. The circulatory system includes the heart, veins, and arteries. They work together to get the job done.

GF 80

Project-Based Inquiry Science

How Are Cells Organized?

20 min.

○ Engage

Have the class recall the activity in the previous Learning Set in which they estimated the size of a cell. Ask students how a microscopic skin cell could possibly cover the entire surface of their body. Guide students to recognize that just as one shingle cannot cover an entire house, it is not possible for a single skin cell to cover your entire body surface. However, just as large numbers of shingles can be fit together to cover a house, large numbers of microscopic skin cells working together can form a surface large enough to cover the body.

In the previous Learning Set you used the microscope to observe cells. You observed that cells are microscopic in size and estimated the size of a cell. How could a microscopic skin cell possibly cover the entire surface of your body?

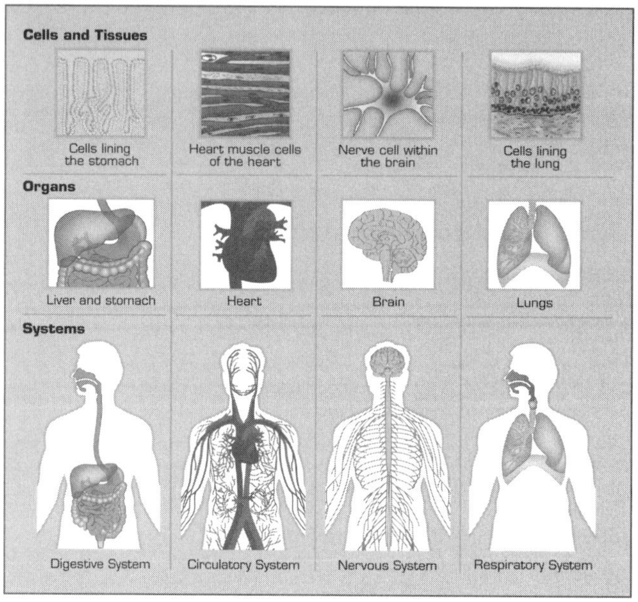

Stop and Think

The levels of organization in a multicellular organism are cells, tissues, organs, and organ systems. Describe each of these levels of organization.

What's the Point?

Your body is made of organ systems. Every system has a different job to do. Systems are made of parts. The parts are organs, tissues, and cells. Every part in a system does a job to make the whole system work. In your body, each system does a different job to keep you healthy. When you know more about how your body works, you can stay healthier. Learning about your body systems will make it easier to understand your body and how to keep it healthy.

△ Guide

Point out that a group of cells with a similar structure and function that work together is called a tissue. (You may want to briefly mention the five tissue types found in animals listed in the *Understanding for Teachers* section for this Learning Set.)

Ask students to imagine what would happen if their legs were made only of skin tissue. Elicit that the skin would be unable to support their body weight and would not be able to cause movement. Ask students what kinds of tissue are found in their legs. As students identify different types of tissue found in the leg, ask them to describe the function that each performs (e.g. skin: cover and protect; muscle: produce movement by contracting; nerve: carry messages from the brain to the muscle to tell it when to contract; blood: carry food and oxygen to the cells; etc.) Guide students to recognize that there are many body parts that require several different kinds of tissue working together to carry out a function. Point out that a structure that has a specific function and is made up of different tissues is called an organ. Ask students if they know of any other organs that perform a specific function, but are made up of several kinds of tissue working together.

Point out that the stomach is an organ. Ask students to point to where they think their stomach is located. Then ask students how they think food gets from their mouth to their stomach. Guide students to recognize that another structure, a tube called the esophagus, carries food from their mouth to their stomach. Point out that a group of organs that have related functions is called an organ system.

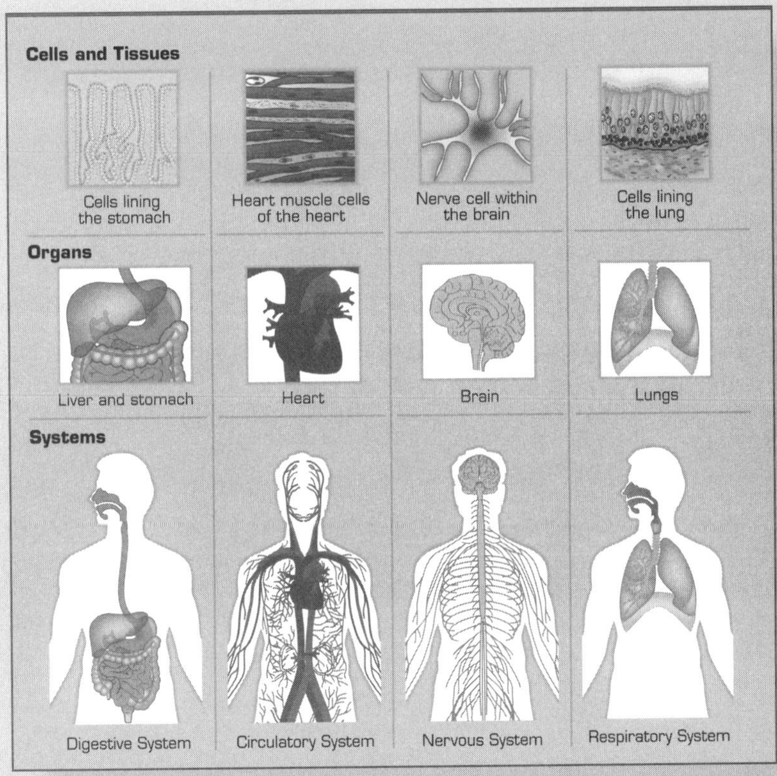

Ask students to examine the examples of organ systems shown in the diagrams in the chart on page 81. Ask students to identify any organs they recognize in each of the systems shown and what related function they think they have in common. If there are organs that the students do not recognize, tell them that they will learn about those organs in later sections of this learning set.

Stop and Think
5 min.

Digestive System | Circulatory System | Nervous System | Respiratory System

Stop and Think

The levels of organization in a multicellular organism are cells, tissues, organs, and organ systems. Describe each of these levels of organization.

What's the Point?

Your body is made of organ systems. Every system has a different job to do. Systems are made of parts. The parts are organs, tissues, and cells. Every part in a system does a job to make the whole system work. In your body, each system does a diff... job to keep yo... ...lthy. When you ...ore

△ Guide

After students have had an opportunity to answer the question, lead a brief class discussion of what students know about the levels of organization in multicellular organisms. Make a class list of students' responses in their own words. Keep this list and refer back to it as students begin to learn about each of the organ systems. Responses should include the following ideas:

- Cells that are similar in structure and function working together as a group are called a tissue.
- Organs are structures that have a specific function and are made up of different tissues.
- A group of organs that work together and have related functions are called an organ system.

Teacher Reflection Questions

- How did the observations of cells that students made in *Learning Set 2* help them better understand the levels of organization in multicellular organisms described in this reading?
- When students examined the four examples of organ systems shown in this activity, which did they seem most familiar with? Least familiar with?

SECTION 3.3 INTRODUCTION

3.3 Investigate

How Often Do You Breathe?

◀ *1 $\frac{1}{2}$ class periods*

A class period is considered to be one 40 to 50 minute class.

Overview

Working with a partner, students investigate the question *How Often Do You Breathe?* by measuring their breathing rate while at rest and after exercise. After collecting and recording data from three trials of each rate, students communicate their results with the class and determine the class average for breaths per minute while at rest and after exercise. Students then consider differences between the class average, their own breathing rate and the average breathing rate of the general population. Students then reflect upon their results and identify cause-and-effect relationships between breathing rate and activity level.

Targeted Concepts, Skills, and Nature of Science	Performance Expectations
Scientists often collaborate and share their findings. Sharing findings makes new information available and helps scientists refine their ideas and build on others' ideas. When another person's or group's idea is used, credit needs to be given.	Students collaborate with a partner to collect breathing rate data. Students then share their breathing rate data with their partner and with the class. The class collaborates to identify possible cause-and-effect relationships between breathing rate and activity level.
Scientists must keep clear, accurate, and descriptive records of what they do so they can share their work with others, consider what they did, why they did it, and what they want to do next.	Students record data for three trials each of breathing rate while at rest and breathing rate after exercise. Students create a table of class data and determine the class average for breathing rate while at rest and breathing rate after exercise.

Targeted Concepts, Skills, and Nature of Science	Performance Expectations
Identify cause-and-effect relationships.	Students compare and contrast average breathing rates while at rest and after exercise to identify cause-and-effect relationships between breathing rate and activity level. Students cite evidence to support the idea that an increase in activity level causes an increase in breathing rate.

Materials	
1 per classroom	Projection of class data table. (Optional: project spreadsheet that calculates the sum of the trials listed in the table.)
1 per student pair	Stopwatch or clock that displays seconds
1 per student	Calculator
1 per student	Worksheet

Activity Setup and Preparation

Check stopwatches to be sure they are working correctly; replace batteries as needed. If this is the first time students are working with stopwatches, demonstrate the correct technique for stopping, starting, and resetting the stopwatch.

You may want to create a spreadsheet for the class data set that calculates the sum (total of all trials) listed in the table, but not the class averages. Errors are common when students add large columns of numbers and the process uses valuable class time. By allowing the spreadsheet to add the numbers, students can focus on calculating class averages. It is important for each student to calculate the class average. You may want to enlist the support of the class' math teacher to review and reinforce this skill.

Students will work in pairs during this section; determine how you will assign student pairs.

Homework Options

Reflection

- **Science Content:** The last time you had a cold and stuffy nose, how was your breathing affected? *(Students typically report difficulty breathing, noisy breathing, mouth breathing and an increase in breathing rate.)*
- **Science Process:** Repeat the investigation you did in class today with other members of your family and/or friends. Before measuring a person's breathing rate, make a prediction about how that person's breathing rate at rest and after exercise will compare with yours. How do you think each of the following factors might affect a person's breathing rates: age, weight, gender, and smoking status? *(Students typically report increases in breathing rate based on age, weight, and smoking status.)*

Preparation for 3.4

- **Science Content:** Draw a diagram of what you think happens to the air you breathe in. Use arrows to indicate the path of the air through your body. *(Answers will vary, but students typically show air moving in through the mouth or nose, through the neck to the chest cavity. Look for details that indicate students' level of understanding of the respiratory system. For example: Configuration of passages in nose and mouth? Does trachea branch into bronchi? Do students show inhaled and exhaled air following the same or different paths? Is a diaphragm shown?)*

NOTES

SECTION 3.3 IMPLEMENTATION

◀ *2 class periods**

3.3 Investigate

How Often Do You Breathe?

How many times do you breathe in a minute? Do you know how different activities affect how often you breathe? Try this investigation to see if you can figure out how often you breathe. Compare that to how often other people breathe. Then think about how different activities affect how you breathe. You will measure your breathing two times, once while at rest and once after exercise.

Procedure

Materials
- stopwatch
- Breathing Rate data pages

Measure Your Resting Breaths

1. Work with a partner. Let your partner use the stopwatch while you count your breaths.
2. Your partner starts the stopwatch. Count your breaths until your partner tells you to stop. Your partner will tell you to stop after 6 seconds. Do this two more times.
3. Multiply your number of breaths by 10 to determine how many times you breathe in one minute (6 seconds times 10 equals 60 seconds or one minute).

4. Switch roles and repeat Steps 1 to 3.

Measure Your Resting Breaths after Exercise

You are going to repeat the steps you just did, but with exercise as part of the investigation. Rather than measuring your resting breaths, you are going to measure how many times per minute you breathe right after you do a little exercise.

5. Work with your partner again. Stand near your desk and make sure you are a safe distance from everyone else. Jog in place for 30 seconds.

3.3 Investigate

How Often Do You Breathe?

5 min.

Students work in pairs to measure their breathing rate at rest and after exercise. They then share their data with the class, create a class data table and calculate class averages for breathing rate at rest and after exercise. Students engage in a class discussion focusing on cause-and-effect relationships shown by the data.

△ Guide

Have students read the introductory paragraph and then hold a brief discussion in which students share their answers to the questions posed in the paragraph. Make a class list of activities that students think will affect their breathing rate. Guide students to recognize that many different types of activities could be grouped under the general heading of exercise. Take note of non-exercise suggestions for possible investigation at a later time. Then, have students form pairs within their groups according to your instructions.

*A class period is considered to be one 40 to 50 minute class.

Procedure: Measure your resting breaths

30 min.

Students work in pairs to measure three trials of their breathing rate while at rest and calculate their resting breathing rate in breaths per minute.

...figure out how often ... Compare ... people breathe. Then think about how different activities ... you breathe. You will measure your breathing two times, once while at rest and once after exercise.

Procedure

Materials
- stopwatch
- Breathing Rate data pages

Measure Your Resting Breaths

1. Work with a partner. Let your partner use the stopwatch while you count your breaths.
2. Your partner starts the stopwatch. Count your breaths until your partner tells you to stop. Your partner will tell you to stop after 6 seconds. Do this two more times.
3. Multiply your number of breaths by 10 to determine how many times you breathe in one minute (6 seconds times 10 equals 60 seconds or one minute).

4. Switch roles and repeat Steps 1 to 3.

Measure Your Resting Breaths after Exercise

You are going to repeat the steps you just did, but with exercise as part of the investigation. Rather than measuring your resting breaths,
you are going to measure how many times per minute you breathe right after you do a little exercise.

5. Work with your partner again. Stand near your desk and make sure you are a safe distance from everyone else. Jog in place for 30 seconds.

Project-Based Inquiry Science GF 82

△ Guide

Read through steps 1-4 of the procedure with the students and demonstrate it for them. Emphasize that each person will take three measurements of their breathing rate while at rest

Remind students that scientists keep track of their data as they collect it. Ask them to recall how important this was during the liquid sharing experiment in *Learning Set 1.* Also, remind students that scientists take many measurements of the same thing (or trials) to make sure their measurements are accurate. Let students know that in this experiment they will be taking each measurement three times.

⬡ Get Going

Have student pairs perform steps 1-4 of the procedure and record their data for breathing rate while at rest in the data table on their worksheet.

☐ Assess

Circulate among groups to check that students are calculating their resting breathing rate correctly (breaths per 6 seconds x10) and that breathing rate at rest is between 10 and 30 per minute.

Make sure students are staying on task while measuring their breathing rates so that recordings are not altered by further activity.

△ Guide

Read through steps 5-7 of the procedure with the students and tell students that they will now make three measurements of their breathing rate after exercise. Demonstrate to students that when they are exercising, they are to stand a safe distance from each other and jog for 30 seconds in place. Then, they are to sit down and immediately start counting how many breaths they take in six seconds. Remind students that their partner is keeping time for them.

⬡ Get Going

Have student pairs perform steps 5-7 of the procedure and record their data for breathing rate after exercise in the data table on their worksheet. Make sure that all students exercise at a similar rate for the same length of time (30 seconds).

☐ Assess

Circulate among groups checking that students are measuring their breathing rate immediately after exercise and that they are calculating their breathing rate correctly (breaths per 6 seconds x 10).

Make sure students are staying on task while measuring their breathing rates so that measurements may not be altered by further activity.

NOTES

Recording Your Data

20 min.

Students record breathing rate data for three trials while at rest and three trials after exercise. Then, they calculate their average breathing rate while at rest and after exercise.

6. Sit down and begin counting your breaths immediately. Stop when your partner tells you that 6 seconds have passed.

   Record your data. Begin counting your second trial immediately after the first, and your third right after the second. Record all three trials.

7. Multiply your numbers by 10 to determine how many times you breathe in one minute.
8. Switch roles and repeat Steps 4 to 7.

Recording Your Data

Scientists always keep track of their data as they collect it. Use a table like the one below to track the data you collect. Use one table for yourself and one for your partner. Complete the tables so you have a record of your data.

My Breathing Rate Data

Name: ______________________ Date: __________

Trials	My number of breaths in 6 s without exercise	My number of breaths in 1 min without exercise (breaths in 6 s x 10)	My number of breaths in 6 s with exercise	My number of breaths in 1 min with exercise (breaths in 6 s x 10)
1		x 10 =		x 10 =
2		x 10 =		x 10 =
3		x 10 =		x 10 =
Average number of breaths per minute for all trials	Add together the number of breaths per minute for each trial and divide by 3			

Stop and Think

1. An average person breathes 15 to 25 times per minute. This is called a range. Does your number of resting breaths fall within that range? If it does not, why do you think it is different? Does your partner's number of breaths fall in that range? If not, why do you think it didn't?
2. How did your results (yours and your partner's) compare for resting and exercising?
3. Were your resting or exercising results higher or lower?
4. Why do you think they were different?

△ Guide

Ask students to calculate their average rest breathing rate per minute and their average breathing rate per minute after exercise. Remind students how to calculate the average, assist them by giving them the formula for calculating the average of 3 trial values.

$$\text{Average} = \frac{\text{Trial 1} + \text{Trial 2} + \text{Trial 3}}{3}$$

☐ Assess

Ask a few students for their averages. Students usually obtain an average breathing rate while at rest between 15 and 25. Their breathing rate after exercise should be higher.

tables so yo... a record of your data.

Stop and Think

1. An average person breathes 15 to 25 times per minute. This is called a range. Does your number of resting breaths fall within that range? If it does not, why do you think it is different? Does your partner's number of breaths fall in that range? If not, why do you think it didn't?
2. How did your results (yours and your partner's) compare for resting and exercising?
3. Were your resting or exercising results higher or lower?
4. Why do you think they were different?

Stop and Think

5 min.

Hold a brief class discussion focusing on observed variations in breathing rates and possible reasons for those differences.

△ Guide

When students have completed the Stop and Think questions, hold a brief class discussion in which students share their answers and focus on ideas they have about why breathing rates vary and why their breathing rate after exercise was higher than their breathing rate while at rest.

Assess students' responses to the Stop and Think questions by listening for the following information in their responses:

1. An average person breathes 15-25 times per minute. This is called a range. Does your number of resting breaths fall within this range? (Answers will vary, but most students will find that their breathing rate falls within this range.) If it does not, why do you think it is different? (Students may cite a variety of factors including gender, weight, and level of fitness.) Does your partner's number of breaths fall within that range? *(Answers will vary.) If not, why do you think it didn't? (Students may cite a variety of factors including gender, weight, and level of fitness.)*
2. How did your results (yours and your partner's) compare for resting and exercising? *(Answers will vary, but most students find that their breathing rates were similar to that of their partner.)*
3. Were your resting or exercising results higher or lower? *(Students should report that their exercising rate was higher than their resting rate.)*

4. Why do you think they were different? *(Students should suggest that when exercising, muscle cells use more food and oxygen to produce energy than when at rest. In order to replenish the oxygen that was used, the body increases breathing rate.)*

At the end of the discussion, create a brief summary of what students think they have learned about breathing rate.

NOTES

Communicate Your Results

Share your pair's results with the class. Create a class table that includes the average breaths per minute for everyone. The table should look like the one shown below. You might use a similar table to organize all the data for yourself.

Class Breathing Rate Data 3.3

Name: ______________________ Date: __________

	Resting breaths per minute	Exercising breaths per minute
1		
2		
3		
4		
5		
6		
7		
8		
9		
10		
11		
12		
13		
14		
15		
16		
17		
18		
19		
20		
21		
22		
23		
24		
25		
26		
27		
28		
29		
30		
Class average number of breaths per minute	Add together the number of breaths per minute for each student and divide by the number of students	

Analyze Your Data

Now that you see everyone's data, you can compare your own information to the class information. When scientists have collected a set of data like this, they analyze it to see what information is important to answering their question. Your question is, *How often do you breathe?* You answered that question for yourself and your partner, but a bigger set of data helps to answer the question in a larger group or population.

1. What was the range of average number of breaths per minute in your class at rest and after exercising?
2. What was the class average for the number of breaths per minute at rest and after exercising? Your teacher will help you use a calculator to figure that out.
3. Did the average for resting fall within the range of 15 to 25 breaths per minute? Was it at the high end, closer to 25 breaths, or at the low end, closer to 15 breaths?
4. How did your at-rest and after-exercising breaths per minute compare to other people's?
5. How reliable do you think the data are? Why?

Communicate Your Results

10 min.

Students share their data with the class, record all data on a class table, and calculate class averages for breathing rate while resting and after exercise.

△ Guide

Have students read the instructions and construct a class data table based on the example shown on page 84. Then, ask student pairs to share their results with the class and record the data on their class table while you record it on a projected class table or in a spreadsheet.

You should calculate the sum of all values or use a spreadsheet to calculate the values. The total number of values is the number of students in the class that reported data.

Tell students that you will give them the sum of all their values, but that each person will calculate the class average on their own. Remind students that the average is found by using the formula:

$$\text{Average} = \frac{\text{sum of all values}}{\text{total number of values}}$$

☐ Assess

Have students compare their calculated class average for breathing rate while at rest and breathing rate after exercise with yours. If calculated results do not agree, have students check their calculations and identify error(s).

Analyze Your Data

10 min.

Students compare their individual results with the class results. Then they compare and contrast answers to the section question, How Often Do You Breathe?, *for individuals and large groups.*

organize all the

yourself.

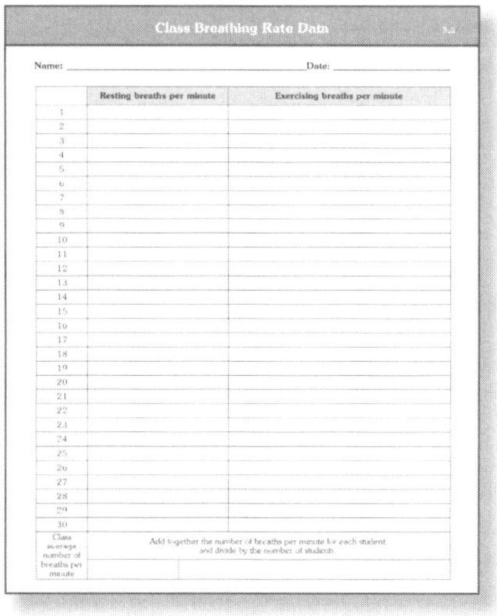

Analyze Your Data

Now that you see everyone's data, you can compare your own information to the class information. When scientists have collected a set of data like this, they analyze it to see what information is important to answering their question. Your question is, *How often do you breathe?* You answered that question for yourself and your partner, but a bigger set of data helps to answer the question in a larger group or population.

1. What was the range of average number of breaths per minute in your class at rest and after exercising?
2. What was the class average for the number of breaths per minute at rest and after exercising? Your teacher will help you use a calculator to figure that out.
3. Did the average for resting fall within the range of 15 to 25 breaths per minute? Was it at the high end, closer to 25 breaths, or at the low end, closer to 15 breaths?
4. How did your at-rest and after-exercising breaths per minute compare to other people's?
5. How reliable do you think the data are? Why?

△ Guide

Read the introductory paragraph together with the class and then have students answer the questions independently. Then, lead a class discussion focusing on the variability of breathing rate. Guide students to recognize that breathing rate may vary from one trial to the next and from one individual to the next. Draw students' attention to the fact that the class averages encompass a range of results. Discuss possible reasons for differences between individual results, class averages, and averages for the general population. Guide students to recognize that, on average, breathing rate is higher after exercise than while at rest.

Reflect

Answer the following questions. Be prepared to discuss your answers with the class.

1. The average resting rate was probably lower than the average exercising rate. Why do you think that is?
2. What was the high number for resting breathing rate? What was the low number? What was the class average number for these rates? Which is the best number to use when you want to know about the population, the high, the low, or the average number of breaths? Why did you choose that number?
3. Why do you think you would need to breathe more often when you are exercising than when you are resting?

What's the Point?

In this investigation, you measured how many times you breathe in one minute when you are sitting still. You compared your at-rest breathing rate to the class average. Then you compared your class average to the average for the general population. Next, you compared the at-rest breathing rate to the after-exercising breathing rate. Your exercising breathing rate was faster than your resting rate. You took more breaths per minute after you exercised than when you were at rest.

Breathing rate increases during exercise because the body needs more oxygen than when at rest.

Reflect

10 min.

Students consider variations in the breathing rate data collected during this investigation and suggest possible reasons for those variations.

△ Guide and Assess

When students have completed the *Reflect* questions, hold a brief class discussion in which students share their answers.

Assess students' responses to the *Reflect* questions by listening for the following information:

1. Student responses will vary, but should convey the idea that the resting rate was probably lower than the active rate because you don't need as much oxygen when you are resting.

2. On the chalkboard, list the highest and the lowest breathing rate while at rest. Next, list the class average. Every measurement contains some error. Errors may result in a measurement larger than the actual value, or can result in a measurement that is smaller than the actual value. When averaging, measurements larger than the actual value are canceled out by measurements smaller than the actual value. As a result, the average (mean) is likely to be closest to the actual value. Therefore, the best number to use is the average number of breaths.

NOTE: If students bring up median or mode, you may want to mention that the mode is the value that appears most frequently in a data set. This value does not provide much information unless a large percentage of the measurements are the same. In this case it may be used to help describe a distribution of values. The median is a measure of central tendency. It is the point that divides the set of data in half so that half the data points lie above the median and half the data points lie below the median.

3. Student answers will vary, but should connect faster breathing rate with increased need for oxygen during exercise.

Teacher Reflection Questions

- In this section, students measured two breathing rates: while at rest and after exercise. How well do you feel that students understood the importance of using class averages rather than individual data points when comparing the two breathing rates? What evidence do you have?
- How well do you feel your students understood the connection between using averages and minimizing errors in measurement?
- What factors negatively or positively affected the success of this investigation? What specifically might have been changed to improve the investigation?
- Good inquiry science connects each investigation with previous and future investigations. How did you help students see the connections between this investigation and previous ones on organ systems and to the next investigation on the respiratory system?

SECTION 3.4 INTRODUCTION

3.4 Investigate

How Does the Respiratory System Work?

◄ *2 class periods*

A class period is considered to be one 40 to 50 minute class.

Overview

In this investigation, students begin by reading about the structure and function of the respiratory system. Students then build a working model of the respiratory system and use the model to simulate breathing. Building on their breathing experiences in the last investigation, students connect the parts of the model to the parts of the respiratory system and investigate how the various parts of the respiratory system work together in the process of breathing. Students then read about the effects of various diseases on the functioning of the respiratory system. Information about respiratory-system diseases is then added to the *Communicable-Disease Information Table* and the *Project Board* is updated.

Targeted Concepts, Skills, and Nature of Science	Performance Expectations
Scientists often collaborate and share their findings. Sharing findings makes new information available and helps scientists refine their ideas and build on others' ideas. When another person's or group's idea is used, credit needs to be given.	Students collaborate with a partner to collect breathing rate data. Students then share their breathing rate data with their partner and with the class. The class collaborates to identify possible cause-and-effect relationships between breathing rate and activity level.
Scientists must keep clear, accurate, and descriptive records of what they do so they can share their work with others, consider what they did, why they did it, and what they want to do next.	Students record data for three trials each of breathing rate while at rest and breathing rate after exercise. Students create a table of class data and determine the class average for breathing rate while at rest and breathing rate after exercise.

Targeted Concepts, Skills, and Nature of Science	Performance Expectations
Models are a representation of something in the world. Simulations use a model to imitate, or act out, real-life situations.	Students successfully build a working model of the respiratory system. Students correctly identify the part of the respiratory system that corresponds to each part of the model. Students use the model to simulate the process of breathing and correctly describe the steps in the process of breathing.
The respiratory system delivers oxygen to the blood and removes carbon dioxide from the blood.	Students correctly identify the exchange of the gases oxygen and carbon dioxide in blood as the primary function of the respiratory system.
Disease breaks down the structures or functions of an organism. Some diseases are the result of failures of a body system. Other diseases are the result of damage by infection from other organisms.	Students describe routes of entry for microbes that cause respiratory illness and their effects on the respiratory system. Students identify black lung and lung cancer as a non-communicable disease.
Identify structure and function relationships in organisms.	Students successfully describe the relationship between the structure of the organs of the respiratory system and their function in the system. For example, the trachea's tubelike structure held open by cartilage rings allows for the free movement of air to and from the lungs.

Materials	
1 per student	Straw
1 per group	Balloon
2 per group	Rubber band
1 per group	2 liter plastic bottle with bottom cut off
1 per group	Cotton balls or lump of modeling clay
1 per group	Plastic wrap or balloon with knot tied in neck end and cut in half.

Activity Setup and Preparation

This investigation requires one empty 2-L plastic bottle for each group of students. You may want to collect these, or ask students to bring them in, well ahead of time so that you can prepare them for use.

Prepare one 2-L plastic bottle for each group by cutting off the bottom so the bottle has a cylindrical shape. See the diagram on page 87 of the student edition.

You may want to build an example of the model to show students how the model should be constructed. In order to obtain the best results, every effort must be made to prevent air from entering the plastic bottle through any opening except the straw leading to the balloon. There are three locations that must be made airtight:

1) The mouth of the bottle through which the straw enters. Modeling clay forms a tighter seal than cotton. Wet cotton creates a better seal than dry cotton.

2) The attachment of the balloon to the straw. If the rubber band collapses the straw when made tight, try sealing the balloon to the straw with adhesive tape. Using a larger balloon allows the balloon to show a greater degree of inflation before the elasticity of the balloon begins to counteract the air pressure inside the balloon. If small balloons are used, inflate and deflate several times to stretch the balloon and decrease the amount of air pressure needed to inflate it.

3) The attachment of the plastic sheet to the bottle. Here again, if the rubber band does not form a tight enough seal, use adhesive tape. Best results are obtained if the plastic sheet is attached to the bottle and pushed upward so it bulges into the bottle before sealing the mouth of the bottle with clay or cotton. This also provides some slack so the plastic can be gripped more easily when pulling down to inflate the balloon.

NOTE: When setting up the model, leave enough slack in the plastic sheet so that you can create an upward bulge. Push upward on the bulge while sealing the opening around the straw. If the balloon still doesn't inflate when you pull downward on the plastic sheet, check the seals and then try tightening the plastic sheet. If it still does not work, discuss with students that this is a flaw of the model and that the way our body works is that the diaphragm muscle moves up and down which fills our lungs with air. When the muscle relaxes, the air is let out of our lungs. Assist students in tightening the plastic covering until the lungs expand.

Homework Options

Reflection

- **Science Content:** Why is it possible to breathe through your mouth, but it is better to breathe through your nose? *(Being able to breathe through mouth or nose enables you to get the oxygen needed to stay alive if one of the passageways is blocked. It is better to breathe through your nose because the nasal hairs and mucus trap particles and germs preventing them from entering the lungs and the nasal passages moisten and warm the air before it enters the lungs.)*
- **Science Content:** The lungs are located in the upper portion of the chest cavity, or thorax. Why is it difficult to breathe if a belt is tightened around your waist? *(Tightening a belt keeps the abdominal organs from moving out of the way when the diaphragm moves downward. As a result, it is more difficult to increase the air space in the thorax so that air will move into the lungs.)*
- **Science Process:** In this investigation, you counted the number of breaths taken in 6 seconds and then multiplied it by10 to determine the number of breaths per minute. How might your results be affected if you counted the number of breaths taken in 30 seconds and then multiplied by two? *(Since it is unlikely that you completed exactly a whole number of breaths in 6 seconds, there was some error in the number you multiplied by 10. That error was also multiplied by 10. If you counted breaths for 30 seconds there would also be some error, but that error would only be multiplied by 2 when converting to breaths per minute, so there would be less error in your result.)*

Preparation for 3.5

- **Science Content:** How does oxygen taken into your lungs get to a cell in your big toe that needs oxygen? *(In the lungs, oxygen enters the blood. The circulatory system transports blood from the lungs to the big toe.)*
- **Science Process:** When you visit the doctor for a physical, what observations does the doctor make to check the functioning of your circulatory system? What instruments does the doctor use to make these observations? *(Routine observations include heart rate and blood pressure. Heart rate is generally measured with a timer and stethoscope. Blood pressure is measured with a sphygmomanometer.)*

SECTION 3.4 IMPLEMENTATION

◀ $1\frac{1}{2}$ *class periods**

3.4 Investigate

How Does the Respiratory System Work?

When you are sick, a medical person may ask you to describe how you feel. He or she is asking you to help identify the symptoms of your disease. If you have a cold, you may say that your nose is stuffy. You may also say that you are coughing. The medical person will think about which diseases affect your body that way. He or she will think about diseases of the **respiratory system**. That is because your nostrils, which are stuffed up, are part of that system. That person may listen to your lungs because they are also part of your respiratory system.

respiratory system: the organ system that delivers oxygen to the blood and removes carbon dioxide, a waste gas, from the blood. It includes the nostrils, trachea, and lungs.

trachea (windpipe): a tube that carries air to the lungs.

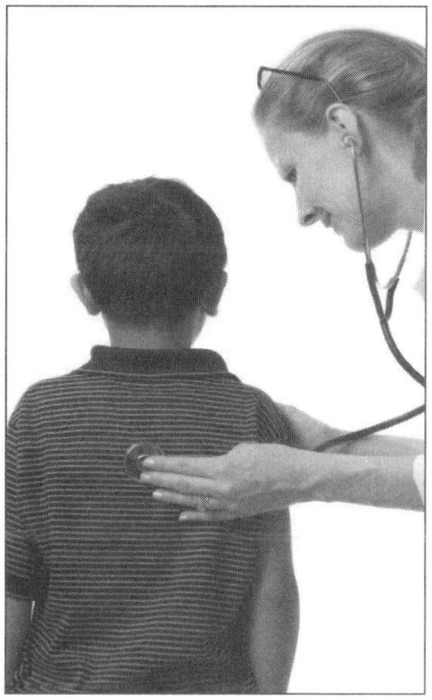

Like all other systems in the body, the respiratory system is made up of particular organs that have particular functions. Some of the organs of the respiratory system include the nose, the **trachea** or windpipe, and the lungs. The trachea is the tube that takes air from your nostrils to your lungs. The function (job) of the respiratory system is to deliver oxygen to the body and remove carbon dioxide from the body. You need oxygen to live. Carbon dioxide is a waste gas that comes from normal body processes.

By listening to the sounds your lungs make as you breathe, a medical person can assess the health of your respiratory system.

Project-Based Inquiry Science GF 86

3.4 Investigation

How Does the Respiratory System Work?

5 min.

Students read about some of the parts of the respiratory system and their functions.

△ Guide

Read the introduction together with the class. After students have completed the reading, briefly discuss the structure and function of the respiratory system. Emphasize that the purpose of the respiratory system is to deliver oxygen to the blood and remove carbon dioxide (a waste gas) from the blood. Ask students how they can connect this with their investigation of breathing rate in the previous section. Students should recognize the connection between an increased need for oxygen during exercise and breathing more rapidly. Point out that because they cannot see the respiratory system in action because it is inside their body, they will build a model of the respiratory system and study how the model works.

*A class period is considered to be one 40 to 50 minute class.

Build and Run Your Model

25 min.

Students construct a working model of the respiratory system and use it to simulate breathing.

It is hard to imagine how your respiratory system works because you cannot see it. Because you cannot see it, building a model is a good idea. You can build a model of the respiratory system with very common materials. When you build it, you will be able to see how the organs of the respiratory system work together to help you breathe.

Build and Run Your Model

1. First, explore what happens when you breathe in and out. Place one hand on your ribs. Place your other hand just under your ribs. Take four or five deep breaths. Observe what happens under each of your hands as you breathe.
2. Answer the following questions:
   a) What do you feel under each of your hands?
   b) What do your lungs do under your hand when you breathe?
   c) Your other hand is just under your ribs. How does this part of you move when you breathe?
3. Using the materials, build your model to match the diagram. Attach a balloon to one end of the straw. Make sure that the fit is tight. You may wish to use a rubber band to make a tight fit.
4. Place the "lung and windpipe" (balloon and part of the straw) into the mouth of the bottle. Secure the straw with a lump of clay.
5. Stretch the plastic wrap tightly over the bottom of the bottle. Secure it with a rubber band.
6. After you have built your model, run your model and answer the following questions. Be prepared to share your answers with the class.
   a) Pull on the plastic wrap at the bottom of the model. What happens to the balloon inside the model? How is this like inhaling?
   b) Push the plastic wrap up at the bottom of the model. What happens to the balloon inside the model? How is this like exhaling?

Materials
- straw
- balloon
- 2 rubber bands
- 2-L plastic bottle with bottom cut off
- lump of modeling clay
- plastic wrap

Do not let anyone use the straw you breathed into.

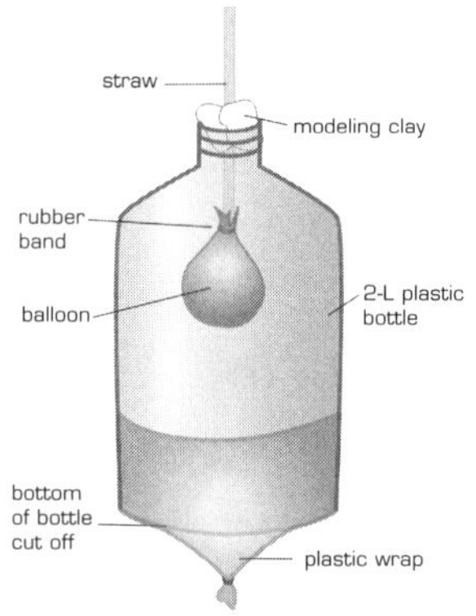

△ Guide

In their groups, have students read through the procedure. Then ask students to describe what they will be doing before, during, and after constructing the model.

⬡ Get Going

Ask students to do the first two steps in the procedure and then hold a brief class discussion of what they observed.

△ Guide

Ask students to describe the movements they felt with their hands while breathing. Make a chart listing the location of their hands in one column and what they felt at each location in the other. The chart may look something like this:

Location of hand	feeling/movement
On ribs over lungs	My chest rose and fell, so I think the lungs expand and contract.
Just under ribs	It feels like it is going up and down as if something is rising up when I exhale and moving down when I inhale pushing out my belly.

⬡ Get Going

Distribute the materials needed to construct the model and caution students not to share straws. Emphasize that each student should use their own straw when breathing into the model. If you made a sample model, show the students what the completed model should look like. Then, have student groups build their model according to the instructions in steps 3-5 of the procedure.

△ Guide and Assess

Guide students through the construction of the model respiratory system. While students are constructing their model, circulate among the groups and check that students are creating a tight seal where the balloon attaches to the straw (between the straw and the mouth of the bottle and between the plastic sheet and the bottle).

Assess the functioning of the models built by groups. If groups are having difficulty getting their model to work properly, troubleshoot their model and help them identify places that need adjustment. After students have constructed their model, ask them to compare their model with the diagram shown in the student text and note any similarities and differences.

⬡ Get Going

Have students carry out step 6 of the procedure, record their observations and write down their answers to the questions.

△ Guide and Assess

When students have completed the questions in step 6, hold a brief class discussion in which students share their answers. Guide students to make connections between the behavior of the model and the observations of breathing made in steps 1 and 2.

Assess students' responses to the questions in step 6 by listening for the following information in their responses:

a) Breathe into the straw. *What happens to the rest of the materials when air enters the model?* Watch all the parts very carefully. Sample response: When air enters the model the balloon fills up and the plastic wrap on the bottom bulges slightly. This is like when someone is being resuscitated. Air is being pushed into the lungs because the diaphragm is not moving.

b) Pull on the plastic wrap and balloon at the bottom of the model. *What happens to the balloon inside the model? How is this like breathing?* Sample response: Air will move from an area of higher pressure to an area of lower pressure until the pressure in both areas is the same. Pulling the plastic wrap or half balloon downward increases the volume inside the model, which causes a decrease in the air pressure inside the model. Air then moves from the area of higher pressure outside the model to the area of lower pressure inside the model through the straw and fills the balloon until the pressure is equalized. Pushing the plastic wrap upward decreases the volume inside the model, which causes an increase in air pressure. Since the air pressure inside the model is now higher than it is outside the model, air moves out of the balloon through the straw until the pressure is equalized. When you inhale, the diaphragm contracts and moves downward; this increases the volume of your chest and causes air pressure inside the chest to decrease. The greater air pressure outside the body causes air to enter through your mouth/nose, move down the trachea and into the lungs. When you exhale, your diaphragm relaxes and moves upward decreasing the volume inside your chest. This increases the air pressure inside the chest, causing air to move out of your lungs, up the trachea and leave through the mouth/nose.

Connect the Model to the Respiratory System

In a good model, the parts work a lot like the real parts they represent. In your respiratory system model, the parts work a lot like parts of the respiratory system. The different parts of the model you made in class are similar to the real structures or organs they are modeling.

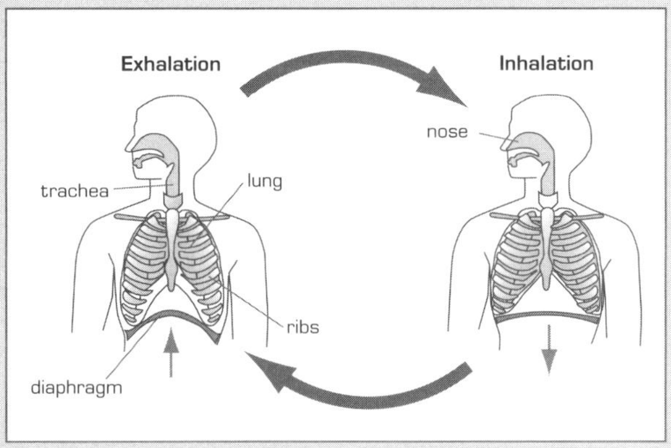

diaphragm: a band of muscle that regulates pressure in the chest cavity.

The **diaphragm** is a band of muscle. It is involved in the breathing process. Look at the labeled pictures of how the respiratory system works. Find the diaphragm. When the diaphragm contracts, it flattens (goes down). The volume of the chest expands. Air moves into the lungs. Try it. Think about what is happening. Remember back to your respiratory-system model.

This process is similar to what happened when you pulled on the balloon in your model of the respiratory system. When you pulled the balloon down, the bottle cavity became bigger and air entered the bottle.

The balloon in the bottle represented the lungs. A balloon is a good model because air can be used to fill the balloon similarly to the way air can fill the lungs. The balloon can hold air and expand just like your lungs do.

You do not usually concentrate so much on breathing. However, if you try, you can control how you breathe. Singers, athletes, and people who

Connect the Model to the Respiratory System

10 min.

Students consider which parts of the respiratory system are represented by each of the parts of the model. They also make connections between the functioning of the model and the functioning of the respiratory system.

△ Guide

Ask students to examine the model and identify the part of the respiratory system represented by each part of the model. If possible, project an image of the *Exhalation and Inhalation* diagram in the student text that can be referred to during the reading. Read this section together with your students, stopping from time to time to assess their understanding of the events that occur during exhalation and inhalation. Emphasize the role of the diaphragm in breathing. Guide students to recognize the connection between movement of the diaphragm, changes in air pressure and the movement of air into and out of the respiratory system.

Ask students to describe how the diaphragm, lungs, trachea and mouth function together as a system.

NOTE: This is an opportune time to conclude the first session of this investigation.

NOTES

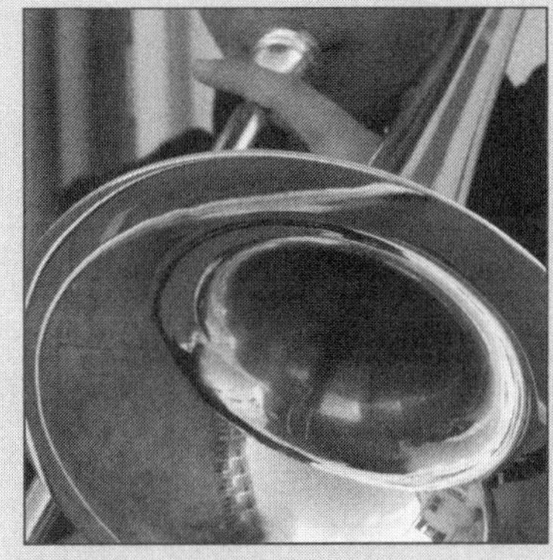

play certain kinds of instruments learn how to control their breathing to help them perform better.

The diaphragm, lungs, trachea, and nose function together as a system called the respiratory system. The same process happens every time you breathe.

Producing a pleasing sound on any brass instrument depends on the proper use and control of breathing.

Reflect

With your partner, look at the diagram you followed to build your model. Answer the following questions. Be prepared to share your answers with the class.

1. For each part of the model,
   a) identify the part of the respiratory system it represents (include nose, trachea, lungs, and diaphragm).
   b) discuss why the materials used are appropriate.
2. In what ways is the model you built a good model of the respiratory system?
3. In what ways is the model different from your respiratory system?
4. Now, use the model to help you make predictions about staying healthy. Using the model, predict how bacteria or viruses might enter the respiratory system to make you sick.

Reflect

10 min.

Students consider the strengths and weaknesses of the respiratory system model they have constructed.

△ Guide and Assess

Have students record their answers to the three *Reflect* questions. Next, facilitate a brief class discussion to check that students have correctly correlated the parts of the respiratory system with the parts of the model they built.

Use a projection of an unlabeled model. Have students fill in the name of the respiratory system part that each part of the model represents. Ask the class to assess whether the part has been correctly labeled and be sure that the class comes to a consensus on the answer.

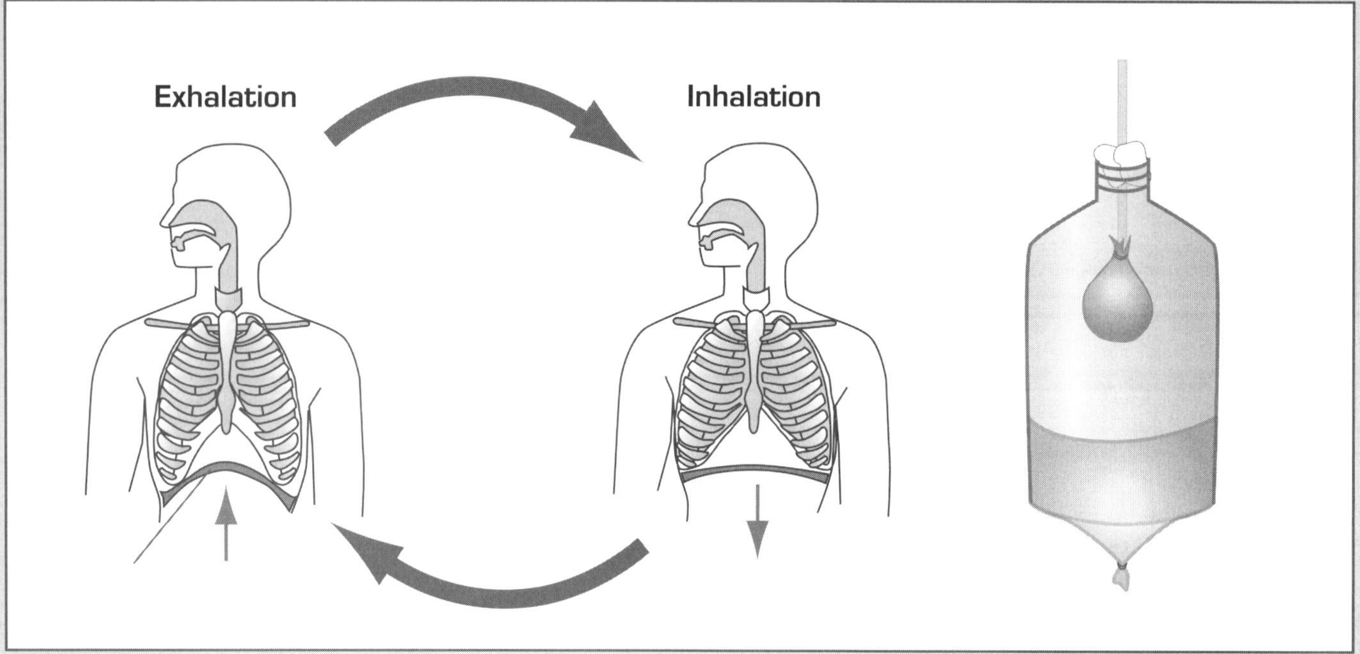

Guide students to recognize that since models are only representations, there will be similarities and differences between a model and the thing it actually represents. Ask students to describe similarities and differences between each part of the model and the part of the respiratory system it represents. Ask students to identify the strengths and weaknesses of the materials used in the model as representations of parts of the respiratory system.

NOTES

Make a chart listing strengths and weaknesses of the model of the respiratory system they built. Have groups present their ideas for each part of the table and encourage the class to discuss those ideas and agree on any suggested changes. Assess students' understanding of the structure and function of the respiratory system by listening for the following ideas:

Strengths of Model	Weaknesses of Model
The balloon is a good representation of the lungs because it inflates, is stretchy, and holds air well.	The balloon is not a good representation of the lungs because the inside of the lung is not a single, large opening as in the balloon. Instead, the lung consists of millions of clusters of smaller balloon-like structures called alveoli.
The straw is a good representation of the trachea because it is like a tube that connects the outside world to the lungs inside our body	The straw is not a good representation of the trachea because it kinks when bent, unlike the trachea which has cartilage rings that keep it from collapsing when bent. The straw is also a single tube while the trachea branches off into bronchi and bronchioles.
The plastic wrap on the bottom is a good model of the diaphragm because it is flexible and moves up and down causing the lungs (balloon) to deflate or inflate.	The plastic wrap does not simulate the active role the diaphragm plays in assisting our lungs in deflating and inflating.

Challenge students to use the model to make predictions about how microbes might enter the respiratory system and make you sick. (You might consider suspending some baby powder in the air near the model to represent microbes.) Assess whether student responses refer to microbes entering the respiratory through the mouth or nose with inhaled air.

NOTES

Beyond the Balloon Model: Diseases that Impact Lung Function

10 min.

Students read about ways in which the respiratory system prevents microbes from entering the body and several non-communicable lung diseases.

Beyond the Balloon Model: Diseases That Impact Lung Function

Many different communicable diseases affect the lungs. Bacteria cause some of the diseases. Viruses cause others. Many bacteria and viruses are in the air. Because you breathe air, it is easy for them to enter your body. There is not much you can do to stop breathing in air. Sometimes though, you get microbes on your hands and pass them to your nose or mouth by putting your hands near those places. The bacteria take the free ride to your nose or mouth and enter your body. Sometimes this can make you sick.

Some germs like flu viruses live in wet and warm places like your respiratory system. The first place germs that affect your respiratory system can enter is your nose. If they invade your nose, you can get a runny, stuffy nose and sneeze. If they get further into your respiratory system, the germs can also affect your lungs. You can get a cough, a fever, and chest pain.

mucus: sticky, wet material in your nose and other organs.

Covering your nose and mouth with a tissue when sneezing can help prevent the spread of germs.

Parts of your respiratory system also help you stay healthy. Many microbes that enter your body do not make you sick, because the **mucus** (sticky, wet material) found in your nostrils and throughout your respiratory system protects you by trapping some harmful particles from the air.

Not all respiratory diseases are caused by infectious agents. Many lung diseases are not communicable. When someone talks about having a "lung disease," they may be referring to a disease caused by something other than an infectious agent. Some

△ Guide

Read through this section together with the students. Emphasize how the respiratory system can act as an entry route for microbes that cause communicable disease. Then, have students think back to the gastrointestinal illness case study they investigated in the previous Learning Set and recall the conditions that are favorable to the growth of bacteria. Guide students to recognize that the warm, moist conditions found throughout the respiratory system are favorable to the growth of bacteria.

Ask students to describe how mucus helps prevent microbes from entering the body and then from making you sick.

After reading the section on lung cancer and black lung, encourage students to share any experiences they have had with other lung diseases that are not communicable. Some students may be familiar with emphysema or know someone who suffers from asthma. Others may bring up lung transplants or people who are living with only one lung. Allow students to discuss their ideas and experiences, but guide them to understand correctly the impact of the conditions they describe on the healthy functioning of the body.

NOTES

Stop and Think

5 min.

Students consider the impact diseases have on the functioning of the respiratory system and update the Communicable-Disease Information Table.

lung diseases are caused by things in the environment. People who smoke are more likely to get lung cancer and other lung diseases than those who do not smoke. A disease called black lung disease was once common among miners, who breathed a lot of dirty air in mines.

Stop and Think

1. What is one disease you have learned about so far that impacts the respiratory system and is caused by bacteria?
2. What is one disease that impacts the respiratory system and is caused by a virus?
3. Go back to the *Communicable-Disease Information Table* you made in *Learning Set 2* and fill in parts of it that you know more about now.

Update the *Project Board*

Your respiratory system investigations may have helped you answer some of the questions on the *Project Board*. The question that guides this *Learning Set* is *What happens to you when you get sick?* You now know about the respiratory system and how it is necessary for breathing. However, bacteria and viruses can also enter your body through this system. What information about the respiratory system helps you answer some of your questions about disease and how disease spreads?

Add what you have learned to the *Project Board*. Be sure to include the evidence for your statements in the *What is our evidence?* column. Look back at what you wrote for the *What do we think we know?* column from the beginning of the Unit. What information from this section of the Unit can you use to support what you thought you knew before? How did the information about the respiratory system change what you thought you knew? Add this information to the *What are we learning?* column. If you have new questions you had not thought about before, add them to the *What do we need to investigate?* column.

△ Guide and Assess

When students have completed the *Stop and Think* questions, hold a brief class discussion focusing on diseases that involve the respiratory system. Review what is contained in each column on the *Communicable-Disease Information Table* and ask students what information can now be added to it.

Assess students' responses to the *Stop and Think* questions by listening for the following information in their responses:

1) Pneumonia
2) Influenza, common cold
3) Students should add information about the parts of the respiratory system that are affected by various diseases. For example, pneumonia affects the lungs; the common cold affects the nose and trachea.

Update the *Project Board*

5 min.

Students add new information and supporting evidence to the Project Board.

Have students read the paragraphs on updating the *Project Board* in their text. Point out that the focus here is on adding information about the respiratory system to the *Project Board.*

Lead a class discussion focusing on what students think should be added to the *What are we learning?* and *What is our evidence?* columns of the Project Board now that they have investigated the respiratory system. During the discussion, guide students to identify what they have learned about the respiratory system that helps answer the question *What happens to you when you get sick?*. This information should be recorded on the *Project Board.*

Guide students by asking:

- What have we learned about the respiratory system?
- Where should we put the information about the respiratory system?

In continuing the discussion about the respiratory system, guide students to associate specific evidence with their claims to answer the question of the section.

For example:

Claim: Bacteria can enter the lungs with air that is breathed in. If the bacteria grow in the lungs, they can cause pneumonia.

Evidence: From our readings, we learned that pneumonia is caused by bacteria and those bacteria can enter the body with air that is inhaled. Since the respiratory system is warm and moist, it provides favorable conditions for the growth of bacteria. When bacteria grow in the lungs they damage lung tissue and the body gets less oxygen when you breathe.

Claim: Diseases can enter the body through the respiratory system.

Evidence: From our readings and by observing the model of the respiratory system, we learned that germs enter the body by way of our nose or mouth when air is inhaled.

Claim: The respiratory system helps protect our body from germs.

Evidence: From our readings, we learned that mucus is found throughout the respiratory system and because mucus is sticky and wet, germs get trapped in mucus.

Hold a discussion focusing on how students think what they have learned will help them answer the *Big Question*, How can you prevent your good friends from getting sick? Record students' ideas in the What does it mean for the challenge or question? Column of the *Project Board*.

Teacher Reflection Questions

- In this investigation, students received new information concerning the respiratory system that needed to be integrated with information from previous investigations and readings about organ systems, respiratory illnesses, and disease spread. How did you assist students with information synthesis?
- When students updated the *Project Board* and made connections between the respiratory system and the *Big Question*, how did you encourage students to rethink, reorganize, and refine their oral and written ideas?
- How well do you feel that students understood the concepts and skills that were presented in this investigation? What evidence do you have?
- During the readings and subsequent discussions, did you provide sufficient opportunity and time for students to work independently, in pairs, and in small groups?

SECTION 3.5 INTRODUCTION

3.5 Investigate

How Fast Does Your Heart Beat?

◀ *2 class periods*

A class period is considered to be one 40 to 50 minute class.

Overview

In this section, students investigate the question *How Fast Does Your Heart Beat?* by measuring their heart rate while at rest and after exercise. After collecting and recording data from three trials of each rate, students communicate their results with the class and determine the class average for heart rate in beats per minute while at rest and after exercise. Students then consider differences between the class average, their own heart rate, and the average heart rate of the general population. Students then reflect upon their results and identify cause-and-effect relationships between heart rate and activity level.

Targeted Concepts, Skills, and Nature of Science	Performance Expectations
Scientists often collaborate and share their findings. Sharing findings makes new information available and helps scientists refine their ideas and build on others' ideas. When another person's or group's idea is used, credit needs to be given.	Students collaborate with a partner to collect heart rate data. Students then share their heart rate data with their partner and with the class. The class collaborates to identify possible cause-and-effect relationships between heart rate and activity level.
Scientists must keep clear, accurate,and descriptive records of what they do so they can share their work with others, consider what they did, why they did it, and what they want to do next.	Students record data for three trials each of heart rate while at rest and breathing rate after exercise. Students create a table of class data and determine the class average for heart rate while at rest and heart rate after exercise.
Identify pulse points and pulse rates.	Students locate their external maxillary pulse point and measure their pulse rate while at rest and after exercise.

Targeted Concepts, Skills, and Nature of Science	Performance Expectations
Identify cause-and-effect relationships.	Students compare and contrast average heart rates while at rest and after exercise to identify cause-and-effect relationships between heart rate and activity level. Students cite evidence to support the idea that an increase in activity level causes an increase in heart rate.

Materials	
1 per classroom	Projection of class data table. (Optional: electronic spreadsheet that calculates the sum of the trials listed in the table.)
1 per student pair	Stopwatch or clock that displays seconds
1 per student	Calculator
1 per student	Worksheet

Activity Setup and Preparation

Check stopwatches to be sure they are working correctly; replace batteries as needed. If this is the first time students are working with stopwatches, demonstrate the correct technique for stopping, starting, and resetting the stopwatch.

You may want to create a spreadsheet for the class data set that calculates the sum (total of all trials) listed in the table, but not the class averages. Errors are common when students add large columns of numbers and the process uses valuable class time. By allowing the spreadsheet to add the numbers, students can focus on calculating class averages. It is important for each student to calculate the class average. You may want to enlist the support of the class' math teacher to review and reinforce this skill.

Students will work in pairs in this activity; determine how you will assign student pairs.

Homework Options

Reflection

- **Science Content:** You may have heard the expression "My heart was pounding!" Write about an experience you had that caused your heartbeat to become so forceful that you noticed it. *(Students typically report experiences in which they were surprised or fearful.)*
- **Science Content:** In this investigation, you measured heart rate by finding your pulse along the side of your neck. At what other locations on the outside of the body could a pulse be felt? *(Answers will vary, but should correspond to locations shown on the diagram of common pulse point in the student text.)*
- **Science Process:** Repeat the investigation you did in class today with other members of your family and/or friends. Before measuring a person's heart rate, make a prediction about how that person's heart rate at rest and after exercise will compare with yours. Other than activity level, what factors do you think might have an effect on a person's heart rate? *(Students typically report factors based on age, weight, and smoking status.)*

Preparation for 3.6

- **Science Content:** Draw a diagram of what you think the inside of your heart looks like. Use arrows to show the path of blood through the heart. *(Answers will vary, students typically show blood simply entering one side of the heart and leaving the other. Less frequently, students show chambers in the heart and a connection between the heart and lungs. Look for details that indicate students' level of understanding of heart structure such as: chambers in the heart, valves, connections between heart and lungs, and correct placement of arrows indicating blood flow.)*

NOTES

SECTION 3.5 IMPLEMENTATION

◀ *2 class periods**

3.5 Investigate

How Fast Does Your Heart Beat?

The respiratory system moves air into your body. In the lungs, the carbon dioxide made by your cells is exchanged for oxygen from the air. But how does the oxygen you need to live move from your lungs to the rest of your body? The oxygen your lungs extract from air is carried around your body by your blood. The **circulatory system** moves blood around your body so all the parts of your body can get oxygen. It also carries nutrients (food) and other chemicals to the cells of the body. It carries waste away from the cells.

The **heart** is the organ responsible for pumping blood through the body. The rate at which your heart beats depends on many factors. In this investigation, you are going to look at how exercise affects your heart rate and think about why.

circulatory system: an organ system that carries nutrients and other chemicals to the cells of the body and carries away waste; includes the heart, arteries and veins.

heart: the organ responsible for pumping blood through the body.

Materials
- stopwatch
- Heart Rate Data page
- blank page

Procedure

1. The circulatory system moves blood with oxygen all around your body. Think about how the oxygen might get from your lungs, where you breathe it into your body, to your big toe. Using an outline of the body, draw the path you think blood with oxygen moves through you to get to your big toe.
2. Use your hands to feel your heartbeat. Try putting your hand on your chest. Can you feel your heart there?
3. Take your index (pointing) and middle fingers, and press them to the side of the middle of your neck and just under your jaw. Move your finger around on your neck until you can find your pulse. This is just like feeling your heartbeat.

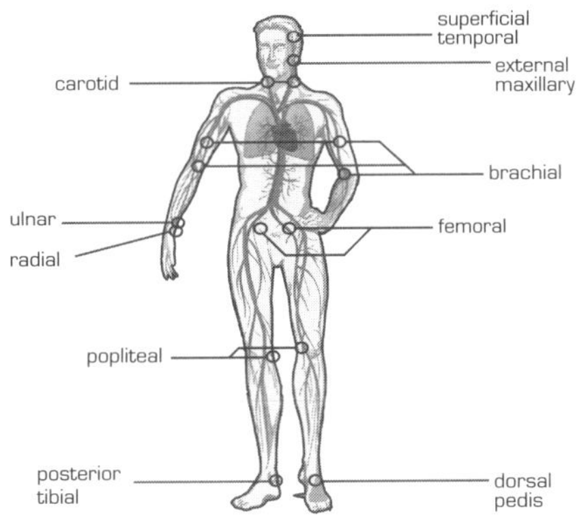

The pulse can be measured anywhere an artery passes close to the skin. The diagram shows several common pulse points.

GF 93 GOOD FRIENDS and GERMS

3.5 Investigation

How Fast Does Your Heart Beat?

5 min.

Students work in pairs to measure their heart rate at rest and after exercise. Then, they share their data with the class, create a class data table, and calculate class averages for heart rate at rest and after exercise. Students engage in a class discussion focusing on cause-and-effect relationships shown by the data.

△ Guide

Have students read the introductory paragraph and then hold a brief discussion focusing on the connection between the respiratory system and the circulatory system. Emphasize the exchange of gases with the bloodstream that occurs in the lungs. Guide students to recognize that the circulatory system transports materials to all cells of the body and that the heart's pumping action provides the force needed to move blood through the system.

*A class period is considered to be one 40 to 50 minute class.

Procedure

30 min.

Students work in pairs to measure three trials of each other's heart rate while at rest and heart rate after exercise.

Have students form pairs within their groups according to your instructions.

• blank page

Procedure

1. The circulatory system moves blood with oxygen all around your body. Think about how the oxygen might get from your lungs, where you breathe it into your body, to your big toe. Using an outline of the body, draw the path you think blood with oxygen moves through you to get to your big toe.
2. Use your hands to feel your heartbeat. Try putting your hand on your chest. Can you feel your heart there?
3. Take your index (pointing) and middle fingers, and press them to the side of the middle of your neck and just under your jaw. Move your finger around on your neck until you can find your pulse. This is just like feeling your heartbeat.

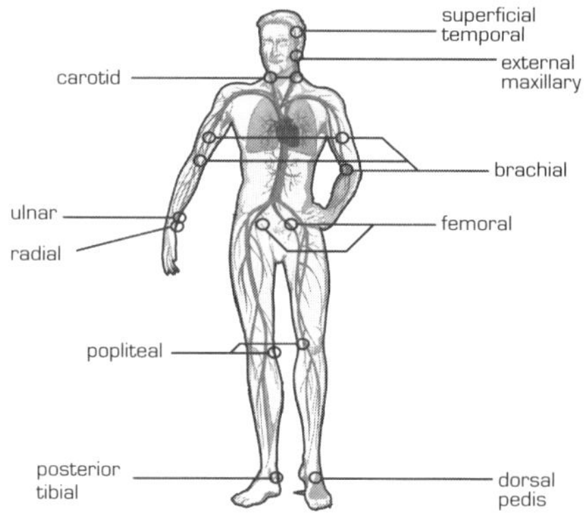

The pulse can be measured anywhere an artery passes close to the skin. The diagram shows several common pulse points.

△ Guide

Read through the Procedure with the students and draw their attention to the diagram showing the location of pulse points in the body. Guide students in locating their external maxillary pulse point by demonstrating how to hold their fingers against their neck, just below the jaw. Have the students locate their pulse, providing assistance where needed.

Once all students have located their pulse, remind students that scientists keep track of their data as they collect it. Therefore, they should record their heart rate immediately after measuring it. Also remind students that scientists do many measurements of the same thing (or trials) to make sure their measurements are accurate. Emphasize that each person will take three measurements of their heart rate while at rest and three measurements of their heart rate after exercise.

Demonstrate to students how, when they are exercising, they are to stand at a safe distance from everyone else and jog for 30 seconds in place. Point out that after exercising, they are to sit down and immediately start counting how many times their heart beats in six seconds. Remind students that their partner is keeping time for them.

⬡ Get Going

Have student pairs perform steps 1-4 of the procedure and record their data for heart rate while at rest in the data table on their worksheet. Before student pairs continue, remind them to stand at a safe distance from others while jogging in place. Have student pairs perform steps 5-7 of the procedure and record their data for heart rate after exercise in the data table on their worksheet. Make sure that all students exercise at a similar rate for

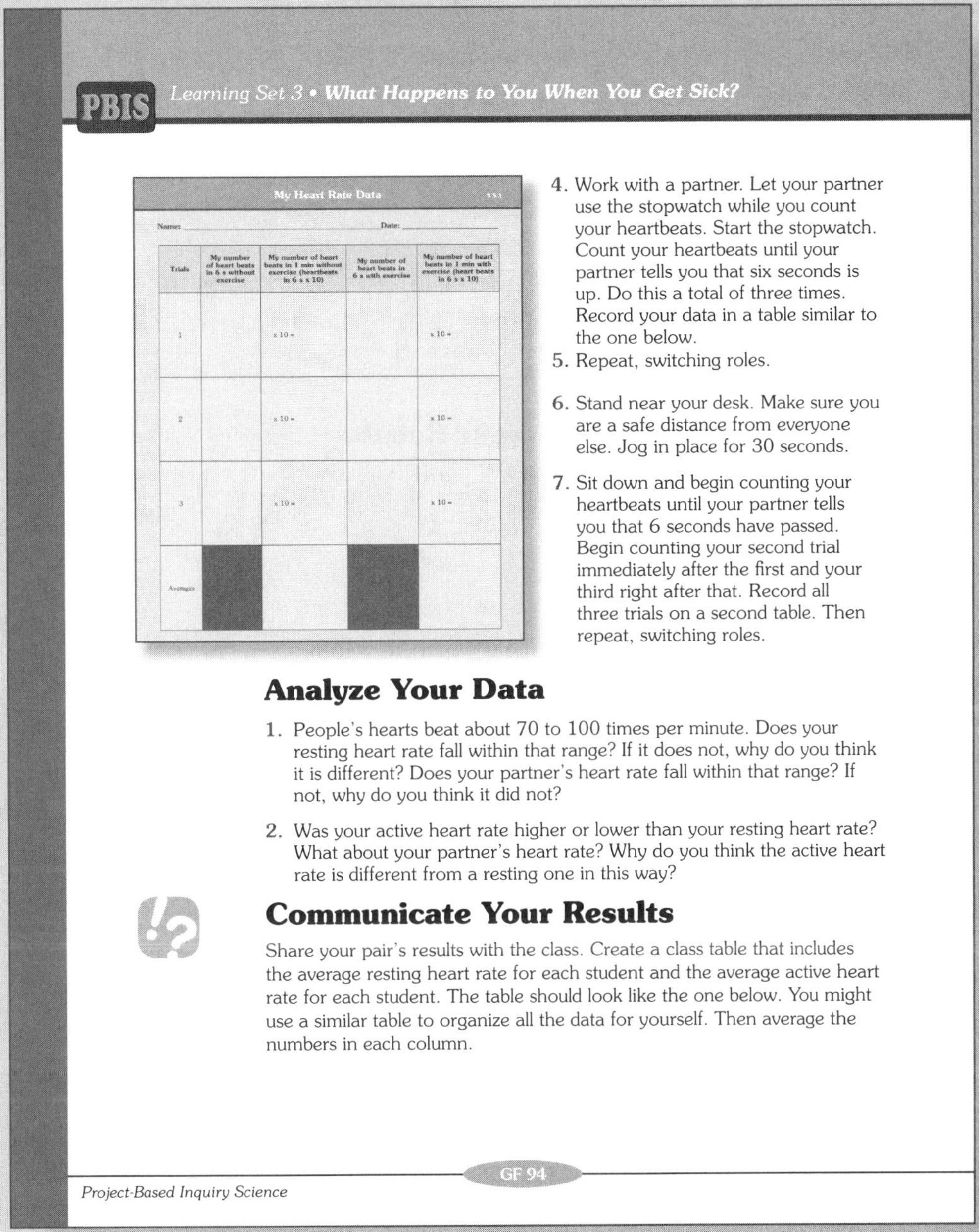

My Heart Rate Data

Name: ______ Date: ______

Trials	My number of heart beats in 6 s without exercise	My number of heart beats in 1 min without exercise (heartbeats in 6 s x 10)	My number of heart beats in 6 s with exercise	My number of heart beats in 1 min with exercise (heart beats in 6 s x 10)
1		x 10 =		x 10 =
2		x 10 =		x 10 =
3		x 10 =		x 10 =
Averages				

4. Work with a partner. Let your partner use the stopwatch while you count your heartbeats. Start the stopwatch. Count your heartbeats until your partner tells you that six seconds is up. Do this a total of three times. Record your data in a table similar to the one below.
5. Repeat, switching roles.
6. Stand near your desk. Make sure you are a safe distance from everyone else. Jog in place for 30 seconds.
7. Sit down and begin counting your heartbeats until your partner tells you that 6 seconds have passed. Begin counting your second trial immediately after the first and your third right after that. Record all three trials on a second table. Then repeat, switching roles.

Analyze Your Data

1. People's hearts beat about 70 to 100 times per minute. Does your resting heart rate fall within that range? If it does not, why do you think it is different? Does your partner's heart rate fall within that range? If not, why do you think it did not?
2. Was your active heart rate higher or lower than your resting heart rate? What about your partner's heart rate? Why do you think the active heart rate is different from a resting one in this way?

Communicate Your Results

Share your pair's results with the class. Create a class table that includes the average resting heart rate for each student and the average active heart rate for each student. The table should look like the one below. You might use a similar table to organize all the data for yourself. Then average the numbers in each column.

the same length of time (30 seconds). (Optional: Jogging in time to music with a strong beat helps students maintain a consistent rate of exercise.)

☐ Assess

Circulate among groups checking that students are calculating their heart rate correctly (heart beats per 6 seconds x 10) and that heart rate at rest is between 85 and 95 per minute.

Make sure students are staying on task while measuring their heart rates so that measurements are not altered by further activity.

Analyze Your Data

5 min.

Students record heart rate data for three trials while at rest and three trials after exercise. Then, they calculate their average heart rate while at rest and after exercise. Hold a brief class discussion focusing on observed variations in heart rates and possible reasons for those differences.

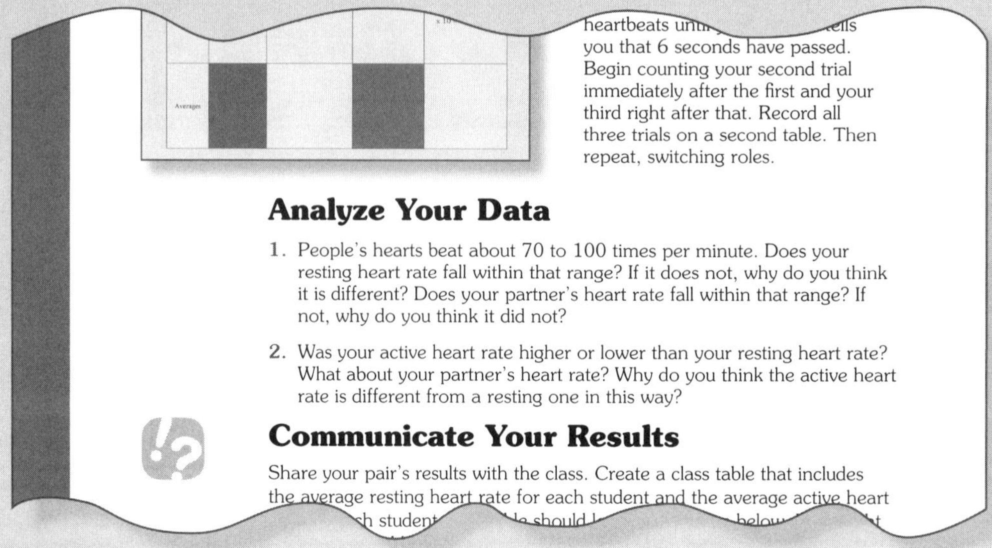

you that 6 seconds have passed. Begin counting your second trial immediately after the first and your third right after that. Record all three trials on a second table. Then repeat, switching roles.

Analyze Your Data

1. People's hearts beat about 70 to 100 times per minute. Does your resting heart rate fall within that range? If it does not, why do you think it is different? Does your partner's heart rate fall within that range? If not, why do you think it did not?
2. Was your active heart rate higher or lower than your resting heart rate? What about your partner's heart rate? Why do you think the active heart rate is different from a resting one in this way?

Communicate Your Results

Share your pair's results with the class. Create a class table that includes the average resting heart rate for each student and the average active heart

△ Guide

Ask students to calculate their average rest heart rate per minute and their average heart rate per minute after exercise. Remind students how to calculate the average, assist them by giving them the formula for calculating the average of three trial values.

$$\text{Average} = \frac{\text{Trial 1} + \text{Trial 2} + \text{Trial 3}}{3}$$

☐ Assess

Ask a few students for their averages. Students usually obtain an average heart rate while at rest between 70 and 100. Their heart rate after exercise should be higher.

△ Guide

When students have completed the *Analyze Your Data* questions, hold a brief class discussion in which students share their answers and focus on ideas they have about why heart rates vary and why their heart rate after exercise was higher than their heart rate while at rest.

Assess students' responses to the *Analyze Your Data* questions by listening for the following information in their responses:

1. People's hearts beat about 70 to 100 times per minute. Does your resting heart rate fall within that range? *(Answers will vary, but most students will find that their heart rate falls within this range.)* If it does not, why do you think it is different? *(Students may cite*

a variety of factors including gender, weight, and level of fitness.) Does your partner's heart rate fall within that range? *(Answers will vary.)* If not, why do you think it did not? *(Students may cite a variety of factors including gender, weight, and level of fitness.)*

2. Was your active heart rate higher or lower than your resting heart rate? *(Students should report that their exercising rate was higher than their resting rate.)* What about your partner's heart rate? *(Answers will vary, but most students find that their heart rates were similar to those of their partner.)* Why do you think the active heart rate is different from a resting one in this way? *(Students should suggest that when exercising, muscle cells use more food and oxygen to produce energy than when at rest. In order to replenish the oxygen that was used, the body increases heart rate so blood reaches cells more rapidly.)*

At the end of the discussion, create a brief summary of what students think they have learned about heart rate.

NOTE: This is an opportune time to conclude the first session of this investigation.

> resting heart rate fall within that range? If it does not, why do you think it is different? Does your partner's heart rate fall within that range? If not, why do you think it did not?
>
> 2. Was your active heart rate higher or lower than your resting heart rate? What about your partner's heart rate? Why do you think the active heart rate is different from a resting one in this way?
>
> **Communicate Your Results**
>
> Share your pair's results with the class. Create a class table that includes the average resting heart rate for each student and the average active heart rate for each student. The table should look like the one below. You might use a similar table to organize all the data for yourself. Then average the numbers in each column.

Communicate Your Results

15 min.

Students share their data with the class, record all data on a class table, and calculate class averages for heart rate while resting and after exercise.

△ Guide

Have students read the instructions and construct a class data table based on the example shown in the student text. Ask student pairs to share their results with the class and record the data on their class table while you record it on a projected class table or in a spreadsheet.

You should calculate the sum of all values or use a spreadsheet to calculate the values. The total number of values is the number of students in the class that reported data.

Tell students that you will give them the sum of all their values, but that each person will calculate the class average on their own. Remind students that the average is found by using the formula:

$$\text{Average} = \frac{\text{sum of all values}}{\text{total number of values}}$$

☐ Assess

Have students compare their calculated class average for resting heart rate and active heart rate. If calculated results do not agree, have students check their calculations and identify error(s).

NOTES

Analyze Your Data

Now that you see everyone's data, you can compare your own information to the class information.

1. How did your resting beats per minute compare to that of others in the class?
2. How did your active beats per minute compare with those of the rest of the class?
3. What was the range of resting heartbeats per minute in your class? What was the range of active heartbeats per minute?
4. What was the average resting heartbeats per minute in your class? What was the average active heartbeats per minute?
5. Did the average resting heart rate fall within the range of 85–95 beats per minute? How reliable do you think the data are? Why?
6. You can hold your breath. For a certain amount of time, you can decide not to breathe. You cannot, however, control your heart rate. You cannot tell your heart when to beat or stop it from beating. You can make it go faster, however, and you can make it go slower when it is beating faster. How would you do that?

Class Heart Rate Data

Name: ______________________ Date: ____________

Trials	Resting beats per minute	Resting beats per minute
1		x 10 =
2		x 10 =
3		x 10 =
Averages		

Analyze Your Data

15 min.

Now that students have recorded data from everyone in the class, hold a class discussion focusing on observed variations in heart rates and possible reasons for those differences.

△ Guide

When students have completed the *Analyze Your Data* questions, hold a brief class discussion in which students share their answers and focus on ideas they have about why heart rates vary and why their heart rate after exercise was higher than their heart rate while at rest.

Assess students' responses to the *Analyze Your Data* questions by listening for the following information in their responses:

1. How did your resting beats per minute compare to that of others in the class? *(Answers will vary, but students should support their answers with data from the class data table.)*
2. How did your active beats per minute compare with those of rest of the class? *(Answers will vary, but students should support their answers with data from the class data table.)*
3. What was the range of resting heartbeats per minute in your class? What was the range of active heartbeats per minute? *(Students should correctly identify the range as the highest and lowest number of heartbeats while at rest and after exercise recorded on the class data table.)*
4. What was the average resting heartbeats per minute in your class? What were the average active heartbeats per minute? *(Students should correctly cite the average number of heartbeats while at rest and after exercise recorded on the class data table.)*
5. Did the average resting heart rate fall within a range of 85-95 beats per minute? *(Students' responses should reflect the information recorded in the class data table.)* How reliable do you think the data are? *(Answers will vary, but might include references to errors in counting heartbeats, rounding off the number of heartbeats in 6 seconds, calculating averages, and errors due to small class population compared to general population.)* Why? *(Pulse is sometimes difficult to locate and feel consistently; small errors in counting are multiplied when calculating beats per minute, errors may be made in adding up large numbers of reported heartbeats when calculating averages; the class is not representative of the general population.)*
6. You can hold your breath. For a certain amount of time you can decide not to breathe. You cannot, however, control your heart rate. You cannot tell your heart when to beat or stop it from beating. You can make it go faster however, and you can make it go slower when it is high. How would you do that? *(You can make your heart rate increase by exercising and decrease by resting. Some students may also recognize that engaging in activities that generate fear or surprise can also increase heart rate and that a return to a safe, calm environment can decrease heart rate.)*

Reflect

Your question for this section was *How fast does your heart beat?* Answer the following questions. Be prepared to discuss your answers with the class.

1. The resting heart rates were probably lower than the active heart rates. Why do you think that is? Why would your heart need to beat more often when you are exercising than when you are resting?
2. What was the high number for the resting heart rate? What was the low number? What was the class average number for these rates? Which is the best number to use when you want to know about the population—the high, the low, or the average? Why did you choose that number?

What's the Point?

When you exercise, your heart beats faster and moves more blood around your body. The blood carries oxygen around to all the places that need it. This includes your lungs, heart, and other muscles. When you exercise, your body works harder, so it needs more oxygen. If you don't get more oxygen to your muscles, you will not be able to exercise anymore. Your muscles will stop working. However, your circulatory system moves more than just oxygen around your body. In the next sections, you will learn more about the importance of the circulatory system in getting you sick and keeping you from getting sick.

When you exercise, your heart rate increases to adapt to the body's need for oxygen.

Reflect

10 min.

Students consider variations in the heart rate data collected during this investigation and suggest possible reasons for those variations.

△ Guide and Assess

When students have completed answering the *Reflect* questions, hold a class discussion focusing on possible explanations for the observed variations in heart rate.

Assess students' responses to the *Reflect* questions by listening for the following information in their responses:

1) The resting heart rates were probably lower than the active heart rates. Why do you think that is? (*Student responses will vary, but should convey the idea that the resting rate was probably lower than the active rate because you don't need as much oxygen when you are resting.*) Why would your heart need to beat more often when you are exercising than when you are resting? *(Student responses should convey the idea that when exercising, muscles use oxygen at a faster rate. Therefore, the heart must beat faster to increase the rate of blood flow so that oxygen is delivered to the muscles more rapidly.)*

2) What was the high number for the resting heart rate? What was the low number? *(Students should correctly report the highest and lowest numbers recorded on the class data table.)* What was the class average number for these rates? *(Students should correctly report the class average recorded on the class data table.)* Which is the best number to use when you want to know about the population – the high, the low or the average? *(Students should identify the average as the best number to use.)* Why did you use that number? *(Students' responses should refer to the class data table and point out that a greater number of students reported heart rates that were closer to the class average than they were to the highest or lowest numbers. Therefore, the average heart rate will be closest to the heart rate of the greatest number of people in a population.)*

Teacher Reflection Questions

- In this section, students gathered data about their heart rate. What evidence did students use to support their claims concerning active and resting heart rates? If they had difficulty with the idea of using evidence, what will you do to help students in the next section?
- How did you connect what students learned about the circulatory system in this investigation with what they learned about the respiratory system in previous investigations? How might you connect this investigation with the next one in which students simulate blood flow through the circulatory system?
- How are you encouraging students to listen and respond to the remarks of their peers during large and small group discussions?
- What factors negatively or positively affected the success of this investigation?

SECTION 3.6 INTRODUCTION

3.6 Investigate

How Does the Circulatory System Work?

◀ *2 class periods*

A class period is considered to be one 40 to 50 minute class.

Overview

In this section, students assume the roles representing the heart, lungs, big toe and blood to construct a model of the circulatory system by arranging themselves around the classroom. They then simulate the flow of blood by moving along imaginary paths between the heart, lungs, and big toe to simulate gas exchange by exchanging cards representing oxygen and carbon dioxide. After running a simulation of blood flow and gas exchange, students connect the model they formed to the circulatory system. Students then extend their understanding by reading about the parts of the circulatory system, the structure and functioning of the heart, changes in the appearance of blood due to gas exchange, and the role of the circulatory system in spreading disease. Students reflect on the strengths and weaknesses of the model, what they now understand about blood flow through the body, they make connections with their previous investigations of breathing rate and heart rate, and add new information to their *Communicable-Disease Information Table.* After discussing how what they have learned relates to the guiding question of this *Learning Set, What happens to you when you get sick?,* students update the *Project Board.*

Targeted Concepts, Skills, and Nature of Science	Performance Expectations
Scientists often collaborate and share their findings. Sharing findings makes new information available and helps scientists refine their ideas and build on others' ideas. When another person's or group's idea is used, credit needs to be given.	Students collaborate to create a model of the circulatory system. The class collaborates to simulate blood flow while at rest and while exercising.

Targeted Concepts, Skills, and Nature of Science	Performance Expectations
Scientists must keep clear, accurate, and descriptive records of what they do so they can share their work with others, consider what they did, why they did it, and what they want to do next.	Students draw a sketch of the model they created and what they now understand about the flow of blood through the body.
Scientific questions are directed toward objects and events that can be described, explained, or predicted by scientific investigations.	Students should be able to write a question to drive the bacteria investigation that meets the criteria for a good question.
Models are a representation of something in the world. Simulations use a model to imitate, or act out, real-life situations.	Students create a model of the circulatory system by assuming the roles of heart, lungs, big toe, and blood cells and arrange themselves to represent the locations of these structures in the body. Students simulate blood flow and gas exchange by moving along imaginary paths through the model and exchanging cards representing oxygen and carbon dioxide.
An organ system is a group of body organs, each of which performs a specific function, that work together to carry on one or more life functions.	Students identify the organs comprising the circulatory system and describe how they work together to transport food, oxygen to cells and remove wastes from cells and eliminate them from body.
Tissues, organs, and organ systems help to provide all cells with nutrients, oxygen, and waste removal.	Students use the simulation they carried out as a context for describing how the circulatory system provides all cells with nutrients, oxygen, and waste removal.
The circulatory system moves substances to and from cells in response to changing demands.	Students create a sketch showing how the circulatory system moves substances to and from cells via the bloodstream.
Disease breaks down the structures or functions of an organism. Some diseases are the result of failures of a body system. Other diseases are the result of damage by infection from other organisms.	Students describe how the circulatory system can spread disease-causing microbes that infect the blood to all parts of the body.

Targeted Concepts, Skills, and Nature of Science	Performance Expectations
Identify structure and function relationships in organisms	Students describe how the structure of the heart enables it to pump blood and how the structure of blood vessels is suited to the transport of fluids.

Materials	
100 per classroom	Red cards marked "oxygen"
100 per classroom	Blue cards marked "carbon dioxide"
1 per classroom	Stopwatch or clock that shows seconds
1 per classroom	Optional: projections of images in student text
1 per classroom	Optional: masking tape for laying out "blood vessels" on floor
1 per student	Optional: role identifiers (i.e. name badges, 5" x 7" index cards worn around the neck on a string.)

Activity Setup and Preparation

Decide how you want to configure the model of the circulatory system in your classroom. You may want to move desks out of the way, or you may want to keep the room as it is and have the students move around obstacles that represent other parts of the body. You may want to have students wear name tags labeled with their role. If you have movable desks, you may want to use desks to create stations representing the heart, lungs, and big toe. Blood vessels connecting the heart, lungs, and big toe can be represented by laying out boundaries on the floor with masking tape.

Homework Options

Reflection

- **Science Content:** How might a blood clot in your upper leg affect your body? *(The clot would block the flow of blood to cells beyond the clot. Student responses should indicate that without the food, oxygen, and waste removal provided by blood flow, cells would stop functioning. The loss of function at the cellular level should then be connected to loss of large-scale body functions. For example, if the blood vessel supplied muscle cells, the lack of food and oxygen and the buildup of wastes would cause the muscle cells to stop functioning. If the muscles in the lower leg and foot stop functioning, you could no longer produce movement in those body parts.)*
- **Science Content:** Red blood cells are shaped like tiny rounded disks. How does this shape help a blood cell do its job? *(The disk-like shape of blood cells allows them to slip past one another easily as they move through blood vessels. The lack of sharp corners or edges makes it less likely that they will get caught in the twists and turns made by blood vessels and cause blockages. The disk-like shape also has a greater surface area through which gas exchange can occur than some other shapes.)*
- **Science Content:** What is the difference between an artery, a vein, and a capillary? *(Arteries carry blood away from the heart, veins carry blood toward the heart, and capillaries are tiny blood vessels that connect arteries to veins.)*

Preparation for 3.7

- **Science Content:** Cells need both oxygen and food in order to produce the energy needed to carry out life functions. In this section, you read that oxygen and food are carried to cells in your blood. How do you think food gets into your blood? *(Students often express ideas such as "food is absorbed into the blood in the stomach." Students' responses to this question will provide you with a good sense of their current understanding of digestion and help you identify ideas that need clarification during the next investigation.)*

SECTION 3.6 IMPLEMENTATION

◀ *2 class periods**

3.6 Investigate

How Does the Circulatory System Work?

In class, you took your pulse to measure how fast your heart was pumping blood. When you took your resting pulse, your heart was working at its normal rate. When you took your active pulse, your heart had been working much harder.

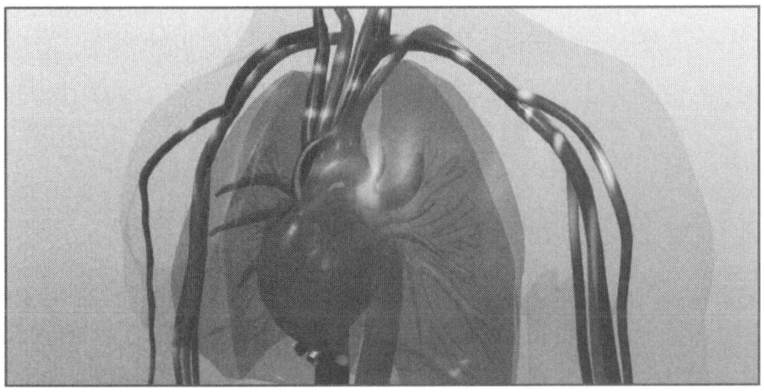

This model of the heart and lungs can help you visualize (see) the parts of the circulatory system and respiratory system that work together to move blood, oxygen, **nutrients** *(the useable substances in food), and wastes through your body.*

nutrients: the useable substances in food.

Your heart is one organ in your circulatory system. It is responsible for pumping blood through your body. One of its functions is to get oxygen to the body cells and carry carbon dioxide away. However, no organ system works on its own. The lungs are part of the respiratory system. They bring oxygen into the body and remove carbon dioxide from the body. The two systems work together to make your body run smoothly.

It is difficult to see what is going on in your body. That is why models are important. You have used models before to see how germs spread and how diseases are transmitted. Now you will model the circulatory system and take part in a simulation to see the path that blood takes around your body. You will see parts of your circulatory system and parts of your respiratory system working together. You need both of these systems to move oxygen and carbon dioxide around your body.

GF 97 GOOD FRIENDS and GERMS

3.6 Investigate

How Does the Circulatory System Work?

5 min.

Students read about the relationship between the respiratory and circulatory systems and the role of each system plays in the exchange of gases that takes place in the lungs.

△ Guide

Have students read the introduction and hold a brief discussion reviewing the connection between the respiratory system and the circulatory system. Emphasize the exchange of gases with the bloodstream that occurs in the lungs. Guide students to recognize that the circulatory system and respiratory system work together to provide oxygen to, and to remove the waste gas carbon dioxide from, all cells of the body. Next, draw students' attention to the model of the heart and lungs shown in the diagram. Ask

*A class period is considered to be one 40 to 50 minute class.

them to identify the parts of the respiratory system and circulatory system shown in the diagram and to describe how they are interconnected.

Emphasize that scientists often use models to help them better understand something that cannot be easily observed, like the workings of organs inside the body. Inform the class that they will be creating a model of their circulatory system and that they will need to pay special attention to what the parts of the model represent in our circulatory system. Point out that they will then use the model to simulate blood flow through the body.

NOTES

Build and Run Your Model for Getting Oxygen to Your Big Toe

Materials
- **100 red cards marked "oxygen"**
- **100 blue cards marked "carbon dioxide"**

1. Arrange yourselves as follows:
   - **Heart (two students)**

     One student will be the left side of the heart (Heart 1).

     One student will be the right side of the heart (Heart 2).

     These two students will stand in the middle of the room.
   - **Lungs (two students)**

     One student will collect blue cards (Lung 1).

     One student will pass out red cards (Lung 2).

     These two students will stand about 1.5 m (about 5 ft) from the heart, as shown in the diagram.
   - **Big toe (two students)**

     One student will collect red cards (Toe 1).

     One student will pass out blue cards (Toe 2).

     These two students will stand about 5 m (about 15 ft) at an angle from the lungs.
   - **Red blood cells (all other students)**

     All other students are red blood cells.

     These students will line up behind the lungs.

     They will receive a red card from Lung 2.

     They will pass a red card to Toe 1.

     They will receive a blue card from Toe 2.

     They will pass a blue card to Lung 1.

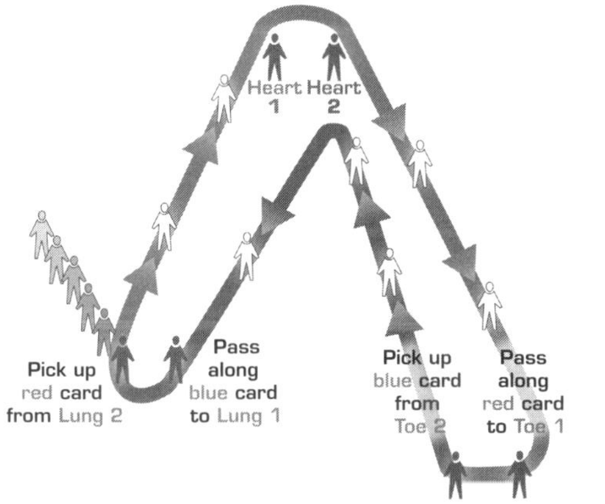

Build and Run Your Model for Getting Oxygen to Your Big Toe

30 min.

Students create a model of the interrelationship between the circulatory and respiratory systems by assuming the roles of the heart, lungs, big toe, and blood cells and arranging themselves to represent the location of these parts in the body. Then, they simulate the flow of blood through the circulatory system by moving from location to location and simulate the exchange of gases by exchanging cards representing oxygen and carbon dioxide.

△ Guide

Read through the entire procedure with the students before creating the model. Assign roles to students and give the red "oxygen" cards to the "Lung 1" student and the blue "carbon dioxide" cards to the "Toe 2" student. Arrange the students as described in the student text. Students representing red blood cells should be lined up and ready to pick up a red card from "Lung 2" before you begin circulating:

2. When everyone is in place, the heart students will begin to chant "pump, pump."
3. Each red blood cell will begin by receiving a red (oxygen) card from Lung 2.
4. They will walk toward Heart 1. Heart 1 will redirect them toward the big toe.
5. Each red blood cell will give their red (oxygen) card to Toe 1.
6. Each red blood cell will then receive a blue (carbon dioxide) card from Toe 2. The red blood cell will then go back to the heart.
7. Each red blood cell will pass Heart 2 and clap his or her hands. The "blood" is being pumped back to the lungs.
8. At Lung 1, pass off the blue card (carbon dioxide). Take a new red card (oxygen) from Lung 2 and continue the process.
9. After you have practiced the procedure, try to complete one cycle in 15 s. Your teacher will keep track of the time.
10. Answer the following questions:
    a) The time it takes for your blood to complete one round is 15 s. How close were you able to come to this time?
    b) Your active heart rate is three times faster than this. In what time would you have to complete one round to simulate this heart rate?
11. Try to simulate the blood flow when you are exercising.

blood vessels: tube-like parts of the circulatory system that transport blood through the body.

Connect the Model to the Circulatory System

In your simulation, the red blood cells that are carrying the oxygen cards move through imaginary paths to get to the big toe. In your body, these paths are called **blood vessels**. At the big toe, they trade in their oxygen cards for carbon dioxide cards. Why does that happen? The blood that travels from the lungs through the heart brings oxygen and other nutrients to your body tissues. On the return trip, the blood brings

The main job of the red blood cells is to carry oxygen through the body. Red blood cells look like disks, as shown in the drawing.

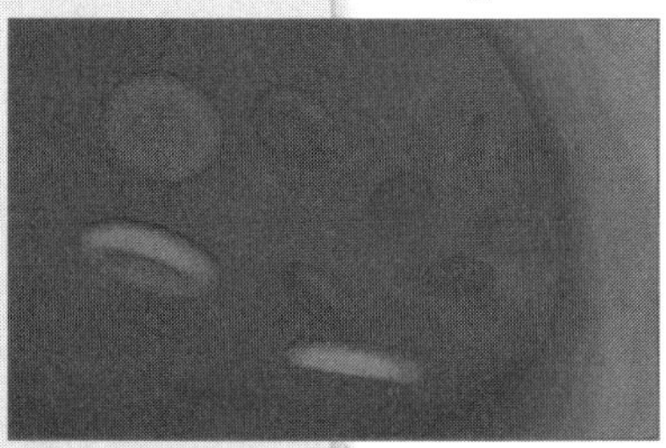

○ Get Going

Try a few practice runs of the simulation so that students understand how they are to move and exchange cards. Then hold an official run in which students try to complete the simulation in 15 seconds. Run a second trial of the 15 second simulation. Measure and record the actual time it takes the class to complete the simulation.

Have students answer the questions in step 10 and write down their answers. Guide students through the calculation of the time in which the simulation would have to be completed in order to represent active heart rate. (You may want to have students answer the questions while they are still part of the model since they will do one more simulation after step 10. If not then, you may have to remind students to write down their answers when they return to their seats.)

☐ Assess

Assess students' responses to the questions in step 10 by listening for the following information:

- Student responses should refer to the actual time it took the class to complete the simulation.
- If your active heart rate is three times faster than your rest heart rate, then you would have to complete one round in 1/3 the time, or 5 s.

△ Guide

Remind students that models are only representations. Therefore, there will be similarities and differences between any model and the real system it represents. Ask the class to briefly share some of their initial ideas about how the model they created is similar or different from the actual systems it represents. Create a list of student ideas.

Connect the Model to the Circulatory System

5 min.

Students consider how their actions while running the model are similar and different from the functioning of actual body systems during circulation.

Have students read this section independently or in small groups. After students complete the reading, hold a brief class discussion focusing on the process of gas exchange that occurred during the simulation. Once you have determined that students grasp the mechanism by which gas exchange occurs and its importance to normal cell function, guide students to recognize the shortcomings of the model created in this investigation. Emphasize that one difference between the model and the real circulatory system is that the model heart had two chambers and the real heart has four chambers. Then, ask students if they want to change or add to their list of similarities/differences between the model and the circulatory and respiratory systems.

NOTE: This is an opportune time to conclude the first session of this investigation.

Beyond the Model: The Circulatory System

20 min.

Students read about the structure and function of the heart.

vein: a blood vessel that carries blood back to the heart.

artery: a blood vessel that carries blood away from the heart.

heart chamber: one of four parts–right atrium, right ventricle, left atrium, and left ventricle–of a heart.

atria (singular, atrium): heart chambers that receive blood from the veins.

ventricles: muscular heart chambers that pump blood through the arteries.

waste products, like carbon dioxide, out of the lungs and through the heart. When you ran your simulation, the red blood cells carried oxygen cards to the toe tissue, traded the oxygen cards for carbon dioxide cards, and carried the carbon dioxide back to the lungs, where it could leave the body.

The science model used in this investigation was just like all other models. It has parts that represent the real thing well and other parts that do not represent the real thing well. For example, you tried to get the time it took red blood cells to go from the lungs and back again to be the same as the time it takes blood to flow through the body. This is one way your model is like your body. One way that your model does not mirror your body is that the heart in your model has only two parts. A real heart has four parts to do the job.

Beyond the Model: The Circulatory System

What Makes the Blood Flow Through Your Body?

At the center of the cirulatory system is the heart, which pumps blood so it can flow around through the body. Blood vessels called **veins** and **arteries** carry the blood around your body. Blood leaves your heart and is pumped to your lungs. In your lungs, the blood picks up oxygen. The blood, rich in oxygen, returns to your heart. From there, it is pumped to the rest of your body.

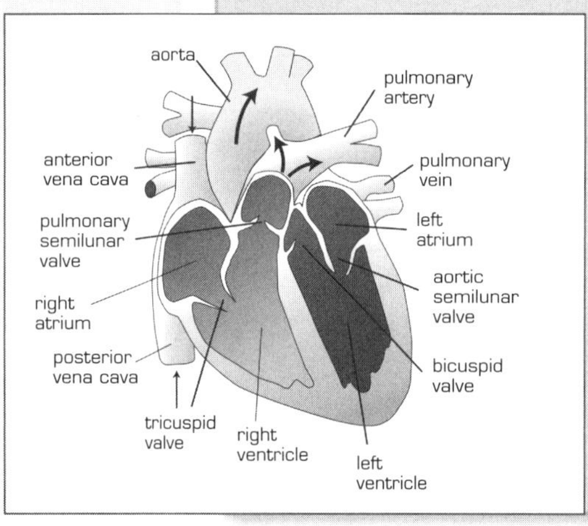

The heart has four parts, or **heart chambers**. The two upper parts of your heart are called **atria** (singular, atrium). They receive the blood from your body (your big toe in your model) or your lungs. The lower chambers are called the **ventricles**. These are the pumping parts of your heart. The muscles in the ventricles are very strong so they can pump blood to distant tissues of your body.

△ Guide

Read this section together with students. You may want to project the image of the heart to serve as a focal point during the reading. Draw students' attention to the definitions of important words printed in the margin. Make sure to stop from time to time to assess students' understanding of the functioning of the heart and to make connections back to the model and the simulation. Focus on connections between the circulatory and respiratory systems and the role of the atria and ventricles in moving blood rather than on memorization of the parts shown in the image of the heart.

TEACHER TALK

"What is the difference between an artery and a vein? (Arteries carry blood away from the heart; veins carry blood back to the heart.)

What parts of the model we created represented arteries? (Heart 1 to Lung 1 and Heart 2 to Toe 1)

What parts represented veins? (Lung 2 to Heart 1 and Toe 2 to Heart 2)

When did Heart 1 act as an atrium? A ventricle? (Atrium: when receiving red blood cells coming from Lung 2; Ventricle: when redirecting them toward Toe 1.)"

Point out that in the model, each heart person represented an atrium and a ventricle, acting as an atrium when taking in the blood cells and then as a ventricle when pushing them or redirecting these blood cells to another part.

NOTES

What Path Does Your Blood Take Through Your Body?

5 min.

Students read about the movement of blood through the body's network of blood vessels.

What Paths Does Your Blood Take Through Your Body?

The model is also different than your circulatory system in another way. In the model, you walked or ran by the shortest route from the lungs to the big toe. In your body, your blood does not choose its route. It moves through a network of veins and arteries that direct it to all the places it has to go.

However, there is a similarity in the path that the blood takes in your body and in your model. In your model, the red blood cells moved in one direction only. None went against the flow. The blood flow in the body is also only one way.

The arteries carry blood away from the heart. Every time your heart beats, blood is pushed from the heart into the arteries. The arteries stretch to accommodate this surge. The **pulse** you felt in your neck is caused by the change in diameter of the arteries near the surface of your body as the blood surges forward.

pulse: the surge of blood in an artery as the blood is pumped by the heart.

Blood from the heart passes into large arteries. From there it passes into smaller and smaller arteries. These eventually lead to extremely tiny blood vessels. These very tiny blood vessels connect the arteries to the veins. The veins are very small at first. However, they become progressively larger as the blood flows back to the heart. The blood returns to the heart through large veins.

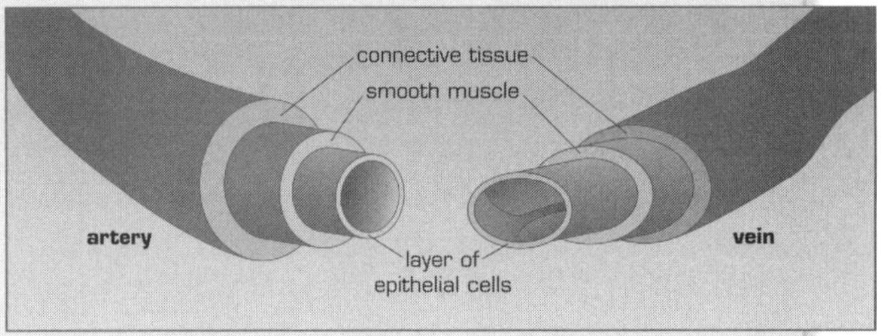

Arteries carry blood away from the heart, while veins carry blood to the heart.

How Does the Circulatory System Contribute to Sickness?

There is one more important way the model is different from your circulatory system. In the model, the blood cells carried oxygen to parts of your body and carried carbon dioxide back to your heart and lungs. In your body, the blood also carries many other things. It carries nutrients (vitamins and minerals) around your body that the cells in your body

△ Guide

Draw students' attention to the question in the heading for this part and encourage them to keep this question in mind as they read. As the class reads, ask students to update their list of ideas about similarities and differences between the pathways of the blood in the model and the real circulatory system. Assess student comprehension of the reading by listening for the following information in their responses:

- Similarities: the blood cells only go in one direction along the passageway (artery or vein)
- Differences: The model did not include branching into different sized passageways such as large arteries branching off into smaller diameter arteries. The model had no indirect routes – the body redirects blood flow based on the needs of cells in various body parts.

Emphasize that the pulse they felt in their neck was caused by the bulging of a large artery due to a pressure buildup when a surge of blood was forced into a smaller artery.

△ Guide

Ask students how they think the circulatory system might contribute to sickness. Make a list of their ideas. As the class reads through this part, ask students to suggest updates to their list of similarities and differences between the model and the real circulatory system and their list of ideas about how the circulatory system might contribute to sickness.

Assess student responses by listening for the following information:

- Model Similarities: the blood cells only go in one direction along the passageway (artery or vein). Blood cells carry oxygen (red cards) and carbon dioxide (blue cards).
- Model Differences: Blood carries nutrients (vitamins and minerals) and it may carry germs (bacteria and viruses), which were not represented.
- Contributions to sickness: blood may carry germs such as the bacteria that cause tetanus.

Emphasize that the nutrients carried by the blood enter the body through the digestive system. Therefore, these two systems also work together to keep the body healthy.

At this point, you may want to run another simulation of the model having students carry different colored cards, to represent either a germ or a nutrient in the blood stream. You may also want to add another pathway. For example, you may want to create a fork in the pathway leading away from Heart 1, with one pathway leading to the toe and another pathway leading to the thumb. Have the path leading away from the thumb then rejoin the path from the toe to Heart 2.

How Does The Circulatory System Contribute to Sickness?

10 min.

Students consider ways in which the circulatory system can contribute to the distribution of disease microbes in the body.

What Color Is Blood?

10 min.

Students read about the difference in appearance between oxygenated and deoxygenated blood.

need to function. You'll learn more about how those nutrients get into the blood when you learn about the digestive system. Your blood can also carry germs (bacteria and viruses) around your body.

Do you remember going to the doctor to get a tetanus shot? Tetanus, also called lockjaw, is a bacterial disease that infects your blood. It causes your jaw to lock up so you can't move it. If you fall down or cut yourself and bleed, the microbes in the environment can get into your bloodstream through your wound. You get tetanus shots to protect you from tetanus bacteria that can get into your body if you cut yourself on something rusty.

What Color Is Blood?

Look at your arm where your elbow bends. The skin there is thin, so you can see the veins and arteries. If you look carefully, you will see that these veins and arteries seem to be different colors. They may be deep red, lighter red, or almost blue. Some people may have told you that the tubes look different because the blood inside them is blue. Some people think that the blood is blue because it does not have oxygen in it. But that is not true. Blood is always red.

Veins are a bluish color because you are looking at them through your skin. The layer of skin on top of veins and arteries affects the color you see. The blood in your arteries has just been through your lungs and is full of oxygen. This makes it bright red. You see that as red through your skin. The blood in your veins is dark red instead of bright red. It has less oxygen because the blood has left its oxygen in other parts of your body. As it is pumped around your body, your blood loses more and more of its oxygen. As it loses oxygen, it becomes darker. When the blood gets very dark, it appears bluish through your skin.

When you donate blood, it is taken from a vein in your arm. The blood that flows through the needle and into the collection bag is very dark red.

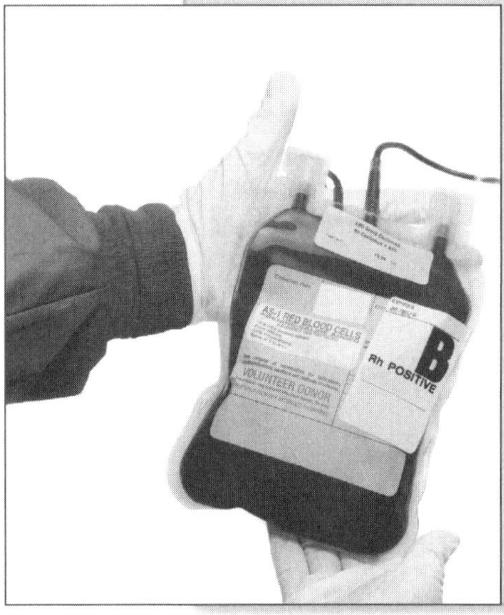

△ Guide

Ask students to describe a situation in which they observed blood directly. Ask students to describe the appearance of blood – What color was it? Was it a thin or a thick fluid? Most students describe blood as a fluid thicker than water and the color of blood as some shade of red. Then have students locate a vein on the inside of their forearm and ask them why they think it appears blue. Make a brief list of students' ideas.

As the class reads through this part, emphasize that all blood is red. Point out that blood is either bright red (carrying oxygen) or dark red (lacking oxygen and carrying carbon dioxide). The veins just appear blue because we are viewing them through the overlying skin cells which filter out much of the red light.

You may want to offer the following explanation: Things appear red because they absorb other colors of light and reflect the red – this is the case of blood. In order to see the blood in our veins, light has to pass through our skin to the blood in our veins and then reflect off the blood and travel back through the skin before it enters our eye. During this process, red light reflected from the blood is absorbed by the cells, so the vein does not appear red.

NOTES

Reflect
5 min.

Students collaborate to make improvements to the model of the circulatory system.

Revise Your Model
5 min.

Reflect

You now know more about the circulatory system than you did when you simulated it earlier. If you could run another circulatory system simulation, what would you change to more accurately model the circulatory system? Be prepared to share your ideas with the class.

Revise your Model

As a class, decide how to better model and simulate the circulatory system.

Update the *Project Board*

Your circulatory system investigations helped you better understand how the system works. The question that guides this section of the Unit is *What happens to you when you get sick?* You now know about the circulatory system and how it is necessary for moving nutrients, oxygen, and waste through your body. However, it can also move bacteria and viruses around your body. What information about the circulatory system helps you answer some of your questions about disease and how disease spreads?

Add what you have learned to the *Project Board*. Review all the information you have already added to the board. Look back at what you wrote in the *What do we think we know?* column in the beginning of the Unit. What information from this section of the Unit can you use to support what you thought you knew? How did the information about the circulatory system change what you thought you knew? Add this information to the *What are we learning?* column. Be sure to include evidence for your statements in the *What is our evidence?* column.

△ Guide

Have students independently consider ways in which they could change the model to make it a more accurate representation of the circulatory system. Have students write down their ideas and be prepared to share them with the class.

Lead a class discussion of students' ideas about how the model could be improved. Guide the class to reach a consensus as to how the model could be improved. You may want to try running the simulation again, but this time incorporate students' suggested improvements to the model.

Have students read this part and emphasize that their focus should be on updating the Project Board with information about the circulatory system which helps answer their questions about disease and disease spread.

Update the *Project Board*

10 min.

Students add new information and supporting evidence to the Project Board.

Lead a class discussion focusing on what students think should be added to the *What are we learning? and What is our evidence?* columns of the *Project Board*, now that they have investigated the circulatory system.

During the discussion, guide students to identify what they have learned about the circulatory system that helps answer the question, *What happens to you when you get sick?* and where that information should be recorded on the *Project Board.*

Guide students by asking:

- What have we learned about the circulatory system?
- Where should we put the information about the circulatory system?

Lead a class discussion of information about the circulatory system that helps answer the question What happens to you when you get sick? Guide students to specifically associate evidence with their claims. For example:

Claim: Bacteria can enter the body through a wound.

Evidence: From our readings, we learned that tetanus is a disease caused by bacteria and that it enters into the circulatory system through an open wound or a puncture and circulates through the body via the circulatory system.

Claim: when disease enters our body through the circulatory system it can be rapidly distributed throughout our body.

Evidence: From our readings and our model we learned that the circulatory system carries important nutrients and oxygen throughout our body and that it does so very rapidly (e.g. 15 s or less). This means that when germs enter our circulatory system they are rapidly distributed throughout our body as they are carried through the circulatory system.

Teacher Reflection Questions

- In this section, students created a model of the circulatory system and ran a simulation of blood circulation and gas exchange. What challenges did students face in setting up the model and running the simulation? What strategies could you share with students that would help them overcome similar challenges in future investigations?
- What were the intended learning outcomes of this investigation? What factors negatively or positively affected students' achievement of the learning outcomes?
- How do you see your role in this investigation? How does that compare with how students may see your role?

NOTES

SECTION 3.7 INTRODUCTION

3.7 Investigate

How Does the Digestive System React to Illness and Help Spread Illness?

◂ *2 class periods*

A class period is considered to be one 40 to 50 minute class.

Overview

In this section, students build a scale model of the digestive system as a story describing the journey of food through the digestive system is read aloud to them. After building their model, students consider and decide if the model does or does not do a good job of representing the organs of the digestive system. Students then read about the ways in which the digestive and circulatory systems work together to provide food to all of the cells in the body, ways in which germs can enter the body via the digestive system and actions that can prevent this from happening. Students also read about non-communicable diseases of the digestive system. Students then update the *Communicable-Disease Information Table* and add information about the digestive system to the *Project Board.* The section concludes with a *More To Learn* reading about the structure and function of the nervous system.

Targeted Concepts, Skills, and Nature of Science	Performance Expectations
Scientists often collaborate and share their findings. Sharing findings makes new information available and helps scientists refine their ideas and build on others' ideas. When another person's or group's idea is used, credit needs to be given.	Students collaborate in pairs to construct a model of the digestive system. Students share their ideas with the class as they consider what they think they have learned and what is their evidence when they fill out the class' *Project Board.* Students will also collaborate as they add new information to the *Communicable-Disease Information Table.*

Targeted Concepts, Skills, and Nature of Science	Performance Expectations
Scientists must keep clear, accurate, and descriptive records of what they do so they can share their work with others, consider what they did, why they did it, and what they want to do next.	Students correctly label each organ in their digestive system model and keep descriptive notes of the role it plays in the digestion of pizza described in the story.
Models are a representation of something in the world. Simulations use a model to imitate, or act out, real-life situations.	Students construct a model of the digestive system and identify the organs of the digestive system represented in the model. Students describe similarities and differences between the model and the actual digestive system.
An organ system is a group of body organs, each of which performs a specific function, that work together to carry on one or more life functions.	Students describe the function of each organ in the digestive system and how it contributes to the digestion and assimilation of food. Students describe how the digestive and circulatory system work together to provide food to all of the cells of the body.
The digestive system is made up of organs that break down food mechanically and chemically into molecules that can be absorbed and transported to cells.	Students identify the organs of the digestive system, trace the path of food through the digestive system, and describe the role played by each organ in the digestion and assimilation of food.
Foods contain substances that serve as fuel and building material for all organisms.	Students identify food as a source of substances needed by the body for energy, the growth or repair of body cells, or to maintain health.
Foods contain a variety of substances that are vital to the survival of an organism, including: carbohydrates, fats, vitamins, proteins, minerals, and water.	Students identify carbohydrates and fats as substances that supply the body with energy and describe ways in which vitamins, minerals, proteins, and water are useful to the body.
Energy in foods is measured in Calories.	Students can define a calorie and distinguish between a calorie and a food Calorie (kcal).
The nervous systems controls and coordinates the body's responses to changes in the environment.	Students identify the brain, spinal cord and network of nerves throughout the body as the major organs of the nervous system.

Targeted Concepts, Skills, and Nature of Science	Performance Expectations
Disease breaks down the structures or functions of an organism. Some diseases are the result of failures of a body system. Other diseases are the result of damage by infection from other organisms.	Students describe how disease affects the digestive system and how disease may enter the body through the digestive system. Students suggest methods to prevent the spread of digestive system illnesses.
Identify structure and function relationships in organisms	Students describe how the structure of a digestive organ is related to its function. For example: the shape of teeth and their ability to grip, tear, grind; the folded inner surface and villi in the small intestines that increases surface area through which assimilation can occur.

Materials	
1 per student	1 large, round balloon
1 per student	8 m (about 25 ft.) of long balloons or rubber tubing
1 per student	Picture of mouth with teeth, side view
1 per student	Tape
1 per student	Construction paper
1 per student	Poster board
1 per classroom	Optional: projections of images in student text

Activity Setup and Preparation

Students will be working in pairs for this investigation because two sets of hands are needed in the model construction process. Decide how you will assign student pairs. (Avoid having students work in larger groups since this will result in some students not playing an active role in the construction of the model.)

Identify a source for, and obtain long balloons in advance. This type of balloon is not as readily available as round balloons are.

Homework Options

Reflection

- **Science Content:** What are the two types of digestion that take place in the mouth? (mechanical digestion by the teeth and chemical digestion by saliva) Why would it be difficult to swallow without saliva? *(The digestive system is shaped like a tube. If food was swallowed dry, it would not slide through the throat and esophagus as easily. Saliva lubricates food, making it easier for it to slide through the esophagus.)*
- **Science Content:** What changes does food undergo on its way to the small intestines? *(Food is physically broken into smaller pieces by the teeth, further broken apart and mixed by the churning action of the stomach, and dissolved by liquid chemicals in the stomach.)*
- **Science Content:** What must happen to the food in the digestive system so that it can get to the other parts of the body? *(Food must be broken down into small molecules that can enter the bloodstream.)*
- **Science Content:** How do you think digestion would be affected if several feet of intestines were removed? *(The effect would be small, since a few feet of intestine represents a small fraction of the whole length of the intestines.)*
- **Science Process:** What activity or process in your home or community is similar to the process of digestion? *(Answers will vary, but should refer to processes that break down large objects into smaller parts – e.g. composting, recycling, and disassembling furniture so it can be more easily moved from one place to another.)*

Preparation for 3.7

- **Science Content:** Think back to a time when you had a cut or scrape that became infected. What changes did you observe in the injury and/or your body when it became infected? *(Redness, tenderness, swelling, pus, and possibly a fever.)*
- **Science Process:** What actions could you take to prevent an injury from becoming infected? *(Clean and disinfect the area surrounding the injury to eliminate bacteria that cause infection. Cover the injury to keep bacteria from entering the body through the injury.)*

SECTION 3.7 IMPLEMENTATION

3.7 Investigate

How Does the Digestive System React to Illness and Help Spread Illness?

The respiratory and circulatory systems work together to move oxygen around your body. You read in the last section that diseases can move into your body and then move around your body through your respiratory and circulatory systems. Sometimes diseases that enter through your mouth and nose end up making your digestive system sick.

What's in Food?

If your digestive system is not working properly, it cannot digest or break down food. Foods contain a variety of substances that are chemically broken down by enzymes into small components. Your body cells can then use these small components. **Carbohydrates** and **fats** are two substances found in food. When broken down, they give your body the energy it needs. The energy in food is measured in **Calories**. A Calorie (C) is the amount of energy in foods. There are 1000 **calories** (c) in 1 Calorie or kilocalorie. (A calorie is the amount of energy needed to raise 1 gram of water 1°C. The more Calories a food has, the more energy it contains. Your body needs a certain number of Calories each day to meet your body's energy needs. **Vitamins**, needed in small amounts, help your body in many ways, such as keeping you from developing certain diseases. **Minerals** such as calcium are needed for strong bones and teeth and for healthy red blood cells. **Proteins** build muscle and body tissue, and water is needed to carry out all the cell functions.

Using common materials, you are going to build a scale model of the digestive system. Remember that one role models can play is to help you see things that are difficult to see. Your digestive system is difficult to see because it is inside you.

carbohydrate: a type of substance found in food that gives the body the energy it needs.

fat: a type of substance found in food that gives the body the energy it needs.

Calorie: the amount of energy in foods. One Calorie is the same as 1 kilocalorie or 1000 calories.

calorie: the amount of energy needed to raise the temperature of one gram of water by 1°C.

vitamin: a type of substance found in food that is needed in small amounts. It helps the body in many ways, such as keeping it from developing certain diseases.

mineral: a type of substance found in food that is needed in small amounts. The mineral calcium is needed for strong bones and teeth and the mineral iron is needed for healthy red blood cells.

protein: a type of substance found in food that builds muscle and body tissue.

GF 105

GOOD FRIENDS and GERMS

*◀ 2 class periods**

3.7 Investigate

How Does the Digestive System React to Illness and Help Spread Illness?

5 min.

Students construct a model of the digestive system while listening to a story describing the journey of a slice of pizza through the digestive system and the changes it undergoes.

△ Guide

Have students read the introduction and remind them that in the last section they investigated the circulatory system and read that it carries nutrients (vitamins and minerals) around the body. Ask students where their body gets nutrients like vitamins and minerals. Students typically respond that the body gets these things from the food they eat. Then ask them how they think something as large as a hamburger, for example, could possibly fit into the tiny blood vessels of the circulatory system.

*A class period is considered to be one 40 to 50 minute class.

Call upon students to briefly share their ideas and make a public list of those ideas. Student responses typically make reference to digestion and the digestive system. Ask students which body parts they think are part of the digestive system and add their ideas to the public list.

What's in Food?

10 min.

The respira... ...circulatory systems work toge... to move oxygen around your body. You read in the last section that diseases can move into your body and then move around your body through your respiratory and circulatory systems. Sometimes diseases that enter through your mouth and nose end up making your digestive system sick.

What's in Food?

If your digestive system is not working properly, it cannot digest or break down food. Foods contain a variety of substances that are chemically broken down by enzymes into small components. Your body cells can then use these small components. **Carbohydrates** and **fats** are two substances found in food. When broken down, they give your body the energy it needs. The energy in food is measured in **Calories**. A Calorie (C) is the amount of energy in foods. There are 1000 **calories** (c) in 1 Calorie or kilocalorie. (A calorie is the amount of energy needed to raise 1 gram of water 1°C. The more Calories a food has, the more energy it contains. Your body needs a certain number of Calories each day to meet your body's energy needs. **Vitamins**, needed in small amounts, help your body in many ways, such as keeping you from developing certain diseases. **Minerals** such as calcium are needed for strong bones and teeth and for healthy red blood cells. **Proteins** build muscle and body tissue, and water is needed to carry out all the cell functions.

Using common materials, you are going to build a scale model of the digestive system. Remember that one role models can play is to help you see things that are difficult to see. Your digestive system is difficult to see ...cause it is insid...

...gives the ...the energy it needs.

Calorie: the amount of energy in foods. One Calorie is the same as 1 kilocalorie or 1000 calories.

calorie: the amount of energy needed to raise the temperature of one gram of water by 1°C.

vitamin: a type of substance found in food that is needed in small amounts. It helps the body in many ways, such as keeping it from developing certain diseases.

mineral: a type of substance found in food that is needed in small amounts. The mineral calcium is needed for strong bones and teeth and the mineral iron is needed for healthy red blood cells.

prote... ...ype of ...

△ Guide and Assess

Read the *What's in Food?* section together with the class, stopping where appropriate to emphasize new vocabulary and assess comprehension. Guide students to recognize the distinction between a calorie and a food Calorie (kcal). You may want to have the class carry out one or two simple calculations involving calories and Calories. For example, a single potato chip contains about 15 Calories? How many calories does it contain? 15,000 (15C x 1000 c/C = 15,000 c)

Ask students to give examples from the reading of ways in which the body uses the nutrients that are identified by bold text. You may want to elaborate on some of the ways in which the various nutrients are utilized by the body, or assign students to research this topic online for homework.

Assess student responses for the following: carbohydrates and fats provide the body with energy; vitamins prevent the body from developing certain diseases; minerals are needed to construct strong bones and teeth, and to keep red blood cells healthy; proteins are used to build muscle and body tissue; water is needed to carry out cell functions.

Then have students read the remainder of the introduction and tell them that they are now going to construct a model of the digestive system. Assign pairs and distribute the materials needed to construct the model.

NOTES

Procedure

20 min.

Students start to simulate and model the pathways of digestion.

Procedure

Someone will read the story below. As you listen to the story, you will use the materials listed to build a model of the digestive system. Consider how the materials mirror the actual parts of your digestive system.

Materials
- 1 round balloon
- 8.5 m (about 25 ft) of long balloons or rubber tube
- construction paper
- mouth and Teeth page
- tape
- poster board

Imagine sitting with your friends at a pizza parlor. You have just ordered your favorite pizza and you are waiting patiently. You notice that your mouth has started to water. You are thinking about the pizza, and your mouth makes more saliva. Before you even begin eating, you have started the digestion process. An enzyme in saliva chemically breaks down carbohydrates into smaller parts (sugars and starches).

1. Take the *Mouth and Teeth* page and write the word "saliva" in the mouth.

When the pizza comes to the table, you dive into eating it. You take a big bite with your front teeth and tear off a piece. Where do you chew the pizza? You probably chew with your back teeth. Molars, your back teeth, are excellent for breaking the food into much smaller pieces and mashing it. Along with the saliva in your mouth, your teeth make it possible for you to swallow the food and begin the digestion process.

esophagus: the tube that carries food from the mouth to the stomach.

2. On the *Mouth and Teeth* page, write "tear" on the teeth in the mouth that help you tear food. Label the teeth in the mouth that help you chew food into smaller pieces with the word "chew." Tape the *Mouth and Teeth* page to the poster board.

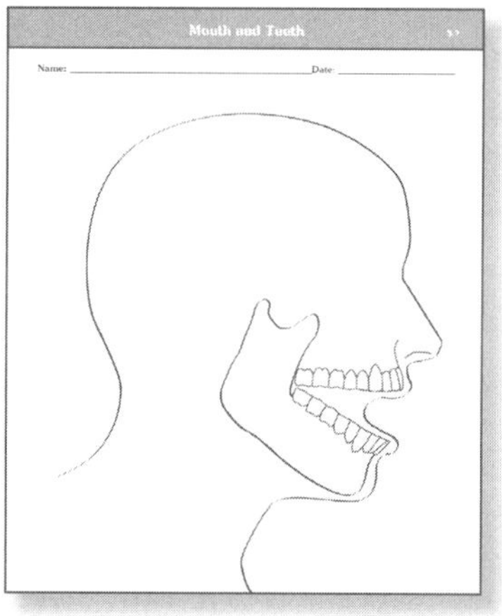

The pizza pieces, now much smaller, travel down your ***esophagus****. The esophagus is the tube that leads to your stomach. This tube pushes the food into your stomach. It takes about 10 seconds for food to get pushed down this tube.*

3. Take one long balloon and attach it to the picture of the head at the back of the mouth.

As the pizza enters your stomach, at the end of your esophagus, your stomach begins to churn. The churning is one way your stomach makes the food you have swallowed even smaller. Your stomach has special cells that make enzymes. These enzymes are sent into your stomach when you eat food. They break

△ Guide and Assess

Let the class know that you will be reading aloud a description of the journey of a bite of pizza through the digestive system. Tell students that as you read about what happens to the pizza in each part of the digestive system, they will add a representation of that part to the model.

Have students read the first paragraph and then ask them to discuss with their partner what part of the digestive system they think each of the materials might be used to represent in the model. Tell students that they *the proteins in the food into smaller substances. Your food gets mashed by the churning of your stomach and digested by the enzymes that are mixed with it.*

Your stomach works on the food for a while. Your stomach stretches to fit more and more food. It is similar to a balloon. When it needs to hold food, it can stretch. When it doesn't need to hold food, it is much smaller. A small opening at the bottom of the stomach remains closed as your food is digested. When the food has been digested enough, the small hole that was closed opens, so the food can leave. The digested food moves into your small intestine.

4. Use the round balloon to represent your stomach. Decide if the stomach is full or empty before you attach it. If you want a full stomach, blow up the balloon and tie a knot to keep in the air. Then attach the balloon to the end of the esophagus with tape.

Your small intestine is the next organ in the digestive system. Food moves from your stomach into your small intestine, where more enzymes break down carbohydrates and proteins into smaller substances. Fats are also broken down into smaller substances in the small intestine. The goal of all the parts of your digestive system is to make the food as small as possible so your body will be able to use it.

villi (singular, villus): tiny, finger-like structures that protrude from the inside surface of the small intestine.

*The small intestine is very long. It can be up to 7 m (about 23 ft) long, but it is only 2.5 cm (about 1 in.) in diameter. This tube squeezes the food along. Tiny finger-like structures, called **villi**, cover the inside wall of the small intestine. These structures allow the digested parts of carbohydrates, proteins, and fats to be carried out into the bloodstream. Other nutrients like vitamins and minerals also pass through the villi. The blood then carries these nutrients to all parts of the body.*

will construct the model by putting materials representing organs on the poster board in a position that corresponds to the position of that organ in the digestive system.

Read the story one step at a time. Before each, step ask student pairs to follow the instructions for building that part of the model. Circulate among groups and assess whether they have correctly placed the organ on their model.

5. Take 7 m (23 ft) of balloons and tape them to the stomach, near the bottom. The balloons represent the small intestine. Lay the intestines on your poster and carefully arrange the balloons (intestines) so they do not get knotted up.

Your body does not use all the bits of food. Some of the food moves on to the large intestine. This is a 1.5 m long (about 5 ft) tube that is attached to the end of your small intestine.

*The small bits of food and water that enter the large intestine are called waste. When the waste enters your large intestine, it is full of water and other liquids. Your large intestine removes the water and other liquids from the waste. What is left is called **feces**. The feces are pushed out of your body through your **anus**. These are the wastes that you flush down the toilet.*

*Both your small and large intestines are organs in your digestive system. They are filled with bacteria that help with the digestion process. These bacteria are called **enteric bacteria**. Enteric means "in the intestines." Almost all of the bacteria in your intestines are good for you. They help you digest food. They also allow your intestines to absorb different vitamins and minerals.*

feces: waste that is produced in the digestive system.

anus: the bottom opening of the digestive tract.

enteric bacteria: bacteria that live in the digestive tract.

Did You Know?

One "famous" bacterium that lives in your intestines is called *E. coli. Escherichia coli* is named after the bacteriologist Theodor Escherich, who first discovered it. You read earlier that *E. coli* can cause gastrointestinal illness. However, many of the *E. coli* bacteria that live in your intestine will not make you sick.

6. Use 1.5 m (5 ft) of balloons to represent your large intestine. Attach them to the end of the small intestine. Again, arrange them on the poster so that they are not twisted or knotted.

7. Label each organ in your model. Make sure you know which parts are the esophagus, the stomach, the small intestine, and the large intestine.

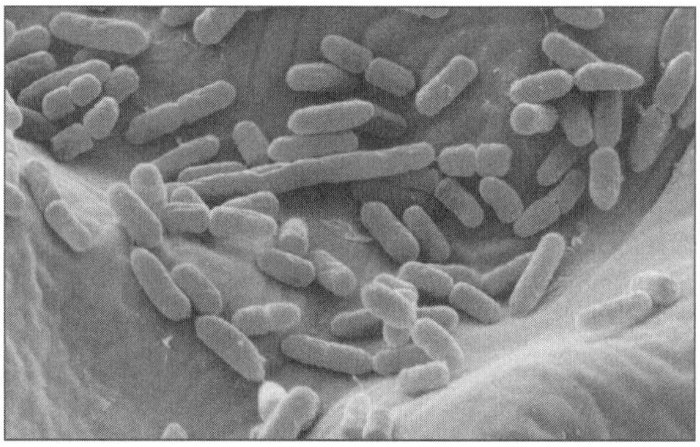

E.coli bacteria on a lettuce leaf.

When students get to attaching the small intestines (7 m of long balloons) and the large intestines (1.5 m of long balloons), make sure students understand that they will have to first tie the balloons together (end to end) to make a string of balloons that is the correct length.

When you get to step 8, make sure students label each of the organs in their model. Assess whether students have correctly labeled the organs of the digestive system in their model. You might consider having nearby pairs examine each others' models and discuss any differences they observe.

Reflect

In this activity, you built a model of the digestive system. This model helped you see what you cannot normally see.

1. You labeled the parts of the digestive system with their names. Now record the function (job) of each part on your poster.
2. In the model, some of the parts were good representations of the organs of your digestive system. Like all models, however, some of the parts do not represent the system well.
   a) Choose one part of this model that does a good job of representing an organ in the digestive system. Describe why you think so.
   b) Choose one part of this model that does not do as good a job of representing an organ in the digestive system. Describe why you think so.

Beyond the Model: Your Digestive System and Circulatory System Work Together

Every cell of every system in your body relies on the digestive system for nutrients. You learned that the circulatory system works with the respiratory system. The circulatory system also works with the digestive system. Your blood absorbs the food that is broken down by the digestive system. Your blood vessels then transport the food to all the cells of the body.

Other systems in your body also work with the digestive system. Muscles and bones (the skeletal and muscular systems) are used to take in food. For example, muscles in your face help you tear and chew food. The nervous system and endocrine system, which are other systems in your body, regulate how the digestive system functions.

How do people get digestive-system diseases? How do they spread them to others? Sometimes, digestive diseases can come from your hands. You may touch something with harmful bacteria or a virus on it. If you then hold your fingers near your face or put them in your mouth, the microbes can enter your digestive system through your mouth.

Symptoms of digestive-system illnesses are vomiting and diarrhea. These are ways the body reacts to rid your body of these microbes so you will get well. When you are sick with a digestive disease, the waste products from your body can have in them the bacteria or viruses that made you sick. Like *E. coli*, these bacteria live in your digestive

Reflect

10 min.

Students connect the model to the real digestive system and consider specific similarities and differences.

△ Guide and Assess

Have students record the function of each part of the digestive system on their poster.

Ask students to collaborate in their pairs to choose one part of the model that did a good job of representing an organ of the digestive system and one that did not. Have students write a description of what they thought were the strengths and weaknesses of the parts they chose.

Ask students to briefly share their responses and lead a brief class discussion focusing on what students perceive as the strengths and weaknesses of the part of the model in representing the digestive system.

Create a list of students' ideas about the strengths and weaknesses of the model, which you should assess and edit during the class discussion. Listen for the following information in student responses:

Part of model	Strengths of model	Shortcomings of model
Mouth (image)	Shows teeth and saliva.	Cannot actually show how food is initially broken down in mouth.
Esophagus (long balloon or tube)	Tube-like shape is similar to that of the esophagus. Connects the mouth to the stomach.	Cannot physically push down the food as the real esophagus.
Stomach (round balloon)	Can expand or stretch when full like the stomach.	Does not churn or create chemicals that real stomach does to break food down further. Does not have a valve at the bottom (gateway to small intestines)
Small intestines (7 m of balloon tied together)	Same length as small intestines and has a similar shape (long, skinny, hollow, expanding).	Since the balloons are tied together, they do not form a continuous open tube like the small intestines. Does not contain villi. Doesn't contain juices that break down food
Large intestines (1.5 m of balloon tied together)	Shape, length, and expandable like the large intestine.	Not a continuous open tube. Much narrower than actual large intestine and doesn't absorb liquids.

NOTE: This is an opportune time to conclude the first session of this investigation.

system. Every year, many people around the world become sick from communicable diseases of the digestive system. Managing the spread of these types of diseases is an important health issue. They are removed in your waste products when you use the bathroom. These bacteria or viruses can get on your hands. If you do not wash your hands, you do not get rid of the bacteria or viruses. They can then be spread directly to other people or by getting on surfaces. (Remember the glow-powder investigation?) Suppose someone who didn't wash his or her hands after using the toilet touches a doorknob. Then if you touch it, you could get the bacteria or virus on your hands. If you put your fingers near your face or in your mouth, you may get the disease.

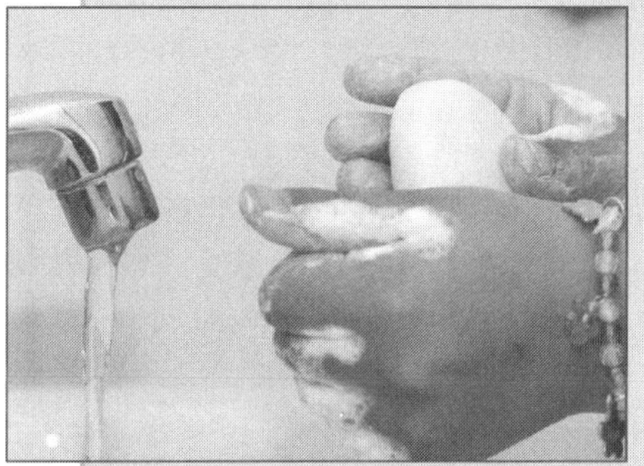

If not handled properly, chicken is one type of meat that can make people sick. Wash your hands thoroughly before and after handling raw chicken.

Use a different plate for cooked meat than you used for raw meat.

Every year, many people around the world become sick from communicable diseases of the digestive system. Managing the spread of these types of diseases is an important health issue.

Some types of digestive-system communicable diseases start in food. When harmful microbes, often bacteria, are in food you eat, they enter your body. Bacteria that need warmth and wetness reproduce in your body. When there are a lot of them, they can make you sick. That's why it is important to keep meat and milk products and some other foods cold in the refrigerator.

It is also important to cook meat well. Meat can have bacteria on it from when it was

Beyond the Model: Your Digestive System and Circulatory System Work Together

20 min.

Students read about the interrelationships between the digestive and circulatory systems, ways in which germs can enter the body via the digestive system, actions that can prevent this from happening and non-communicable diseases of the digestive system.

△ Guide and Assess

Refer students to their initial ideas about connections between the digestive system and circulatory system. Ask students how they think the nutrients (vitamins and minerals) get into the blood stream, now that they have learned more about the digestive system. Assess student responses by listening for references to absorption through the villi in the small intestines.

processed. Cooking the meat well kills the bacteria. But then the meat can get more bacteria on it from the air. Suppose you leave meat that has bacteria on it out of the refrigerator for a long time. Those bacteria can live and rapidly reproduce in the warm air. If you eat the meat, the bacteria will make you sick. A refrigerator is cold enough so bacteria reproduce very slowly. Food kept in the refrigerator is less likely to have large amounts of harmful bacteria in it.

Some digestive-system diseases are not communicable. One disease of the digestive system that is not communicable is gallstones, a gallbladder disease. The gallbladder is a small, pear-shaped organ on your right side, under your liver. It stores a chemical called bile. Bile is made in your liver. Your digestive system uses bile to help digest fats. Sometimes substances in the gallbladder form gallstones. Gallstones can cause severe pain, nausea, and vomiting. But gallstones are not communicable; you cannot develop gallstones from being with someone who has them.

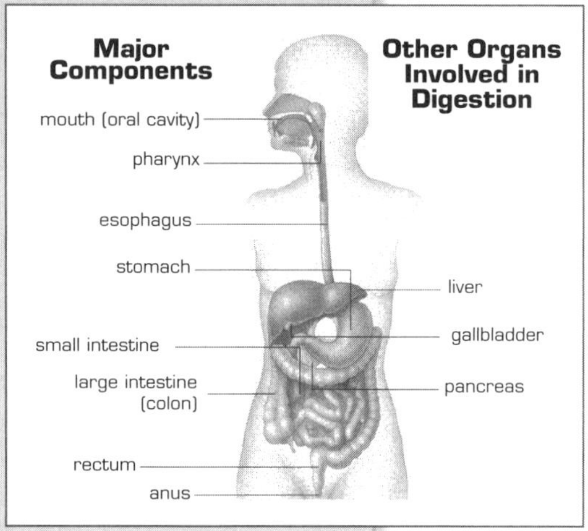

The digestive system includes many organs. They work together to help your body use the food you eat and help you grow and stay healthy. If one organ does not function properly, the whole digestive system cannot function properly.

Stop and Think

1. Go back to the *Communicable-Disease Information Table* you made in *Learning Set 2*. Fill in parts of it that you now know more about.
2. What are some good ways to reduce the chances of spreading digestive-system communicable diseases? Follow the path of disease-causing

Get Going

Read the first paragraph with students and remind them that it is through the villi in the small intestines that nutrients enter the blood stream.

Before reading the second paragraph, ask students how they think a person might get a digestive disease and how it might spread from a person that has the disease to someone that does not. Suggest that they think back to what they learned in *Learning Set 1* when they examined the case study. Make a class list of their ideas.

Have students complete the reading.

bacteria through the digestive system and decide what you would suggest to someone to prevent the spread of disease.

3. Choose one of the organs and describe what you think would happen to the digestive system if the organ stopped working correctly.

Update the *Project Board*

Your digestive system model building and reading may help you add more to the *Project Board*. The question for this section is *What happens to you when you get sick?* You now know about the digestive system and how it reacts to bacteria and viruses. You even have some ideas about how the digestive system moves bacteria and viruses out of your body and increases the chances that you could spread illness to someone else. What information about the digestive system do you think you should add to the *Project Board*?

Add what you have learned to the *Project Board*. Be sure to include evidence for your statements in the *What is our evidence?* column. Look back at what you wrote for the *What do we think we know?* column from the beginning of the Unit. What information from this section of the Unit can you use to support what you thought you knew before? How did the information about the digestive system change what you thought you knew before? Add this information to the *What are we learning?* column.

What's the Point?

The digestive system is responsible for digesting or breaking down food. Food enters your digestive system through your mouth. The saliva in your mouth contains an enzyme that chemically breaks down carbohydrates into sugars and starches. Along with the saliva in your mouth, your teeth tear and mash food so it possible for you to swallow food and begin the digestion process. Next, smaller pieces of food travel down a tube called the esophagus that connects to the stomach. Once food enters the stomach, it begins to churn the food and release enzymes that break down proteins into smaller substances. The next digestive organ that food travels into is the small intestine where more enzymes break down carbohydrates and proteins into smaller substances. Finger-like structures called villi cover the inside wall of the small intestine. They allow digested parts of carbohydrates, proteins, and fats as well as vitamins and minerals to be carried out into the bloodstream to all parts of the body. Lastly, the undigested parts of food move into the large intestine and are pushed out of the body. The large intestine also removes water from the wastes.

Stop and Think

10 min.

△ Guide

Ask student groups to read through the *Stop and Think* questions and to write down their answers.

For question 1, lead a brief class discussion of what new information this investigation added to what they already knew about gastrointestinal disease. Have students add this new information to their *Communicable-Disease Information Table.*

Student responses may vary. Assess students' *Communicable-Disease Information Table* for the following information:

Name of disease	Cause (bacterial or viral)	How does it spread?	Which body systems does it affect?	What are the symptoms?	How is it treated? Is there a vaccine?	Other notes
Gastrointestinal infections	bacteria or virus (or parasite)	by dirty hands, contaminated food or water, pets, and dirty surfaces (like the bathroom).	digestive system (stomach/ intestines)	diarrhea and vomiting	rest, drinking lots of fluids	can enter the body through mouth or nose

Hold a brief class discussion of students' responses to question 2 focusing on ways to reduce the chances of spreading digestive system communicable diseases. Guide students to connect methods of prevention with the route taken by disease-causing bacteria into and through the digestive system.

Assess students' answers by listening for the following information:

- The disease enters through the mouth or nose and then works its way down the esophagus, into the stomach, and into the small and large intestines. It causes diarrhea and/or vomiting.
- Some ways to prevent the spread of digestive system disease from one person to another are:
    - wash hands properly to remove bacteria,
    - clean kitchen and bathroom surfaces,
    - refrigerate food to minimize bacterial growth,
    - cook meats well to kill any bacteria that are present, and
    - wash hands before you eat and after you've handled a pet.

For question 3, hold a class discussion of possible consequences of loss of functioning in one of the organs of the digestive system. Guide students to make connections between the function of a specific part, and the effects that loss of function in that part would have on other organs in the system.

Student answers will vary, some examples of student responses are given below:

Organ	Causes of loss of function	Consequences of loss of function
mouth	bad teeth	Cannot properly bite or chew food; large pieces cannot fit into esophagus; must have food pureed first.
esophagus	blockage by cancer or scar tissue due to acid reflux	Food cannot reach stomach; body lacks nutrients.
stomach	Ulcers; too much or too little acid; cancer invades and blocks	Unable to break down food properly; blood leaks into digestive system through ulcers, food and bacteria leak out of stomach through perforated ulcer and infect surrounding organs, loss of nutrients.
small intestines	Infection, blockage by cancer, kink in tube	Unable to absorb nutrients leads to a loss of nutrients absorbed by the body.
large intestines	Infection, blockage by cancer	Difficulty absorbing water runny stools, diarrhea too much water absorbed, hard stool, constipation

suggest to someone ... spread of di...

3. Choose one of the organs and describe what you think would happen to the digestive system if the organ stopped working correctly.

Update the *Project Board*

Your digestive system model building and reading may help you add more to the *Project Board*. The question for this section is *What happens to you when you get sick?* You now know about the digestive system and how it reacts to bacteria and viruses. You even have some ideas about how the digestive system moves bacteria and viruses out of your body and increases the chances that you could spread illness to someone else. What information about the digestive system do you think you should add to the *Project Board*?

Add what you have learned to the *Project Board*. Be sure to include evidence for your statements in the *What is our evidence?* column. Look back at what you wrote for the *What do we think we know?* column from the beginning of the Unit. What information from this section of the Unit can you use to support what you thought you knew before? How did the information about the digestive system change what you thought you knew before? Add this information to the *What are we learning?* column.

What's the Point?

The digestive system is responsible for digesting or breaking down food. Food enters your digestive system through your mouth. The saliva in your mouth contains an enzyme that chemically breaks down carbohydrates ... and starch ... with the s... mouth, yo...

Update the *Project Board*

10 min.

Students add new information and supporting evidence to the Project Board.

△ Guide

Have students read this section and emphasize that their focus should be on updating the Project Board with information about the digestive system which helps answer the for *Learning Set 3* question, *What happens to you when you get sick?*

Lead a class discussion focusing on what students think should be added to the – What are we learning? and What is our evidence? columns of the Project Board now that they have investigated the digestive system.

Guide students by asking:

- What have we learned about the digestive system?
- Where should we put the information about the digestive system?

Lead a class discussion of information about the digestive system that helps answer the question What happens to you when you get sick? Guide students to specifically associate evidence with their claims. For example:

Claim: A bacteria could enter my body through my nose or mouth and affect my digestive system.

Evidence: From our readings we learned that many germs like warm, moist environments such as those found in our mouth or near our nose. These are entry points into our digestive system.

Claim: Diarrhea and vomiting are symptoms of disease affecting the digestive system.

Evidence: From our readings, we learned that the digestive system contains the stomach which is responsible for vomiting if sick, and the intestines, which are responsible for diarrhea when sick.

Claim: When disease enters our body through the digestive system it can cause a decrease in nutrients available to the rest of our body systems.

Evidence: From our readings, we learned that the small intestines is responsible for transferring nutrients from the digestive system to the circulatory system, which then carries these nutrients throughout the body. If our digestive system is sick and we are vomiting or have diarrhea, then our body is unable to absorb many of the nutrients we are ingesting.

Add what you have learne... *Project Board*. Be s... ...ude evidence for your statements in the *What is our evidence?* column. Look back at what you wrote for the *What do we think we know?* column from the beginning of the Unit. What information from this section of the Unit can you use to support what you thought you knew before? How did the information about the digestive system change what you thought you knew before? Add this information to the *What are we learning?* column.

What's the Point?

The digestive system is responsible for digesting or breaking down food. Food enters your digestive system through your mouth. The saliva in your mouth contains an enzyme that chemically breaks down carbohydrates into sugars and starches. Along with the saliva in your mouth, your teeth tear and mash food so it possible for you to swallow food and begin the digestion process. Next, smaller pieces of food travel down a tube called the esophagus that connects to the stomach. Once food enters the stomach, it begins to churn the food and release enzymes that break down proteins into smaller substances. The next digestive organ that food travels into is the small intestine where more enzymes break down carbohydrates and proteins into smaller substances. Finger-like structures called villi cover the inside wall of the small intestine. They allow digested parts of carbohydrates, proteins, and fats as well as vitamins and minerals to be carried out into the bloodstream to all parts of the body. Lastly, the undigested parts of food move into the large intestine and are pushed out of the body. The large intestine also removes water from the wastes.

GF 112

Project-Based Inquiry Science

What's the Point?

5 min.

Teacher Reflection Questions

- In this investigation, student pairs alternated between listening to a narrative and constructing a model of the digestive system. What are the benefits to learning of this strategy?
- The *Beyond the Model* reading selection made reference to the glow powder investigation in *Learning Set 1* and the gastrointestinal illness case study performed in *Learning Set 2*. Why is it so important for students to make connections with previous investigations during a *Learning Set?*
- What evidence do you have that students learned both the structure of the digestive system and the functional role of each organ in the process of digestion?
- To what extent did your questioning during class discussions foster critical and creative thinking?

NOTES

SECTION 3.8 INTRODUCTION

3.8 Read

How Does Your Body Protect Against Disease?

◀ ***1 class period***

A class period is considered to be one 40 to 50 minute class.

Overview

Students read about the body's wide variety of defenses against disease. The concept of immunity introduced the smallpox case study in *Learning Set 2* is extended and the various methods employed by the body to prevent germs from entering it and to combat germs once they have entered are described. The parts of the immune system and their roles in defending against disease and/or fighting disease are described in some detail. After they have completed the reading, groups presentat one part of the immune system using a poster showing that part of the immune system, its function, and how it works together with other parts of the immune system. The class then collaborates to construct a more complete understanding of how the parts of the immune system work together to keep you healthy. The section concludes with students adding information to the *Communicable-Disease Information Table* and updating the *Project Board.*

Targeted Concepts, Skills, and Nature of Science	Performance Expectations
Scientists often collaborate and share their findings. Sharing findings makes new information available and helps scientists refine their ideas and build on others' ideas. When another person's or group's idea is used, credit needs to be given.	Students collaborate to create a poster of one part of the immune system and make a class presentation describing its function and how it works together with other parts of the immune system. Students share their ideas with the class as they consider what they think they have learned and what is their evidence when they fill out the class *Project Board.* Students will also collaborate as they add new information to the *Communicable-Disease Information Table.*

Targeted Concepts, Skills, and Nature of Science	Performance Expectations
Scientists must keep clear, accurate, and descriptive records of what they do so they can share their work with others, consider what they did, why they did it, and what they want to do next.	Students make detailed notes as they read. Students produce a poster and presentation on one part of the immune system. Students add information along with supporting evidence to the *Project Board* and *Communicable-Disease Information Table*.
An organ system is a group of body organs, each of which performs a specific function, that work together to carry on one or more life functions.	Students make a class presentation in which they identify the parts of the immune system and describe the function of each part. After the class presentations, students collaboratively identify ways in which the organs of the immune system work together.
Disease breaks down the structures or functions of an organism. Some diseases are the result of failures of a body system. Other diseases are the result of damage by infection from other organisms.	Students add information about disease and its effects on the body to the *Communicable-Disease Information Table* and the *Project Board*.
Specialized cells protect the body from infectious disease. The chemicals they produce identify and destroy microbes that enter the body.	Students describe the role of leukocytes, T-cells and B-cells in protecting the body from infectious disease.
Identify structure and function relationships in organisms.	Students describe ways in which the structure of the immune system is related to its ability to prevent and fight against disease.

Materials	
1 per student group	Poster board, materials for making a poster (crayons or colored pencils)
1 per classroom	Project Board
1 per classroom	Optional: projections of images in student text

Activity Setup and Preparation

You may want to make a projection of the image of the lymphatic system that is in the student text to use as a focal point during reading.

Decide how you would like students to keep a record of individual and group work. Make sure the class *Project Board* is posted where everyone can see it.

Homework Options

Reflection

- **Science Content:** What are the body's two types of defenses against disease? *(The first type of defense consists of barriers that block the entry of germs, such as the skin, the linings of the nose and throat, the hairs in the in the nostrils, and acids in the stomach. The second type of defense fights germs once they have entered the body and includes the inflammatory response, white blood cells and antibodies.)*
- **Science Content:** What are some different types of white blood cells and how do they fight disease? *(Some white blood cells hunt down, trap, and suffocate germs, T-cells puncture infected cells causing them to die and then other white blood cells eat the remains, B-cells produce antibodies which are specialized to the shape of a germ and inactivate it.)*
- **Science Process:** Make analogies between parts of your body's immune system and structures, devices or systems in your home, school, or community. *(Student answers will vary, but may include: window screens to nasal hairs; mucus linings to rodent glue traps; walls to skin cells; T-cells to soldiers)*

Preparation for Back to the Big Question

- **Science Content:** What are some connections between organ systems that you studied in this Learning Set? *(Respiratory-Circulatory, Digestive-Circulatory, Lymphatic-Circulatory)*
- **Science Process:** Of all the strategies for preventing disease that you have learned about, which would have the greatest benefit with the least amount of effort? *(Student answers will vary, but may include: washing hands and good sanitation.)*

NOTES

SECTION 3.8 IMPLEMENTATION

3.8 Read

How Does Your Body Protect Against Disease?

The next body system you will learn about is the **immune system**. This body system's function is to protect against disease. The immune system detects bacteria and viruses in your body and fights them off. Sometimes it can do that before you even feel anything. After you've been sick, some cells in your immune system continue to live for many years and remember the bacteria or virus they fought off earlier. Those cells can give you **immunity** from that same bacteria or virus. Immunity means that you will not become sick by that germ again.

immune system: body system that fights disease.

immunity: a condition of being able to resist an infectious disease.

The immune system is very complicated. In all the other systems you have learned about, the parts are connected to each other physically, and parts of the system move substances to other parts of the system. In the immune system, it is different. The immune system has only some parts that are connected to each other and have the specific job of fighting disease. In addition, it depends on other parts of your body. For example, your skin is very important to protecting you from disease. It acts as a barrier to keep bacteria and viruses from entering your body. If germs can be kept from your immune system, it does not have to work as hard.

You will begin learning about how your body defends against disease by thinking about a time when you were sick.

What Does It Mean to Protect Against Disease?

Think back to a time when you had a cold or the flu. What symptoms did you have? Did you have a runny nose, scratchy throat, stomachache, headache, or fever? Maybe you had other symptoms, as well. Why did your body react the way it did?

Your body reacts the way it does because it is trying to protect you. What good does it do to have stuff running out of your nose? When your nose runs, you know your body is fighting off bacteria, viruses, or other germs. Mucus serves a purpose. It traps germs. When you have a cold and you blow your nose, you remove some of the mucus from your body along with the germs trapped by the mucus.

GF 113 GOOD FRIENDS and GERMS

◀ *1 class period**

3.8 Read

How Does Your Body Protect Against Disease?

5 min.

Students read about the immune system and the lymphatic system. The concept of immunity is introduced. The mechanisms by which the body defends itself against germs are described. Students make a poster showing the immune system's role. They update their Communicable-Disease Information Table *and their* Project Board.

△ Guide

Before students begin reading, have students recall that they have investigated four systems so far: the respiratory, the circulatory, the digestive, and the nervous systems. Ask students what their investigations have revealed about how these systems prevent disease from entering the body. Guide students to recollect specific examples introduced in previous investigations such as nose hair, mucus membranes, and stomach acid.

Create a list of students' initial ideas about how the body defends itself against disease-causing microbes. Introduce the skin as a defense structure by asking students what keeps germs from entering their arm, leg, or toe. Point out that the skin is a natural barrier protecting almost every inch of the body from disease and that the skin is also the largest organ in the body.

Let students know that they will be learning about the immune system and ask them what they think the immune system is and what it does. Let them know also that the immune system consists of parts of other systems that act as barriers to prevent disease from entering our bodies (such as mucus or skin) and other parts that actively attack and kill germs in our body.

Emphasize that the immune system is different from some of the other systems they have studied because not all of the parts are physically connected to one another.

Consider alternating the way in which students read different segments of this section. You may want to read some parts aloud to the class and have students read others individually or in groups. Changing the reading strategy keeps students more actively engaged in the reading assignment.

Encourage students to make detailed notes during the reading, perhaps using a graphic organizer to make the logical sequence of ideas more readily apparent. Make clear to students that these notes will be very useful when preparing their class presentations later in the section.

Have students read the introduction and briefly discuss the concept of immunity. Guide students to make connections between immunity and the work of Edward Jenner in the smallpox case study examined in the previous *Learning Set.*

NOTES

sick by

The immune system is very complicated. In all the other systems you have learned about, the parts are connected to each other physically, and parts of the system move substances to other parts of the system. In the immune system, it is different. The immune system has only some parts that are connected to each other and have the specific job of fighting disease. In addition, it depends on other parts of your body. For example, your skin is very important to protecting you from disease. It acts as a barrier to keep bacteria and viruses from entering your body. If germs can be kept from your immune system, it does not have to work as hard.

You will begin learning about how your body defends against disease by thinking about a time when you were sick.

What Does It Mean to Protect Against Disease?

Think back to a time when you had a cold or the flu. What symptoms did you have? Did you have a runny nose, scratchy throat, stomachache, headache, or fever? Maybe you had other symptoms, as well. Why did your body react the way it did?

Your body reacts the way it does because it is trying to protect you. What good does it do to have stuff running out of your nose? When your nose runs, you know your body is fighting off bacteria, viruses, or other germs. Mucus serves a purpose. It traps germs. When you have a cold and you blow your nose, you remove some of the mucus from your body along with the germs trapped by the mucus.

GF 113

GOOD FRIENDS and GERMS

What Does It Mean to Protect Against Disease?

5 min.

△ Guide

Before students read, ask them to describe any symptoms they had the last time they were sick. Ask them why they think their body reacted in this way. Create a class list of students' ideas about what caused their symptoms.

Tell students that it is often the immune system that causes the symptoms we experience. The immune system sends signals throughout the body and each system uses the parts it has to fight disease. For example, the respiratory system makes mucus, which results in a runny nose and congestion; the digestive system makes you vomit to eliminate germs, and so on. Ask students for more ideas about how the body defends against germs.

As students read this section, stop at appropriate points to discuss the information and guide students in making connections to their own experiences and to the ideas suggested during the class discussion they had before they began reading.

How Does Your Body Attack and Kill Germs?

15 min.

filter: to separate.

expel: to force out.

microscopic: so small you can see it only through a microscope.

white blood cells (leukocytes): blood cells whose main job is to protect the body from invading microbes.

lymphatic system: body system that collects fluid lost by the blood and returns it to the circulatory system. It also produces some types of white blood cells.

antibodies: substances that help destroy harmful microbes.

The body has several defenses against germs. Skin keeps germs from getting into your body. Mucus linings in the breathing passages work to keep germs from getting through your nose and into your respiratory system. White blood cells and other substances attack microbes to kill them after they are in your body. Small nodes and glands all over the body help white blood cells develop and reproduce. Other glands **filter** microbes from your body fluids and trap them or **expel** them from your body. Together, these parts form the body's immune defense system.

The immune system defends the body against **microscopic** living things, like bacteria and viruses, that might make you sick. You know that even in a clean and spotless place, there are many bacteria and viruses. They get onto your skin, into the food you eat, into the drinks you consume, and even into the air you breathe. These bacteria and viruses may also get into your body through a cut or scrape on your skin, through your nasal passages, or through the throat. You get sick when the bacteria and viruses that get into your body have a chance to reproduce and multiply. When that happens, glands in your body work overtime to create more **white blood cells**, or **leukocytes**, to kill the germs and to filter the germs from your body fluids.

How Does Your Body Attack and Kill Germs?

The **lymphatic system** is part of the immune system that collects fluid lost by the blood and returns it to the circulatory system. The organs of the lymphatic system include lymph nodes and lymph vessels.

The lymphatic system also produces several different kinds of white blood cells. These are the cells that hunt down and kill diseases. They also protect you from diseases. There are many types of white blood cells. Each type has a special job in protecting you from disease. Some white blood cells wrap themselves around invading bacteria to trap and suffocate them. Some white blood cells, called killer T-cells, recognize when a cell has been invaded by a virus and punctures holes in those cells. When that happens, the infected cell's cytoplasm leaks out, and the cell dies along with the virus that invaded it. Another kind of white blood cell acts as a vacuum cleaner. They eat the remains of the dead cells.

Some white blood cells, called B-cells, produce **antibodies** that are specialized to the shape of invading microbes. The antibodies then attach themselves to the invaders. They might kill the invader or attach themselves in a way that inactivates the invader (keeps it from doing its job). Some

△ Guide

Read this section about the lymphatic system together with students. Consider creating a chart listing the parts of the lymphatic system and their functions. As you complete each paragraph ask students what should be added to the chart. Let students know that later they will have to draw a poster of a part of the lymphatic system and present it during the *Communicate Your Understanding* section.

antibodies keep watch for the kind of invader they specialize in. Remember Edward Jenner? His experiments with vaccination relied on these memory cells. When they recognize a familiar cell, they send out a signal, and other white blood cells come and kill those invading cells. Because of memory cells, the second time your body encounters some bacteria or virus, your body attacks it more quickly. This is how many vaccines work.

The organs in your lymphatic system produce white blood cells, help them develop and reproduce, and carry them around your body. They also filter or trap bacteria, viruses, and dead cells to remove them from your blood.

Bone Marrow

Your **bone marrow** is the soft material in the center of your bones. All your blood cells, including white blood cells, are produced in bone marrow. The long bones in your body produce most of your blood cells.

Thymus

Your **thymus** is an organ in the immune system. The thymus is located high in your chest, beneath your breastbone and between the two parts of your lungs. It is important in the development of some types of white blood cells. Some of the white blood cells produced by your bone marrow are ready to go to work right after they are created. Others need to mature or develop more complicated structures. The thymus helps the kind of white blood cells called T-cells to develop.

Spleen

Your **spleen** is located under your ribs and in front of your lungs on the left side of your body. This organ has many different functions. It filters your blood to remove bacteria or viruses that might make you sick. B-cells are activated in the spleen. Old red blood cells that cannot do their job anymore are destroyed in the spleen. If somebody has to have their spleen removed, other organs in the immune system take over the spleen's functions. But the spleen carries out those functions better than any other organs can.

Lymph Vessels

Your **lymph vessels** carry the different kinds of **lymphocytes** around your body. Lymphocytes are a type of white blood cell that makes antibodies. Both T-cells and B-cells are lymphocytes. Lymph vessels are very much like blood vessels, and they can be found right next to blood vessels in your body.

bone marrow: a soft tissue in the center of bones where most blood cells are produced.

thymus: an organ that is important in the development of a type of white blood cell.

spleen: filters blood by removing harmful bacteria and viruses, activates B-cells, and destroys old red blood cells.

lymph vessels: structures that carry lymphocytes throughout the body.

lymphocyte: a blood cell that protects the body from invading microbes.

Discuss the roles that different types of white blood cells play in attacking and killing germs found in the body. Differentiate between the actions of T-cells, B-cells, and other types of white blood cells. Elaborate on the mechanism by which antibodies fight disease-causing microbes.

Guide students in extending their understanding of vaccines by describing the mechanism by which they help the body develop immunity to a particular disease. Point out that a vaccine introduces a tiny amount of virus (dead or weakened) into the body so that our immune system can build up antibodies against the virus.

How Else Does Your Body Defend Against Illness?

5 min.

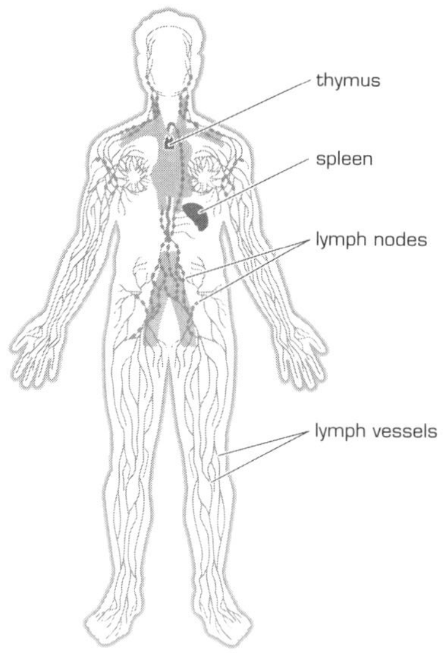

Lymph Nodes

Lymph nodes are all over your body. They can be found along the pathways of your lymph vessels. They are in your neck, under your arms, and in your groin and chest. Like the spleen, the lymph nodes have many functions. One of their functions is to hold large numbers of lymphocytes for when they are needed to fight infection. When you have an infection, the lymph nodes become warm and provide a warm place for lymphocytes to reproduce actively so there are enough available to fight your infection. When lymphocytes reproduce quickly, they can build up in your lymph nodes. This makes the lymph nodes swell. When you have a cold, the lymph nodes in your neck sometimes swell enough so you can feel lumps on either side of your neck. When you are sick and go to the doctor, he or she may feel each side of your neck to see if the lymph nodes are swollen.

The lymph nodes also act as filters, like the spleen.

lymph nodes: structures in the lymphatic system that filter lymph and trap microbes.

barrier: something that obstructs (or blocks).

How Else Does Your Body Defend Against Illness?

If the lymphatic system had to fight every bacteria and virus in the air or on things you touch, it would have too much work to do. Your body helps the lymphatic system in several ways:

- Your body provides **barriers** that keep harmful microbes from getting into your body. Your skin is one of those barriers. The mucus in your nose is another. In fact, you have mucus on the insides of organs in your respiratory system, your digestive system, and other systems. Mucus forms a barrier. A layer of mucus helps keep microbes from getting through to healthy tissue. When some body part with mucus gets irritated, it secretes additional mucus.

△ Guide

Before students begin reading, have them review their initial list of ideas about how the body defends against disease and ask them if they would like to add anything to their list.

Read the remainder of this section together with students and hold a brief class discussion focusing on the inflammatory response. Guide students to understand how increased blood flow provides a favorable environment for white blood cells and an unfavorable environment for invading microbes. Connect the inflammatory response with some of the symptoms of an infection, such as redness, tenderness, vomiting and diarrhea. Then, ask students how they should further edit their list of how the body defends itself against disease.

- Some of your body systems make substances that break down the cell walls of bacteria. The saliva in your mouth does that, and so do your tears.
- All of your cells have an **inflammatory response**. When a cell is invaded by a microbe, it gives off a chemical signal (called a histamine) that increases blood flow to the area. That area becomes red and heated. A heated area of your body provides a good environment for white cells to reproduce and enough heat to kill some of the invading microbes.

- Some body systems have particular inflammatory responses. For example, when your stomach is inflamed, you vomit. When your intestines are inflamed, you get diarrhea. These inflammatory responses, or **inflammations**, expel harmful microbes from your body.

Stop and Think

1. Sometimes when you are very sick, the doctor will take some blood and count the different kinds of cells in your blood. Why do you think he or she might count your white blood cells?
2. The immune system is very complicated and has many parts. The parts work together to keep you healthy. If the parts of this system did not work your body would not be able to fight off infections. Even the smallest infection could make you very sick. Describe in your own words how the parts of the immune system work together to fight disease.

inflammatory response: the body's reaction to invading bacteria or viruses; includes the release of a chemical signal and increased blood flow.

inflammation: a reaction to invading microorganisms; signs of this include redness, heat, swelling and pain.

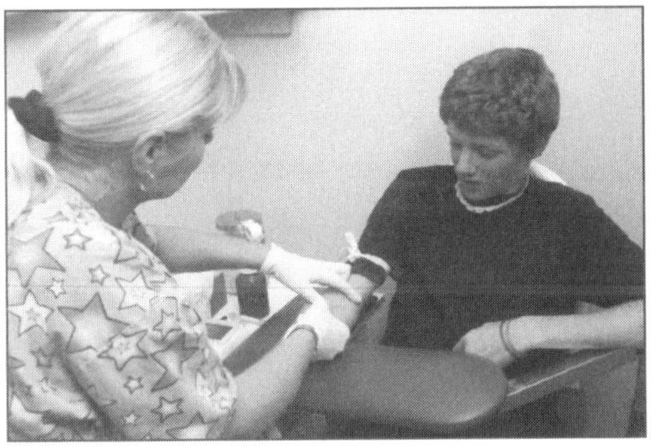

Sometimes when you are sick, a doctor may order a WBC count. This is a blood test to measure the number of white blood cells.

Stop and Think

15 min.

△ Guide

Ask students to read through the *Stop and Think* questions and then write down their answers. Then, call upon two students to share their responses (one for each question) and hold a class discussion of the ideas they share. Students' answers to question 1 will vary, but they should make a connection between the white blood cell count and an increased production of white blood cells (lymphocytes) to attack the disease, explaining why a doctor may count white blood cells when a patient is very sick.

In question 2, the description of immune system should contain the following information:

- Barriers, such as skin and mucus, help to keep unwanted germs out of our body so we do not get sick.
- The lymphatic system fights germs when they get past barriers. The lymphatic system consists of:
  - bone marrow: creates blood, including white blood cells that protect the body from invading germs
  - thymus: organ that helps white blood cells develop into more complicated structures that fight disease in a specific way.
  - spleen: organ that filters blood and removes germs and old blood cells
  - lymph vessels: carry lymphocytes around the body. A lymphocyte is a type of white blood cell.
  - lymph nodes: found all over the body they hold large numbers of lymphocytes, reproducing and dispersing them when needed.)

NOTES

Communicate Your Understanding

Choose one part of the immune system. Make a poster showing the following three things:

- what this part of the immune system does to protect agaist disease
- what other parts of the immune system depend on it
- what other parts of the immune system it helps

Each group will have a chance to show their poster. As you listen to the presentations, take notes about how the different parts of the immune system work together.

After the presentations, work as a class to figure out how the immune system organs work together to keep you healthy.

Go back to the *Communicable-Disease Information Table* you made in *Learning Set 2* and add the new information.

Update the *Project Board*

The question for this section of the Unit is *How does your body protect against disease?* You now know about the immune system and how it is devoted to maintaining good health. What information about the immune system helps you answer some of your questions about disease and how disease spreads?

Add what you learned from the presentations and reading about the immune system to the *Project Board* in the *What are we learning?* column. Be sure to include evidence for your statements on the *Project Board*.

Look back at what you wrote at the beginning of the Unit in the *What do we think we know?* column. What information from learning about the immune system can you use to support what you thought you knew before? How does what you have read about the immune system change what you thought you knew before? Add this information to the *What are we learning?* column.

Communicate Your Understanding

10 min.

△ Guide

Have students read through the directions for this activity. Emphasize that they will be using their poster as the basis for making a presentation to the class.

⬡ Get Going

Provide students with the materials needed and have small groups begin construction of their posters.

Have each group present their poster, describing what they have learned about that part of the immune system and how it helps and is helped by other parts of the immune system.

Then, hold a class discussion focusing on how the different parts of the immune system work together to keep you healthy. You may want to do this by using a diagram of the body and discussing how body systems are affected when germs enter the body. Have students help you draw a path by which germs could try to enter the body and how the body prevents or responds to the invasion.

Display the class' *Communicable-Disease Information Table* and ask them to suggest additions to the table. Suggest adding information to the "other notes" column about what the body does to prevent the disease or to attack and kill disease germs once they enter the body. Students may also include additional information about vaccines..

Update the *Project Board*

10 min.

Students add new information and supporting evidence to the Project Board.

...ch group will have a ch... ...ow their poster. ... the presentations, take notes about how the different parts of the immune system work together.

After the presentations, work as a class to figure out how the immune system organs work together to keep you healthy.

Go back to the *Communicable-Disease Information Table* you made in *Learning Set 2* and add the new information.

Update the *Project Board*

The question for this section of the Unit is *How does your body protect against disease?* You now know about the immune system and how it is devoted to maintaining good health. What information about the immune system helps you answer some of your questions about disease and how disease spreads?

Add what you learned from the presentations and reading about the immune system to the *Project Board* in the *What are we learning?* column. Be sure to include evidence for your statements on the *Project Board*.

Look back at what you wrote at the beginning of the Unit in the *What do we think we know?* column. What information from learning about the immune system can you use to support what you thought you knew before? How does what you have read about the immune system change what you thought you knew before? Add this information to the *What are we learning?* column.

△ Guide

Remind students that the guiding question of this *Learning Set is What happens to you when you get sick?* Then, ask students to look at the class *Project Board* and review the questions they have added to it. Ask them to think about what they have just learned about the immune system and what information they may be able to add to the *Project Board.*

What's the Point?

Sometimes bacteria or viruses make your body ill. When you are sick, your immune system fights against these germs. Sometimes your lymph nodes, especially the ones in your neck, swell up. When you are healthy, they are about the size of a pea or grape, and you can barely feel them. During illness, they can be as big as golf balls.

When your lymph nodes swell, it means that your immune system is doing important work. The immune system includes white blood cells that attack germs. If bacteria or viruses get into the body, the immune system tries to find them and get rid of them before they can reproduce. Lymph nodes contain billions of white blood cells that multiply rapidly to fight invading germs.

Your lymphatic system would have too much to do if your body did not also have other defenses against germs. Those defenses include barriers, secretions, and inflammatory reactions.

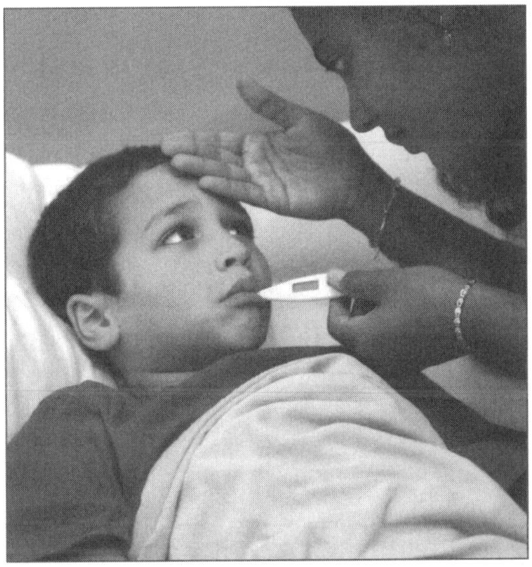

A fever is often a symptom of an infection. Viruses and bacteria have a hard time surviving at temperatures higher than body temperature. A fever is the body's way of fighting an infection. Fever increases blood flow and speeds up the body's defense mechanisms.

Have students read this section and emphasize that their focus should be on updating the *Project Board* with information about the immune system which helps answer the question for this section of the Unit, *How does your body protect against disease?*

Lead a class discussion focusing on what students think should be added to the *What are we learning?* and *What is our evidence?* columns of the *Project Board* now that they have investigated the immune system. Guide students to specifically associate evidence with their claims.

For example:

Claim: A bacteria or virus may be prevented from entering the body by defense mechanisms of the immune system.

Evidence: From our readings, we learned that germs may be stopped from entering the body by barriers such as skin and mucus that act as part of the immune system although these are part of others systems. For example, our lungs and nose (respiratory system) create mucus to trap germs entering our system and flush them out of our system before they can cause harm.

Claim: Diarrhea and vomiting are symptoms that may result from the inflammatory response of the immune system

Evidence: From our readings we learned that the immune system will often cause the body to expel (e.g. vomit, diarrhea) disease microbes entering or growing within our bodies.

Claim: One way to prevent the spread of disease caused by virus is to use a vaccine.

Evidence: We read that vaccines are created by introducing a dead or weakened virus into the body so that the immune system can prepare lymphocytes to fight against the virus.

Teacher Reflection Questions

- This section connects the knowledge of communicable diseases with the role of immune systems. What evidence do you have that students recognize these connections and are seeing what they are learning in the context of the driving question for this Unit, What happens to you when you get sick?
- How did you help students make connections between what they have learned about other body systems and the functioning of the immune system?
- The written questions can be used as a formative assessment to help you monitor student progress. Students' questions and their answers can help you identify what students are learning and what ideas they have that need further clarification. Were the assessment techniques used in this section appropriate to the type and quantity of information presented? Did students complete the assessment under conditions promoting the best possible performance?
- This section involved extensive reading. How did you provide sufficient opportunity and time for the students to work independently, in pairs and in small groups?

BACK TO THE BIG QUESTION INTRODUCTION

Learning Set 3

Back to the Big Question

◀ ***2 class periods***

A class period is considered to be one 40 to 50 minute class.

Overview

Students collaborate in small groups to create transparencies of four different body systems showing organs in their correct size and position in relation to the body. Students then overlay the transparencies, collaborate within their group to identify ways in which the body systems are interconnected and how they interact. Groups then present their findings to the class in a Communicate Your Ideas: *Solution Showcase*. Students conclude the *Learning Set* by updating the *What does it mean for the challenge or question?* and *What do we need to investigate?* columns of the *Project Board*.

Targeted Concepts, Skills, and Nature of Science	Performance Expectations
Scientists often collaborate and share their findings. Sharing findings makes new information available and helps scientists refine their ideas and build on others' ideas. When another person's or group's idea is used, credit needs to be given.	Students collaborate in small groups to create transparencies of four body systems showing organs in their correct size and position in the body. Groups collaborate to present their transparencies to the class and report their ideas about how the organ systems are interrelated during a *Solution Showcase*. Students share their ideas with the class as they consider what they think they have learned and what is their evidence when they fill out the class *Project Board*.

Targeted Concepts, Skills, and Nature of Science	Performance Expectations
Scientists must keep clear, accurate, and descriptive records of what they do so they can share their work with others, consider what they did, why they did it, and what they want to do next.	Students create transparencies of organ systems using information recorded throughout the *Learning Set* and present them to the class during a *Solution Showcase*. Students record their claims, supporting evidence and science knowledge on Create Your Explanation pages. Students add information along with supporting evidence to the *Project Board*.
Explanations are claims supported by evidence, accepted ideas, and facts.	Students complete *Create Your Explanation* pages as they develop explanations for how we get sick with communicable diseases

Materials	
1 per student	Transparency with an outline of the human body, materials for drawing on transparencies such as transparency markers or wax pencils.
1 per classroom	*Project Board*
1 per classroom	Posters of each organ systems
1 per classroom	Optional: projections of organ system images in student text

Activity Setup and Preparation

Display posters of organ systems around the room. Then, reproduce sufficient human body outline transparencies so that each student will have their own copy.

Decide on how organ systems will be assigned within each group. Each student in a group must work on a different organ system.

Homework Options

Reflection

- **Science Content:** What part of creating the transparencies so that they would overlay properly did you find most challenging? *(Students' answers will vary, but students typically report having more difficulty drawing organs to the correct scale than they do in correctly positioning the organs.)*
- **Science Process:** What challenges might a doctor face when operating on different patients that are similar to the challenges you faced in preparing your transparencies? *(The size and exact position of organs varies from person to person. The doctor's challenge is not to inadvertently cut into an organ that is in a slightly different position from what is expected. Modern imaging technology has greatly improved the ability of a doctor to see the position of organs inside the body before beginning surgery.)*

Preparation for Learning Set 4

- **Science Content:** Just as multicellular organism can change over time to adapt to changes in the environment, bacteria and viruses can also change. Suppose your body had developed antibodies to fight a specific virus, but then the virus changed. How might that affect your body's immunity to the virus? *(Since an antibody is specific to a particular virus, if the virus changes the antibody will no longer work. The body would have to create new antibodies to produce immunity to the new virus.)*
- **Science Process:** Most people use antivirus software on their computers. What part of your immune system is similar to an antivirus program? Why do antivirus programs need to be updated frequently with new virus definitions? *(Antivirus programs are similar to antibodies in that they identify and delete or inactivate a computer virus just as an antibody identifies and deletes (kills) or inactivates an actual virus. Antivirus programs need to be updated frequently because new computer viruses are always being created. Since the program identifies computer viruses by looking for information that matches a specific "virus definition," new viruses will not match old definitions and, therefore, will not be recognized.)*

NOTES

BACK TO THE BIG QUESTION IMPLEMENTATION

Learning Set 3

Back to the Big Question

How can you prevent your good friends from getting sick?

Your body is one large system made up of many smaller systems. You explored the systems within your body that make it possible to breathe; to move oxygen, nutrients, and wastes around; and to digest food. You also learned how your body systems protect you from disease. Each of the systems in your body interacts with other systems. They work together to keep your body healthy.

Beginning with cells in *Learning Set 2*, you began to see the wide variety of systems in your body. Each cell is a small but complicated system. Specialized cells perform specialized functions, and those cells cooperate to form tissues. These tissues are then grouped to form organs. Organs are grouped together to make up organ systems. Each type of cell, tissue, organ, and organ system has a distinct structure and set of functions. Only when your body systems are all working together with each other does your body function in a healthy way.

The systems you learned about, from cells to your whole body, show that the job a system can do is related to the way it is built. Lungs help you breathe because they can move in and out to allow air in and push it out. The parts of your heart, the chambers, each have a special job and are built a particular way to do their jobs.

You will finish this *Learning Set* by exploring the ways different body systems work together with each other and then revising your explanations about how we get sick with communicable diseases and your recommendations for staying healthy.

Each group will get a set of four transparencies and special pens for writing on the transparencies. Each will have the outline of a body. Each person in your group will choose a different body system and draw that body system on their transparency.

The hardest part of doing this is that when you lay all of the transparencies your group draws on top of each other, the organs will need to be in the right

△ Guide

Have students read this section and assess students' comprehension of the assigned task by asking students to summarize what it is they are expected to accomplish. Listen for the following in their responses:

- Each person in a group is to draw a different organ system on their transparency.

◀ *1 class period**

Learning Set 3

Back to the Big Question

10 min.

Students reflect on what they have learned about body systems in Learning Set 3. *They look back over the structural hierarchy of living things that they have investigated beginning with the structure of cells in* Learning Set 2 *through the organ systems of the human body they studied in* Learning Set 3. *Students conclude the Learning Set by constructing scale diagrams of organ systems on transparencies that overlay to show structural and functional relationships between them.*

*A class period is considered to be one 40 to 50 minute class.

- Organs should be the correct size and in the correct position in relation to the outline of the body.
- When the organ system transparencies are finished, they will be placed on top of one another.
- Each group will have to show their transparencies to the class and present their solution – a diagram showing all of the organs in their correct location in the body.
- Each group will present their findings in connection with the four questions listed in the *Solution Showcase* section.

Draw students' attention to the organ system posters placed around the room. You may also want to list the page numbers where diagrams of organ systems can be found in the student text.

⬡ Get Going

Assign a different organ system to each member of a group based using your selection method of choice. Then, distribute transparencies and special pens and have groups begin working on their organ system drawings.

☐ Assess

While students are working on their transparencies, walk around the room and evaluate the accuracy of their drawings. Encourage students within a group to collaborate on the sizes and placement of organs by testing the overlays frequently. If you see significant errors, encourage students to examine the posters and diagrams in the student text more carefully.

NOTES

places. This means that, as you are doing this activity, you will have to talk to each other about where the different organs go.

Make sure you include all of the organs in each body system. This might be hard, but around the room are posters showing each system, and there are pictures throughout this book.

Communicate Your Ideas

Solution Showcase

Each group will have a chance to show their transparencies to the class and present their solution. When you show your drawings, report to the class about the following four things:

- What did you find difficult about the activity?
- How did you work with your group to make the systems fit the body?
- What organs or systems were most difficult to identify, draw, or locate?
- How does your drawing help you better realize the complexity of how your body works?

Afterward, as a class, discuss the reasons this activity was so challenging. Make a list of the complexities of the human body and how it works that you find most interesting or surprising.

Explain

It is time now to explain how we get sick with communicable diseases using what you have been reading about body systems. Each group in the class will focus on a different body system and develop claims and explanations to answer some of the following questions:

- How does the body system your group is focusing on keep you healthy?
- How do people get sick through the body system your group is focusing on?

Answer whichever questions are applicable to the body system your group is focusing on. As always, begin by developing a claim to answer one of the

Communicate Your Ideas: Solution Showcase

30 min.

△ Guide

Have groups present their transparencies to the class and report their responses to the four questions of this section.(Using an overhead projector to show the transparencies would greatly enhance this part of the Solution Showcase.) Then, hold a brief class discussion of each group's solution and encourage students to listen and respond to the remarks of their peers in a positive and respectful way. Conclude the *Solution Showcase* with a class discussion of the more challenging aspects of the activity and the surprising complexity of the human body.

NOTE: This is an opportune time to conclude the first session of this activity.

Afterward, as a class, discuss the reasons this activity was so challenging. Make a list of the complexities of the human body and how it works that you find most interesting or surprising.

Explain

It is time now to explain how we get sick with communicable diseases using what you have been reading about body systems. Each group in the class will focus on a different body system and develop claims and explanations to answer some of the following questions:

- How does the body system your group is focusing on keep you healthy?
- How do people get sick through the body system your group is focusing on?

Answer whichever questions are applicable to the body system your group is focusing on. As always, begin by developing a claim to answer one of the

Explain

10 min.

Students create explanations for what happens to you when you get sick – how body systems keep you healthy and how you get sick from communicable diseases through body systems.

⬡ Get Going

Assign a specific organ system to each group. Then, have groups use the Create Your Explanation pages to explain how we get sick with communicable diseases based on what they have investigated and read about body systems. Review the parts of a good explanation with the class: the claim; the evidence; the science knowledge; and a logical statement tying the claim to the evidence. Emphasize that students should use evidence presented during all of the investigations in this Learning Set, because all investigations provided information on how we get sick with communicable diseases.

☐ Assess

Check each group's explanation and ask them what claims they are making and what evidence they are using to support it.

Some possible explanations are:

- Our group has the respiratory system. The respiratory system keeps people healthy by creating mucus so that diseases caused by bacteria and viruses cannot get into the body. Also, it is important to wash our hands and keep our hands away from our face. Washing kills the germs that can get into our bodies. Keeping hands away from our faces helps keep us healthy because it keeps some of the viruses and bacteria that could be on our hands away from the openings to our body. The nose and eyes are places where a lot of germs can get into the respiratory system. From there, they can move throughout that system and cause sickness.

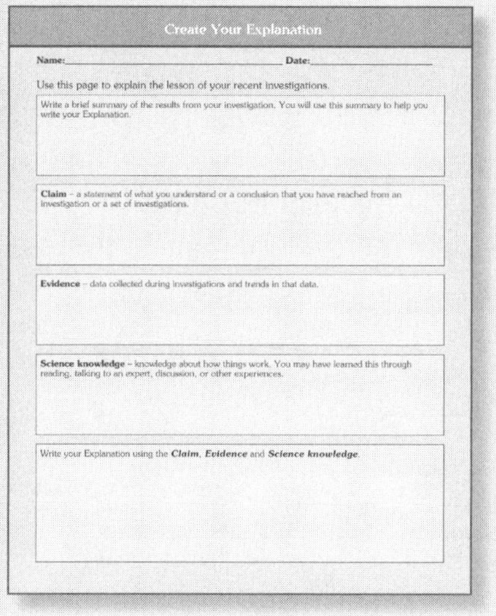

Create Your Explanation

Name:__________ Date:__________

Use this page to explain the lesson of your recent investigations.

Write a brief summary of the results from your investigation. You will use this summary to help you write your Explanation.

Claim – a statement of what you understand or a conclusion that you have reached from an investigation or a set of investigations.

Evidence – data collected during investigations and trends in that data.

Science knowledge – knowledge about how things work. You may have learned this through reading, talking to an expert, discussion, or other experiences.

Write your Explanation using the ***Claim***, ***Evidence*** and ***Science knowledge***.

questions, and record it on a *Create Your Explanation* page. Then record evidence from the simulations you have been running and science knowledge from your readings to support your claim. Develop a statement that ties together your claim, your evidence, and your science knowledge. Make sure that your statement describes enough about how your body system works so that others will be convinced of your claim. Remember that a good explanation can convince someone else that your claim is trustworthy. If your statement doesn't seem convincing, revise your claim so your evidence and your science knowledge support it.

Recommend

Using your claims and explanations about how your assigned body system keeps us healthy and how we get sick through that body system, develop recommendations for staying healthy. Focus on answering the questions below that are relevant to the body system you are assigned.

- How can I keep from getting sick through the body system I am focusing on?
- How can I prevent friends from getting sick when there is an infection in the body system I am focusing on?

Use a *Create Your Explanation* page for each recommendation. Record your recommendation as a claim, and then record evidence and science knowledge that support it. Develop a statement linking your claim to the evidence and science knowledge that support it. Remember that a good explanation can convince someone else that your recommendation is good. If your statement doesn't seem convincing, revise your recommendation and the evidence and science knowledge that support it so that you can make a more convincing statement.

- The circulatory system keeps me healthy by picking up oxygen in the lungs and transporting it to cells and by picking up carbon dioxide from cells and transporting it to the lungs where it is removed from the body when we exhale. The circulatory system also keeps me healthy by transporting food molecules absorbed from the digestive system to cells and by removing waste products from cells and transporting them to the excretory system for removal from the body. The blood also contains white blood cells that attack and kill germs.

- People get sick through their digestive system when they eat food that contains harmful bacteria. The bacteria can spread through the digestive system as food moves through it. Germs can also be absorbed into the blood through the digestive system and get carried throughout the body and cause sickness. That is why it is important to refrigerate food – bacteria do not grow as quickly in cooler temperatures. Also, it is important to cook food thoroughly before eating – the heat of cooking kills any bacteria in or on the food. Washing before eating kills germs on our hands that could be transferred to food and enter our body when we eat.

Recommend

5 min.

Students make recommendations for staying healthy.

knowledge support it.

Recommend

Using your claims and explanations about how your assigned body system keeps us healthy and how we get sick through that body system, develop recommendations for staying healthy. Focus on answering the questions below that are relevant to the body system you are assigned.

- How can I keep from getting sick through the body system I am focusing on?
- How can I prevent friends from getting sick when there is an infection in the body system I am focusing on?

Use a *Create Your Explanation* page for each recommendation. Record your recommendation as a claim, and then record evidence and science knowledge that support it. Develop a statement linking your claim to the evidence and science knowledge that support it. Remember that a good explanation can convince someone else that your recommendation is good. If your statement doesn't seem convincing, revise your recommendation and the evidence and science knowledge that support it so that you can make a more convincing statement.

△ Guide

Remind students that a recommendation is a kind of claim that suggests to someone what to do when certain kinds of situations occur. Then, have students read this section and emphasize that the recommendations they develop must answer the two bulleted questions in the instructions.

⬡ Get Going

Have students record each recommendation on a separate *Create Your Explanation* page and also record any supporting evidence or science knowledge. Next, have students write a logical statement connecting their recommendation with supporting evidence and science knowledge.

Communicate

Share your Explanations and Recommendations

Share your group's claim and explanation with the class. Then share the recommendations you have developed about staying healthy. For each claim and recommendation, tell the class what makes your claim accurate based on your evidence and science knowledge. As you listen, pay special attention to how the other groups are supporting their claims with science knowledge. Ask questions or make suggestions if you think a group's claim or recommendation is not as accurate or complete as it could be or if the group has not supported it well enough with evidence and science knowledge. Make sure you are convinced that each group's claims are accurate. Make sure you are convinced of their recommendations. If you are not convinced, share what you know about the body system they are making claims about, and try to offer a better claim, explanation, or recommendation.

Revise Your Explanations and Recommendations

As a class, work to make all of the claims and explanations as accurate and as convincing as possible. Then work to make all of the recommendations as complete and convincing as possible.

Update the *Project Board*

The final column of the *Project Board* asks you to summarize *What does it mean for the challenge or question?* The question for this Unit is *How can you prevent your good friends from getting sick?* It is time now to look at all the information you recorded on the *Project Board* and think about how what you now know helps you answer this question. What do you know, so far, that will help you answer the *Big Question?* Write your ideas in the last column. Add other questions you need to investigate. Write them in the *What do we need to investigate?* column.

Communicate: Share Your Explanations and Recommendations

10 min.

Groups share their explanations and recommendations and receive peer feedback.

△ Guide

Have students read this section and make sure that groups understand that for every claim and recommendation, they are expected to share the evidence and science knowledge that supports it. Have groups share their explanations and recommendations with the class. Emphasize to the class that as each group makes their presentation, the other groups should suggest ways in which the explanations and recommendations can be made more accurate and more convincing. Hold a class discussion of each group's explanations and recommendations and encourage students to listen and respond to the remarks of their peers in a positive and respectful way.

Revise Your Explanations and Recommendations

5 min.

Groups make revisions based on peer feedback.

you are convinced that each group's claims are accurate. Make sure you are convinced of their recommendations. If you are not convinced, share what you know about the body system they are making claims about, and try to offer a better claim, explanation, or recommendation.

Revise Your Explanations and Recommendations

As a class, work to make all of the claims and explanations as accurate and as convincing as possible. Then work to make all of the recommendations as complete and convincing as possible.

Update the *Project Board*

The final column of the *Project Board* asks you to summarize *What does it mean for the challenge or question?* The question for this Unit is *How can*

△ Guide

After groups have presented their explanations and recommendations to the class and received feedback from their peers, guide the class in working collaboratively to make all of the recommendations as complete and convincing as possible. Make a class list of the edited explanations and recommendations that the class can refer to when updating the *Project Board*.

Update the *Project Board*

10 min.

As a class, work to make all of the claims and explanations as accurate and as convincing as possible. Then work to make all of the recommendations as complete and convincing as possible.

Update the *Project Board*

The final column of the *Project Board* asks you to summarize *What does it mean for the challenge or question?* The question for this Unit is *How can you prevent your good friends from getting sick?* It is time now to look at all the information you recorded on the *Project Board* and think about how what you now know helps you answer this question. What do you know, so far, that will help you answer the *Big Question?* Write your ideas in the last column. Add other questions you need to investigate. Write them in the *What do we need to investigate?* column.

△ Guide

Have students read this section and emphasize that their focus should be on the final column, What does it mean for the challenge or question?, of the *Project Board.* Make it clear that their task is to now consider how the information they have recorded on the *Project Board* throughout this Unit helps answer the *Big Question,How can you prevent your good friends from getting sick?*

Lead a class discussion in which students share their ideas about how what they have learned helps them answer the *Big Question* and record these in the last column of the *Project Board.* Ask students if what they have learned has raised any new questions that need to be investigated. Listen to students' ideas and record them in the *What do we need to investigate?* column.

Teacher Reflection Questions

- How well do you feel students understood the concepts and skills that were presented in this *Learning Set*? What evidence do you have?
- What were the unintended and unanticipated learning outcomes of *Learning Set 3*? What factors negatively or positively affected the success of *Learning Set 3*?
- Specifically, how should what you have learned about teaching and learning as a result of this experience be used to improve your performance in other aspects of your teaching and learning?

NOTES

NOTES

MORE TO LEARN INTRODUCTION

More To Learn

Other Body Systems

◀ ***2 class periods***

A class period is considered to be one 40 to 50 minute class.

Overview

This section provides information about other important human body systems. You may want to assign portions of it as supplementary reading for homework throughout the *Learning Set*, or use it as the basis for a culminating activity at the conclusion of the *Learning Set*. You may want to consider approaching this section using a cooperative learning strategy called "jigsaw." The following describes how this strategy can be used to support student learning of the ideas and concepts presented in the *More to Learn: Other Organ Systems* section.

META NOTES

If this section is done using the jigsaw strategy, allow at least 1 class period for expert groups to read about their assigned organ system and plan how to teach it to the others in their jigsaw team, and one class period for members of the jigsaw team to teach one another about their assigned organ systems.

The Jigsaw Strategy

What is it?

The jigsaw strategy (Aronson, 1978) is designed for cooperative learning. The idea is analogous to a jigsaw puzzle in that "pieces" or topics of study are researched and learned by students across groups and then put together in the form of peer teaching within groups.

Students work in jigsaw teams of three to six to become experts on a particular topic which is based on an overall theme or unit of study. In this case, each student in a jigsaw team is assigned one of the organ systems in the *More to Learn: Other Organ Systems* section. The team members are charged with learning everything they can about their assigned organ system. The students then leave their jigsaw teams to join "expert groups" where they collaborate with all of the other students assigned that organ system in a joint research effort to become team's "expert" for that organ system. The members of the expert group then work on how they will teach the others in the jigsaw teams about their assigned organ system. Then, the jigsaw teams come back together to teach each other what they have learned. Each student listens and takes notes, and at the end of the Unit, is accountable for the information shared throughout the class.

Instructional technology can easily be incorporated into the jigsaw strategy. Research can be accomplished via the internet. Presentations can be developed with various software packages and enhanced with video camera pictures, student voices, music, and moving illustrations from other sources in to the presentations.

Note: The teacher's preparation and planning is key for the success of this project. It can be set up for a couple of days or a couple of weeks, depending on the nature of the topic, the students who will be conducting the research, and the extent of the research that is involved.

How to set it up:

1. Divide class into 4-6 member jigsaw teams and assign each team member a different organ system.
2. Members from all of the teams that were assigned the same organ system leave the group and meet together in an expert group devoted to that organ system. Expert groups are given a variety of resource materials and texts available to explore their organ system. Alternatively, a single reading from the textbook or another source could be used to complete the assignment.
3. Within the expert group, the students collaborate to prepare how they will teach the information to others.
4. Everyone in the expert groups returns to their jigsaw team to teach what they learned to the other team members.
5. Team members listen and take notes as each "expert" teaches them.

All students are assessed on the overall topics/concepts that have been taught in sections within each jigsaw group.

MORE TO LEARN IMPLEMENTATION

More to Learn

Other Body Systems

nervous system: contains the brain, spinal cord, and an enormous number of nerves that carry messages about what is happening outside the body to the brain and messages from the brain to all parts of the body.

Your body has other systems than the ones you just read about. The systems you will read about now are just as important to your health and the way your body functions. All your body systems must work together to carry out the functions of life.

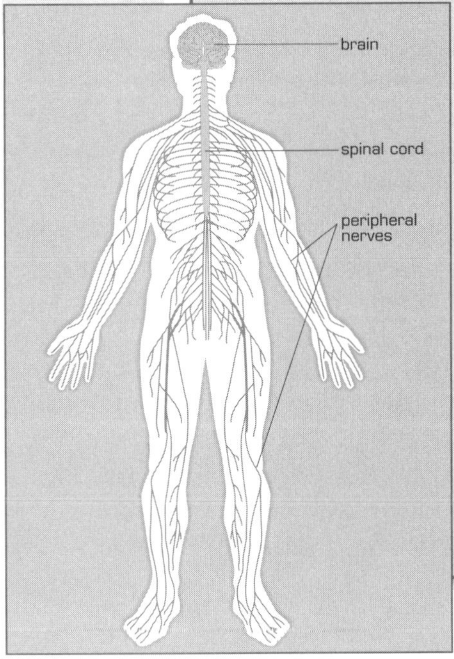

The Nervous System

Without your **nervous system**, you could not smell and taste your favorite foods or watch and listen to your favorite TV shows. Every breath you take and every beat of your heart depend on your nervous system. Your nervous system is an incredible communication system. It makes it possible for you to respond to things around you and to respond to what is taking place inside your body.

Your nervous system is made up of your brain, spinal cord, and an enormous number of nerves that form a network through your body. Nerves are very special cells. They carry messages about what is happening outside your body to your brain. They also carry messages about what is happening inside your body to your brain.

Your brain uses the information from these messages to coordinate how you will react. Then the brain sends messages using different nerve cells to tell your body parts what to do.

Project-Based Inquiry Science GF 124

*◀ 2 class periods**

More to Learn

Other Body Systems

The Nervous System

15 min.

△ Guide

Before reading, ask students to recall the shapes of the nerve cells they observed with the microscope in *Learning Set 2.* Briefly review how the cells had long, thin parts (axons) that bridged distances between nerve cells and allowed them to link together into long chains and networks.

Have students read the *More to Learn* section on the nervous system. Draw students' attention to the diagram of the nervous system and have them locate the brain, spinal cord, and peripheral nerves. Point out the bilateral

*A class period is considered to be one 40 to 50 minute class.

symmetry of the nervous system. Have students physically trace a path with their fingertip on the diagram leading from the brain to the outside of the left knee. Ask students what would happen if the spinal cord was severed just below the ribs during an accident.

Briefly discuss the nervous system's ability to sense information outside and inside the body. Ask students how they think their body obtains information about the outside world and how that information gets into their brain. Guide students to recognize that their body obtains information about the outside world through their sense organs and that the sense organs are connected to the brain via nerves.

Then, challenge students to identify situations in which they have sensed something inside their body. Student responses may include example such as: muscle pain or stiffness, joint pain, feeling coldness in the esophagus or stomach after swallowing ice water, stomachache, sore throat, gas pains, chest pain from coughing, and so on. Ask students how they think that information was obtained and how it got to their brain.

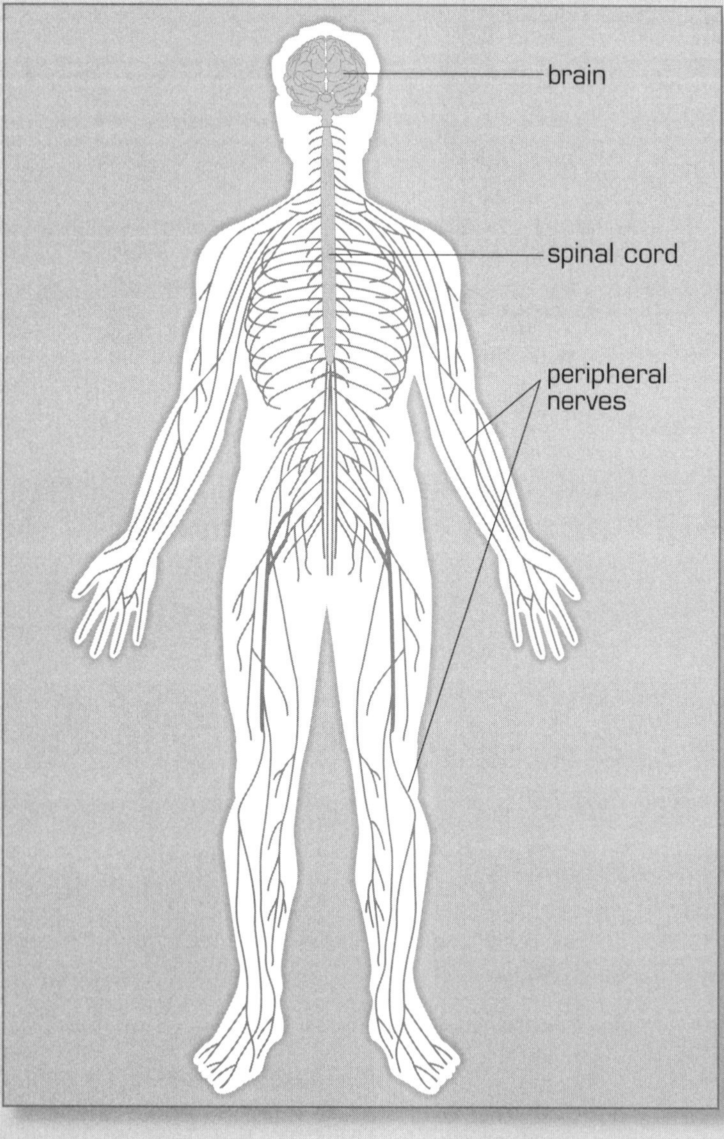

Briefly discuss meningitis and polio, focusing on the effects of these diseases on the nervous system.

- Meningitis outbreaks are frequently reported in newspapers and on television and radio. You may want to perform an internet search on "meningitis outbreak" and have students read an account of a recent outbreak of the disease.
- You may want to display images of the effects of polio on the human body. A powerful historical connection can be made to President Franklin Delano Roosevelt, who contracted the polio virus as an adult and suffered partial paralysis due to nerve damage by the virus.

There are a number of communicable diseases that affect the nervous system. You may have heard of **meningitis**. Meningitis is the inflammation of the lining surrounding the brain (called the **meninges**). Meningitis can be caused by viruses or bacteria. You also read about polio. Polio is a viral disease. The polio virus enters the nervous system and in some cases can cause paralysis. Vaccines have eliminated almost all cases of polio.

The Endocrine System

Your **endocrine system** controls the way other organs work. It is made up of **glands** (specialized tissue that secretes a substance) that produce and release **hormones**. Hormones are chemical messengers that regulate the activities of cells in an organ or group of organs. Hormones produced in organs of the endocrine system are carried through the body by the blood. This allows them to control even cells and organs that are far away from the glands that produce the hormones.

Your body could not operate without an endocrine system. Nearly everything that goes on inside your body is regulated in some way by the organs of your endocrine system. Your endocrine system regulates the amount of sugar in your blood, affects the rate at which food is converted into energy, and controls your growth rate. Without an endocrine system, your cells could not carry out the many functions they perform every day.

A major gland of the endocrine system is the **hypothalamus**, a tiny part of the brain near the middle of your head. The hypothalamus and the nervous system work together to control and regulate your body's responses to the environment, and to regulate growth, development, and reproduction. The hypothalamus is the link between the endocrine system and nervous system. It plays a major role in maintaining homeostasis (the process of keeping a constant internal environment) through the hormones and nerve messages it produces. You read earlier about homeostasis in the section on bacteria.

The endocrine system works by using a process called **feedback**. Feedback can be negative or positive. In negative feedback, when the amount of a particular hormone in the blood reaches a certain level, an endocrine organ sends signals that stop the release of that hormone. Positive feedback works in the opposite way.

meningitis: a viral or bacterial disease that results in inflammation of the lining surrounding the brain.

meninges: the linings (membranes) that cover and protect the brain and the spinal cord

endocrine system: the body system, made up of glands that produce hormones, which regulate all activities in organs and cells and help to maintain homeostasis in the body.

gland: specialized tissue that secretes a substance.

hormone: a chemical messenger that regulates the activities of cells in an organ or group of organs, or affects the activities of all the cells in the body.

hypothalamus: a major gland of the endocrine system that is part of the brain.

feedback: a process that uses nerve messages to turn on or turn off the production of hormones by the endocrine system.

The Endocrine System

15 min.

△ Guide

Have students read this section and point out that the endocrine system works together with the nervous system to carry messages throughout the body. Emphasize that the nervous systems' messages take the form of nerve impulses, while endocrine systems' messages take the form of chemicals called hormones that are carried throughout the body by the blood.

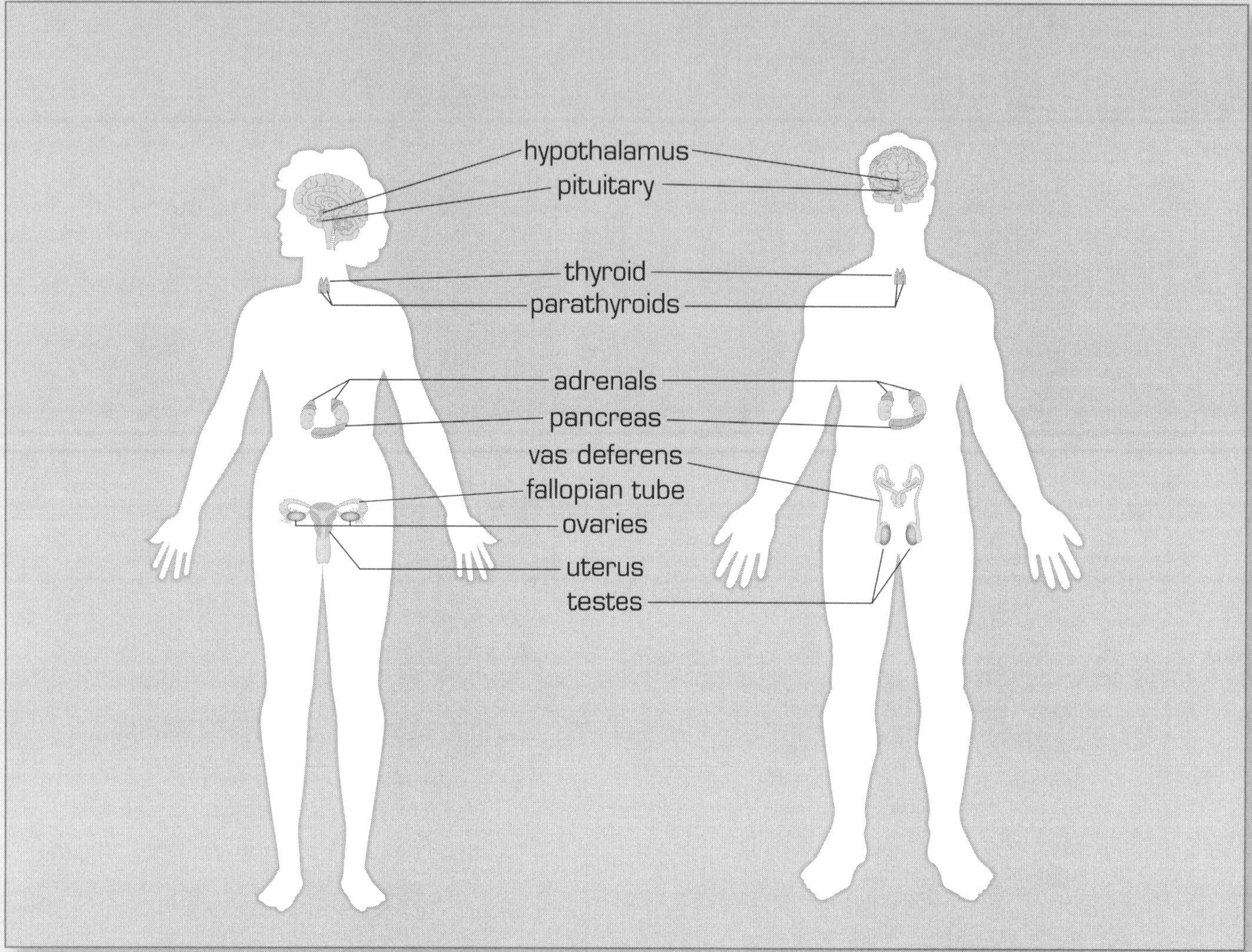

Draw students' attention to the diagram on page 126 and lead a class discussion of the glands that make up the endocrine system. Point out that glands are specialized tissue that produce and release hormones. Emphasize the key role of the hypothalamus as a link between the nervous and endocrine systems. Hold a class discussion of the feedback process and guide students to distinguish between positive and negative feedback. Guide students through specific examples of positive and negative feedback, such as those that regulate blood sugar and growth. Guide students to recognize the effects of diseases that interfere with the functioning of the endocrine system, such as the wasting syndrome associated with AIDS.

AIDS (acquired immune deficiency syndrome): a communicable disease caused by a virus that affects all the body systems.

wasting syndrome: a disease resulting from AIDS and affecting the hormones that metabolize food.

reproductive system: the body system that is specialized to produce sex cells (eggs and sperm) and sex hormones.

ovaries: female reproductive organs that produce eggs.

fallopian tube: a passageway for eggs from each ovary to the uterus.

uterus (or womb): hollow, muscular organ of the female reproductive system in which a fertilized egg develops.

fetus: a developing human from the ninth week of development until birth.

Most diseases of the endocrine system are non-communicable. However, **AIDS** (Acquired Immune Deficiency Syndrome), a communicable disease caused by a virus that affects the entire body, can also affect the endocrine system. AIDS can affect the production of hormones that regulate the metabolism of food. Without the proper hormones, food is not broken down properly, and the body does not receive the nutrition it needs. This condition can lead to **wasting syndrome**, a serious disease that is difficult to treat.

The Reproductive System

The **reproductive system** makes it possible for you to reproduce. Its organs are specialized to produce sex cells (eggs and sperm) and sex hormones. The organs in the reproductive system that produce sex hormones are also part of the endocrine system. The **ovaries** are the female reproductive organs that produce eggs. Each ovary is located near a **fallopian tube** that acts as a passageway for eggs as they travel to the **uterus** (womb). The uterus is a hollow, muscular organ where

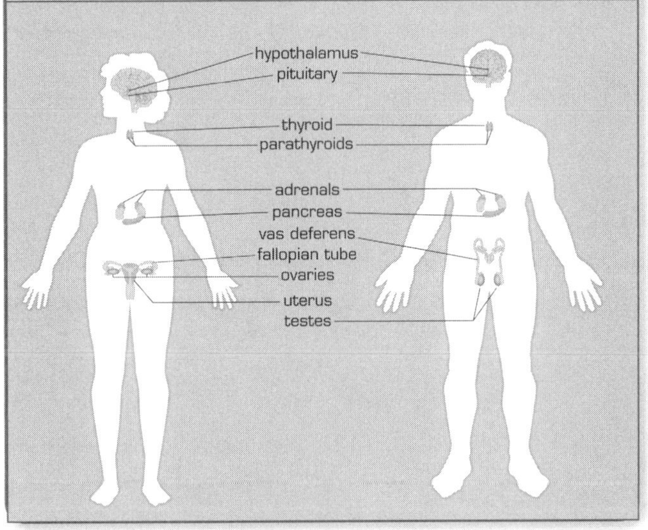

The Reproductive System

15 min.

△ Guide

Have students read this section and guide students to understand that the reproductive system and endocrine system are interrelated. Point out that, like the endocrine system, the reproductive system also contains glands. However, the organs of the reproductive system are specialized to produce cells and hormones associated with sexual reproduction.

Discuss the structure of the organs of the reproductive system and the function of each in the process of sexual reproduction with the class. Emphasize that many of the communicable diseases of the reproductive system are transmitted during sexual activity, thus the term Sexually Transmitted Diseases, or STDs. Point out that AIDS can also be spread by blood transfusions or contact with contaminated body fluids.

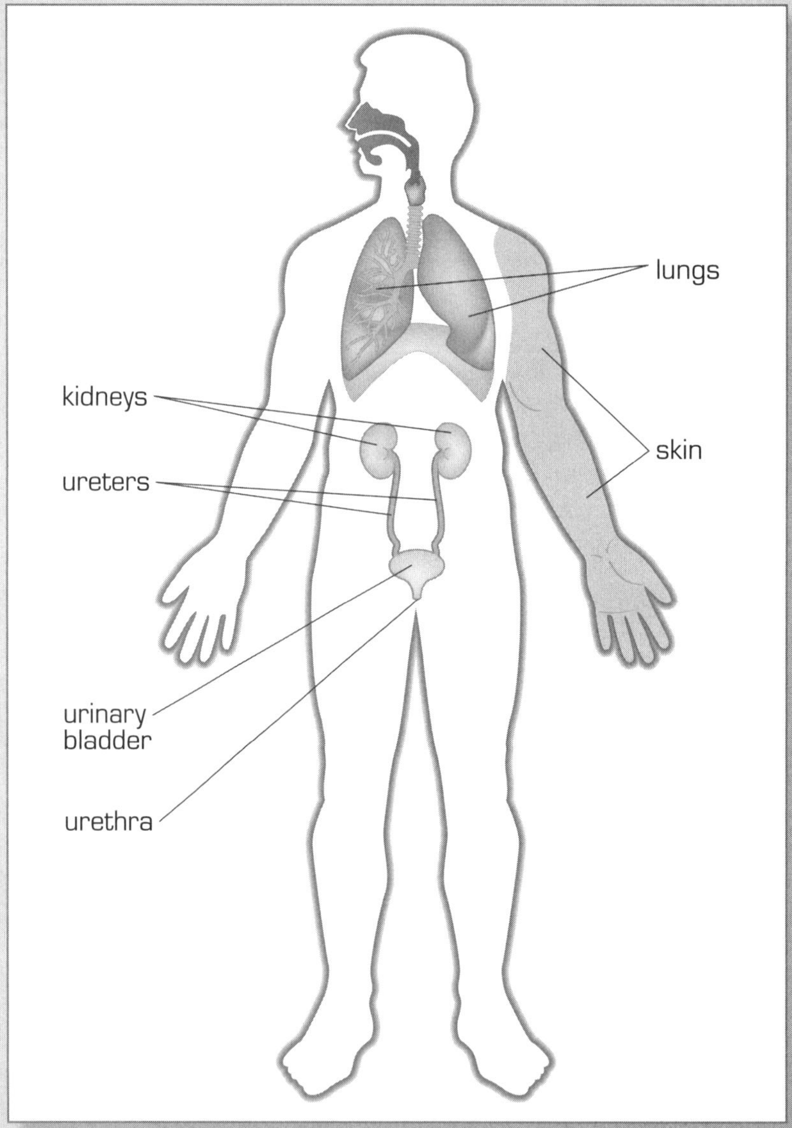

a **fetus** develops. The **testes** are the male reproductive organs that produce sperm. The sperm travel through tubes called **vas deferens** before they leave the body.

Many of the communicable diseases of the reproductive system are sexually transmitted diseases (STDs). One STD is AIDS. AIDS cannot be passed from person to person through casual contact. AIDS is sexually transmitted and also transmitted through blood transfusions or contact with other contaminated body fluids.

testes (singular, testis): male reproductive organs that produce sperm.

vas deferens: a tube that carries sperm from the testes in males.

The Excretory System

Your **excretory system** keeps the chemicals in your body in balance. The food you eat, the water you drink, and the air you breathe provide all the materials your body needs to survive. The materials are broken down inside your body by the digestive system into the proper chemicals that are sent through your blood to your cells. Your cells use these chemicals, converting them into energy and other substances used for growth and survival.

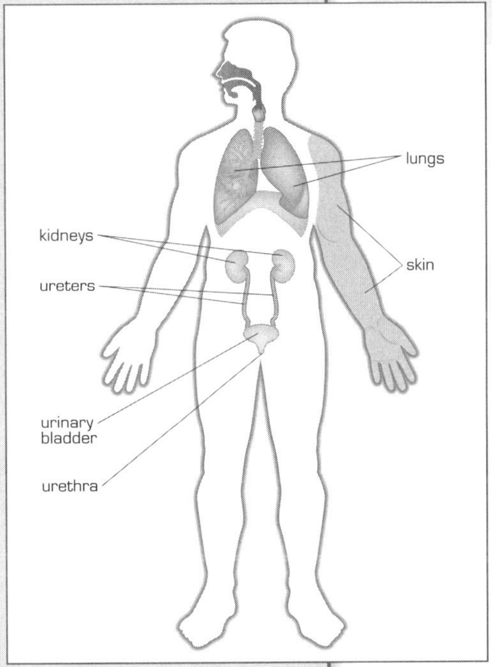

The process of changing food, water, and air into useful chemicals produces waste materials that can't be used by your body. Nitrogen compounds, salts, carbon dioxide, and excess water are all waste products. If these materials remain inside your cells or blood, the build-up of harmful wastes could result in disease and damage your body. It is the job of your excretory system to get rid of the harmful waste materials.

The Excretory System

15 min.

△ Guide

Have students read this section and discuss with the class the body processes that produce waste products. Ask students to recall what they learned about the removal of carbon dioxide from the body when they investigated the respiratory and circulatory systems. Ask students to recall what they learned about what happened to food when they investigated the digestive system. Then ask students to recall that the digestive system breaks down food and other materials into chemicals that cells can use for growth, repairs, and to obtain energy.

Then, guide students to recognize that these processes produce waste materials that can't be used by the body and that if these wastes build up inside cells or the blood, they can have harmful effects. To help students grasp the harm that accumulated body wastes might cause, ask students to imagine some of the effects they might experience if household wastes were not removed from their home.

Draw students' attention to the diagram of the excretory system and have them note that it contains organs of the respiratory system. Remind students of their investigations of the respiratory system and its role in removing the waste gas, carbon dioxide, from the body. Point out that when we exhale, our breath also contains some water. You may want to have students exhale into their cupped hands several times and then feel the moisture that accumulates. Point out that the excretory system also contains organs that are specialized to remove liquid wastes from the body such as the urinary organs and the sweat glands in the skin.

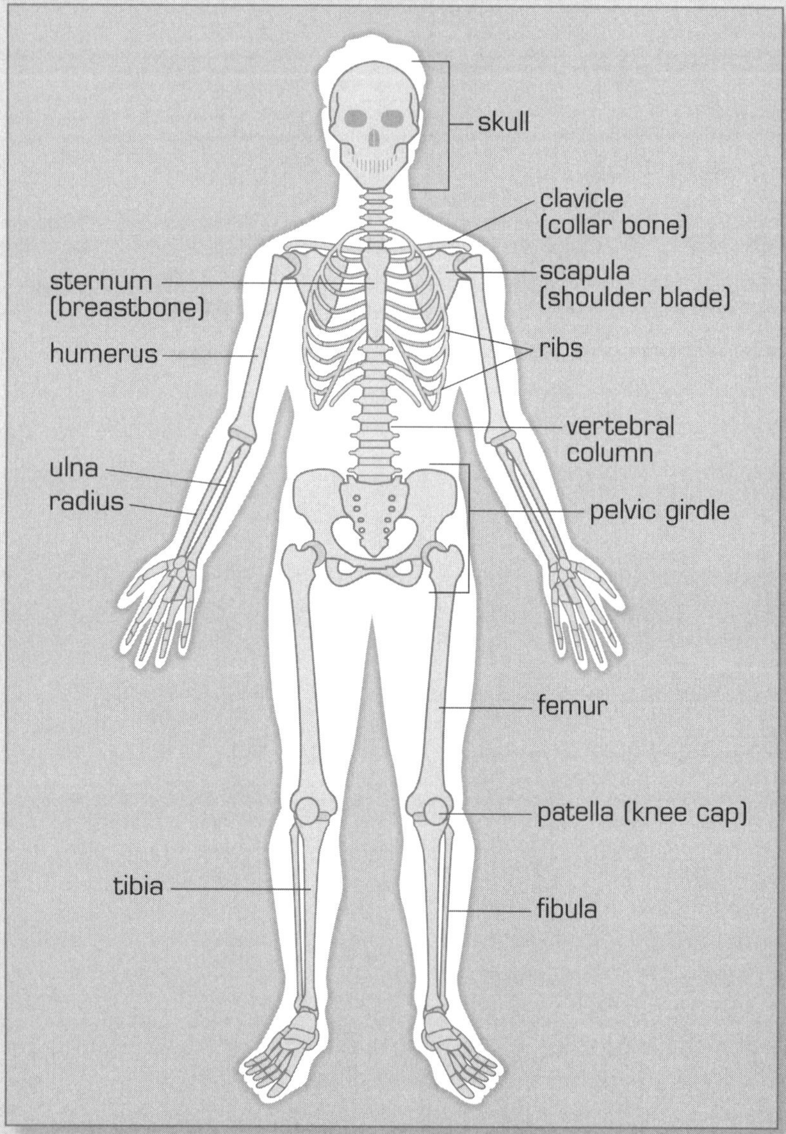

Then, hold a class discussion of the structure and function of the urinary organs focusing on how they remove excess water, salts, and nitrogenous wastes from the blood and eliminate it from the body. Similarly, discuss the structure and function of sweat glands and guide students to recognize how they remove water and excess salt from the blood and eliminate them from the body. Ask students if they have ever had a urinary tract infection, and if so, what its symptoms were. Then, briefly discuss the causes, symptoms and treatment for cystitis described in the reading.

kidney: one of two excretory organs that filter the blood and remove waste products.

urine: a watery fluid produced by the kidneys that contain wastes.

ureter: one of two tubes that carries urine from each of the kidneys to the urinary bladder.

urinary bladder: an organ that stores urine until it is eliminated from the body.

urethra: a tube through which urine flows from the body.

sweat glands: structures in the skin that remove water and salts.

cystitis: an infection of the lining of the urinary bladder, caused by bacteria.

skeletal system: the internal framework of bones that provides shape and support, allows movement, protects organs, produces blood cells, and stores minerals.

marrow: soft tissue inside the bones that makes red blood cells and stores fat.

Your excretory system is made up of several organs. **Kidneys** filter your blood and remove the waste products of metabolism. These waste products are combined with water to form **urine**, which moves through your **ureters** to your **urinary bladder** and out of your body through your **urethra**. Most of the waste is removed through your kidneys. As your kidneys filter the blood, they help maintain homeostasis by regulating the amount of water in your body.

Your lungs and skin also remove some wastes from your body. When you exhale, carbon dioxide and some water are removed from your body. **Sweat glands** in your skin also remove water and some salts. Some wastes need to be broken down before they can be removed. Your liver carries out this function.

One common infectious disease of the excretory system, **cystitis**, is caused by a bacterial infection. Cystitis affects the lining of the urinary bladder and makes elimination of urine painful. Cystitis is easily treated with antibiotics, which you read about earlier in **Learning Set 2**.

The Skeletal System

When you look up at a tall building, you may wonder how it keeps from falling down. The building has an internal structure of steel or wooden beams that act as a framework. The framework is not visible, but it is necessary to keep the building from collapsing. Your body also has an internal framework that you cannot see. This framework is your **skeletal system** and is made up of the bones in your body. Without your bones, you could not stand up, and your body would collapse.

Your skeletal system has many functions other than support. It allows you to move, protects your organs, produces blood cells, and stores minerals and other materials. A special material, called **marrow**, is found inside your bones. One type of marrow makes red blood cells that carry oxygen to the cells in the body. Without bones, not only would your body collapse, but your cells would be unable to get the oxygen they need to survive. The other type of marrow stores fat that can act as an energy reserve.

You probably already know about your bones acting as support and protection for your body, but you probably never thought about the fact that your bones are living tissue. Bones are not only strong, they contain living cells. They grow and develop, just like the other parts of your body. As you grow, new bone tissue grows. Blood vessels and nerves enter your bones through small canals to the living cells inside and allow the

The Skeletal System

15 min.

△ Guide

Have students read this section and emphasize the analogy between the internal framework that supports a building and the role of the skeletal system in the body. (You may want to use supplement your lesson with multimedia resources that support this theme.)

Discuss the role the skeletal system plays in the production of blood cells and the storage of minerals and other materials. Emphasize that bones are living tissue and elaborate on their connection to the circulatory and nervous systems. Point out that like other living tissue in the body, bone tissue is capable of growth and repair.

Briefly discuss the causes of osteomyelitis and point out the role of the circulatory system in the spread of bacteria or fungi to bone tissue. Distinguish between spread of bacteria from person to person and the spread of osteomyelitis within the body.

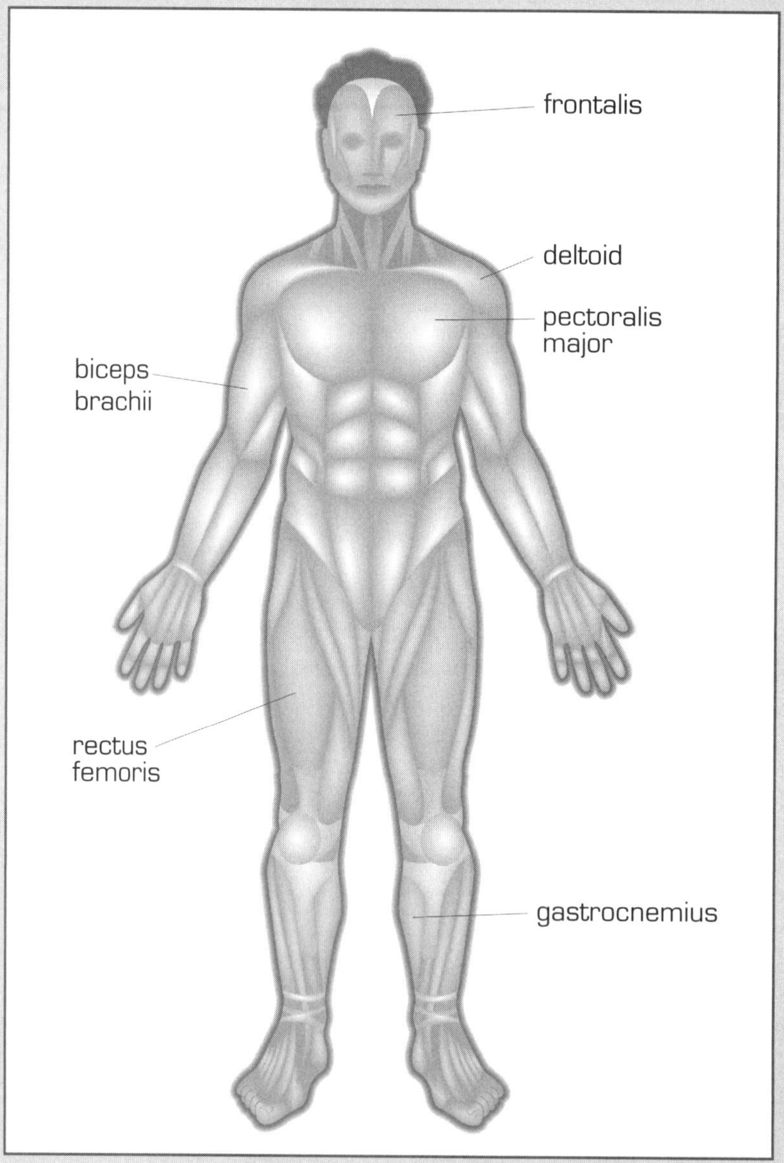

bones to grow. Because the cells of your bones receive nerve impulses and blood that contains nutrients and oxygen, bone can also heal itself when it is broken. New bone tissue grows to fill the gap between the broken ends of the bone.

Osteomyelitis is a bone disease caused by bacteria or fungi that is often in another part of the body and spreads to the bone through the blood. The bone disease is not communicable, but the bacteria that cause it can be passed from person to person

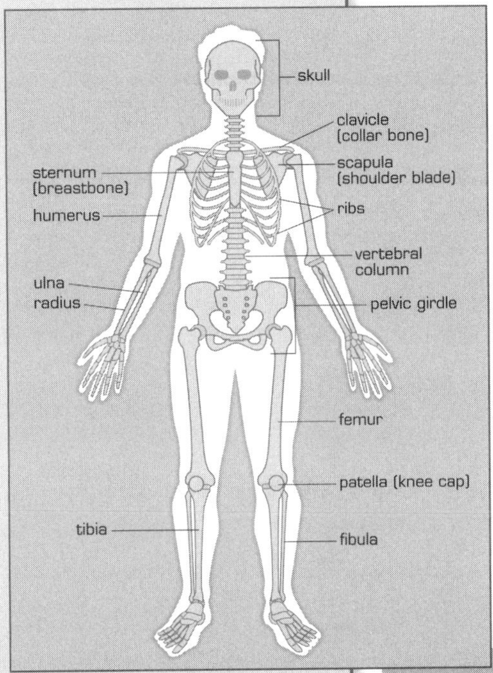

The Muscular System

You may ride a bike, play sports, or walk to school. These are all active movements. But you also breathe, swallow, smile, and sit in a chair. You digest food, and your heart beats. Without a **muscular system,** you would be unable to do any of these things.

Your muscular system is made up of the muscles in your body. Like other body tissues, muscle tissue is living tissue that receives blood and nerve impulses. The muscular system works together with the nervous system, the skeletal system, the circulatory system, and your other body systems to move the parts of your body. Some of the things you do, like riding a bike and walking, are voluntary movements. They are under your conscious control and are controlled by **voluntary muscles**. Other things, like breathing and the beating of your heart, are involuntary. **Involuntary muscles** are not under your conscious control.

osteomyelitis: an infectious bone disease caused by bacteria or fungi.

muscular system: the body system, made up of the muscles, which moves the parts of the body.

The Muscular System

15 min.

△ Guide

Have students read this section and emphasize the role of muscles in all body activities that require movement. Emphasize that muscles are the organs of this body system and that nerve impulses are what trigger muscles to contract and produce movement. Point out the connections between the muscular, circulatory, and nervous systems.

Lead a class discussion of the distinction between voluntary and involuntary muscles. Guide students to recognize that because muscles can only

voluntary muscles: muscles that can be moved by conscious control.

involuntary muscles: muscles that move without conscious control.

Muscles move by contracting; they become shorter and thicker. The muscle cells contract when they receive messages from your nervous system. Because muscle cells can only contract, not extend, they have to work in pairs. When one muscle contracts, the other muscle in the pair relaxes to its original length. The bones of the skeletal system support the muscles. With your skeletal system and your muscular system working together, you are able to move your body.

Polio, a viral disease you read about earlier, affects the muscular system as well as the nervous system. When the polio virus enters the nervous system, the nerve cells can become inflamed. This inflammation leads to destruction of the nerves that supply the muscles. As a result, the muscles become weak and poorly controlled. Finally, the muscles can become paralyzed.

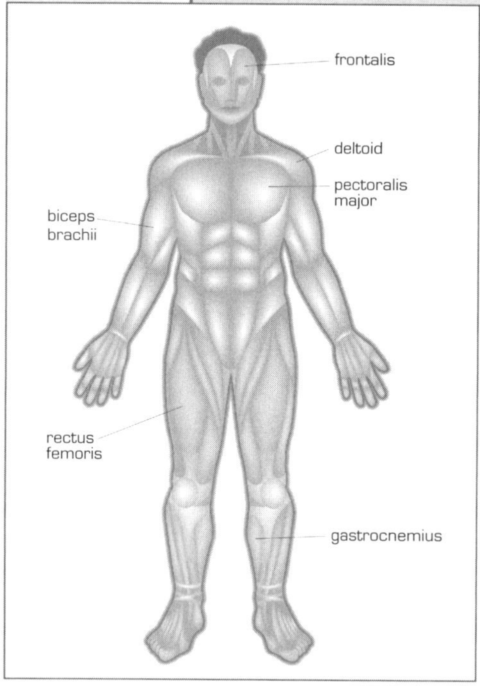

contract, they must work in opposing pairs. Point out how bones provide support and act somewhat like levers on which skeletal muscles can exert force to produce movement. Remind students of the earlier class discussion about polio and point out that although it does not attack muscle cells, by attacking the nerves that supply muscles, it causes the body to lose the ability to control muscles properly. If muscles are not used, they become weak. As the damage to nerves increases, the loss of muscular control becomes more pronounced until there is a total loss of control and the muscle is paralyzed.

LEARNING SET 4 INTRODUCTION

Learning Set 4

How Do Scientists Identify and Stop Disease Outbreaks?

◀ *5 class periods*

A class period is considered to be one 40 to 50 minute class.

Students consider the strategies epidemiologists use to identify and contain disease outbreaks. Students are then challenged to track a disease outbreak, first through a series of simple interactions and then through a more complex series of interactions. Students then read and reflect on some of the difficulties involved with trying to track diseases in today's world.

Overview

In *Learning Set 4*, students begin by reading a brief introduction that connects what they have learned about the causes, symptoms, and prevention of communicable diseases to the spread of bacteria and viruses.

In Section 4.1, students read about three recent disease outbreaks: the SARS outbreak in 2003, the *E.coli* outbreak in 2006, and the Ebola outbreak in 2003. They consider the unpredictability of outbreaks and how this may be related to mutations in bacteria and viruses. Students then think about the question for this *Learning Set: How do scientists identify and stop disease outbreaks?* and collaborate in small groups to consider what they already know about disease outbreaks and why it is important that they be investigated. After groups share their ideas about disease outbreaks, their ideas are added to the *Project Board.*

Section 4.2 is a reading that describes factors that influence a disease outbreak, such as incubation period and quarantine. After reading this section, students consider the effects of a long incubation period on the spread of disease.

Section 4.3 consists of two short investigations in which students engage in a simulation of how epidemiologists identify the source of a disease outbreak. In the first investigation, students create an interaction diagram to help them trace a disease outbreak through a series of simple interactions, and then collaborate as a class to identify the sentinel case. In the second investigation, students are challenged to trace a disease outbreak through a more complex series of interactions. Working in small groups, students collaborate to create an interaction diagram, write a story describing how the interactions might have occurred, and then identify the sentinel case. Groups then share how they identified the sentinel case and the supporting evidence for their conclusions. The section concludes with a class discussion of why it was more difficult to determine the identity of the sentinel case in the second investigation as compared to the first.

In *Section 4.4*, students consider some of the factors that contribute to an outbreak developing into a pandemic, including increased travel, higher populations, poor sanitary conditions. Students then return to the Big Question and create explanations for how diseases spread through and across populations, develop recommendations about helping populations avoid contagious diseases, and update the *Project Board.*

The Unit concludes with a *Solution Showcase* in which each group focuses on a particular communicable disease and develops recommendations about how to avoid spreading that disease to others. Students then update the *Project Board* and the class collaborates to develop a set of recommendations for making their school a healthier place. The class recommendations are then made into persuasive posters that can be displayed throughout the school.

Targeted Concepts, Skills, and Nature of Science	Section
Scientists often collaborate and share their findings. Sharing findings makes new information available and helps scientists refine their ideas and build on others' ideas. When another person's or group's idea is used, credit needs to be given.	All sections
Scientists must keep clear, accurate, and descriptive records of what they do so they can share their work with others; consider what they did, why they did it, and what they want to do next.	All sections
Explanations are claims supported by evidence, accepted ideas, and facts.	BBQ
Models are a representation of something in the world. Simulations use a model to imitate, or act out, real-life situations.	4.3
Disease outbreaks are occurrences of more cases of a particular disease than expected in a given period of time or a given area.	4.1
The sentinel case in a disease outbreak is the first person in the population to become infected with the disease.	4.3
Mutation is a permanent change in genetic material. Mutation may result in changes in the traits of a germ that render existing antibodies ineffective.	4.1

Targeted Concepts, Skills, and Nature of Science	Section
The incubation period of a disease is the time period between initial infection and the onset of symptoms. During this time, a person is contagious but may not realize that they have the disease.	4.2
Quarantine is the isolation of infected people to prevent transmission of a disease to healthy people.	4.2
An epidemic is a rapidly spreading outbreak of an infectious disease. If an epidemic spreads throughout the world it is considered a pandemic.	4.4

NOTES

NOTES

LEARNING SET 4 IMPLEMENTATION

◀ *5 class periods**

Learning Set 4

How Do Scientists Identify and Stop Disease Outbreaks?

You read about many different diseases, the bacteria and viruses that cause them, and their symptoms. You read about how body systems work and how they work together to keep you healthy, to fight disease, and even to spread disease. You simulated how bacteria and viruses can be transmitted to others and read about how diseases can spread. You have seen many examples showing how hard it is sometimes to keep from spreading diseases.

Bacteria and viruses are responsible for most communicable diseases. It is amazing that there are so many different diseases and ways to spread them. One reason for this is that bacteria and viruses change to react to changes in the environment.

When this happens, scientists have to learn how to combat the new microbes and the diseases they cause. Several recent **outbreaks** of new viral diseases have surprised scientists. Each time, however, these new viruses are stopped from spreading further by actions taken by people to prevent their spread.

In this *Learning Set*, you will investigate how scientists identify and stop disease outbreaks.

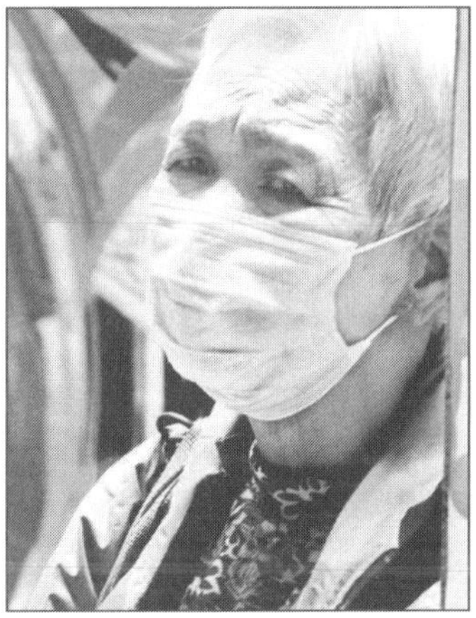

This person is protecting herself from the SARS virus (Severe Acute Respiratory Syndrome) in 2003. What is she doing to prevent getting sick or spreading the disease to someone else?

outbreak: occurrence of more cases of a particular disease than expected in a given period of time or a given area.

Learning Set 4

How Do Scientists Identify and Stop Disease Outbreaks?

20 min.

Students reflect on what they have learned about communicable diseases in previous Learning Sets and consider the wide variety of diseases, methods of transmission, and difficulties faced when trying to prevent the spread of disease. Students then consider the challenges scientists face when outbreaks of new diseases occur, which launches the driving question for Learning Set 4: How do scientists identify and stop disease outbreaks?

△ Guide

Before reading the introduction, briefly review the driving questions of the previous three *Learning Sets* and the *Big Question* for the Unit. Write these questions on the board and hold a brief discussion guiding students to recognize how the smaller questions connect in a progression that leads to being able to answer the *Big Question.* Emphasize the many different kinds of diseases they have studied, their causes, their effects on body systems, and the body's defenses against disease germs. Point out that bacteria and viruses can change, giving rise to new diseases.

*A class period is considered to be one 40 to 50 minute class.

TEACHER TALK

"Before we begin the final Learning Set, it is important for us to consider what we have done so far, why we did it and what we want to do next. What is the *Big Question* that we are trying to answer in this Unit? What are the smaller questions we investigated in each of the first three *Learning Sets*?

Notice the overall progression that our investigations have made from helping us understand how disease spreads, to understanding what causes disease and then to understanding the effects of disease on the body and the body's responses to disease.

How have the answers to each of these questions contributed to answering the *Big Question*?"

Look back over your *Communicable-Disease Information Table.* How many different diseases have we learned about? What causes most of these diseases? What are some of the ways in which bacteria and viruses can be transmitted from person to person? What are some of the problems faced when trying to prevent the spread of disease?"

Read through the introduction with the students. Consider having students read aloud. Emphasize that the smaller question they will be focusing on in this *Learning Set* is *How do scientists identify and stop disease outbreaks?* and that as they pursue answers to this question, they will investigate some of the ways in which epidemiologists fight the spread of disease.

Draw students' attention to the photograph of the woman protecting herself from SARS and ask students to answer the question in the caption *What is she doing to prevent getting sick or spreading the disease to someone else?* Encourage students to elaborate on their thinking and make connections between the use of a face mask and the respiratory system as a possible entry point for disease and a source for droplet transmission.

SECTION 4.1 INTRODUCTION

4.1 Understand the Question

Thinking about Disease Outbreaks

◀ ***1 class period***

A class period is considered to be one 40 to 50 minute class.

Overview

This section is designed to engage students in active reflection about disease outbreaks in preparation for answering the question for this *Learning Set: How do scientists identify and stop disease outbreaks?* Working collaboratively, they identify what they think they know about disease outbreaks, what they are unsure of and what needs to be investigated. Students formulate questions that might help them answer the Learning Set question and share them with their group. Groups evaluate whether their questions meet the criteria for questions: interesting, require several resources to answer, relate to the *Big Question*, and require collecting and using data. Each group then selects their two most interesting questions to share with the class and the *Project Board* is updated.

Targeted Concepts, Skills, and Nature of Science	Performance Expectations
Scientists often collaborate and share their findings. Sharing findings makes new information available and helps scientists refine their ideas and build on others' ideas. When another person's or group's idea is used, credit needs to be given.	Students should share their individual questions with their group and their group's questions with the class as they consider what they think they need to learn and what they need to investigate as they fill out the class *Project Board*.
Scientists must keep clear, accurate, and descriptive records of what they do so they can share their work with others, consider what they did, why they did it, and what they want to do next.	Students participate in a class discussion updating their Project Board, which is an organizing record of their ideas and what they are learning.
Disease outbreaks are occurrences of more cases of a particular disease than expected in a given period of time or a given area.	Students distinguish between an outbreak and the occurrence of an expected number of cases of a disease.

Targeted Concepts, Skills, and Nature of Science	Performance Expectations
Mutation is a permanent change in genetic material. Mutation may result in changes in the traits of a germ that render existing antibodies ineffective.	Students connect the unexpected appearance of new diseases with mutation of bacteria and viruses.

Materials	
1 per class	*Project Board*

Activity Setup and Preparation

This activity begins with students working individually to formulate questions and ends with a class discussion leading to updating the Project Board. Decide how you would like students to keep a record of individual work—will it be in the form of a science notebook, a learning journal, or some other format? Make sure the class Project Board is posted where everyone can see it.

Homework Options

Reflection

- Science Content: Suppose you recovered from *strep throat* (an infection of Streptococcus bacteria) after your body had made antibodies that could inactivate strep. Why would those strep antibodies not protect you if there was a mutation in the strep bacteria? *(Like a lock and key, antibodies are specified to the shape of an invading microbe. If a mutation changes the shape of the microbe, the shape of the existing antibody will no longer match it and the antibody will not be able to attach itself to the microbe and inactivate it.)*
- Science Content: How could a mutation in a disease-causing microbe result in an outbreak of the disease? *(A mutation could result in a new microbe for which no one in a population has antibodies. Without antibodies to fight the disease, any person that came into contact with the disease could become infected and spread the disease to others.)*

- Science Process: Of all the questions that other groups shared with the class, which one do you think best meets all of the criteria for a good question? *(The question should match the criteria for good questions. The question was interesting; it is something I personally want to learn more about. It is possible to find answers to this question by doing an investigation and I can think of some ways in which this might be done.)*

Preparation for 4.2

- **Science Content:** Do research into reports of recent disease outbreaks online or look at the health section of a local newspaper to see which communicable diseases are currently causes for concern. Prepare a short summary to share with the class. The summary should include the name of the disease, its symptoms, how it is being spread and what is being done to stop the outbreak. *(Answers will vary but the summary should include all of the requested information.)*

NOTES

NOTES

SECTION 4.1 IMPLEMENTATION

4.1 Understand the Question

Thinking About Disease Outbreaks

The question for this *Learning Set* is *How do scientists identify and stop disease outbreaks?* It is a good idea to think about what you know about disease outbreaks. It is also important to think about what you are unsure about and what you would like to investigate.

Sometimes, bacterial or viral diseases start, and no one predicted they would happen. When people got *E. coli* from eating contaminated spinach, nobody could have predicted it. That's because nobody knew the spinach was contaminated. The same thing happened in 2003 when the SARS (Severe Acute Respiratory Syndrome) virus made many people around the world very sick. Nobody knew where it came from. When many people get sick unexpectedly, it is called an outbreak. Outbreaks can happen for many reasons. SARS was caused by sudden change, or **mutation**, in a virus. When a virus mutates, its shape changes. Then the antibodies that used to protect against it no longer work.

mutation: a permanent change in the genetic material of a cell.

Get Started

In your group, share your ideas about what you think you know about disease outbreaks. Listen carefully to all the ideas presented. Consider what you know about

- how scientists find out about disease outbreaks, and
- why it is important to investigate outbreaks.

Think of several questions that might help you answer the question for this *Learning Set*. Develop two questions that might help you better understand disease outbreaks. Then share your two questions with your small group. Carefully consider each question and decide if it meets the criteria for a good question. With your group, refine the questions that do not meet the criteria. Choose the two most interesting questions to share with the class. Give your teacher the rest of the questions so they might be used later.

◀ *1 class period**

4.1 Understand the Question

Thinking About Disease Outbreaks

10 mins.

Students consider the question of Learning Set 4: How do scientists identify and stop disease outbreaks?

△ Guide

Read through the introduction with the students. Emphasize that the focus of this *Learning Set* will be the question *How do scientists identify and stop disease outbreaks?* Emphasize that disease outbreaks often occur without warning and elicit the reasons why the two outbreaks described in the reading were unexpected: it was not known that the spinach was contaminated with *E. coli* and SARS was the result of a mutation of a virus.

*A class period is considered to be one 40 to 50 minute class.

Ask students what they think should be done if every day more and more students in their school suddenly begin to experience symptoms such as sudden severe stomach cramps and diarrhea followed by bloody stools. Make a list of students' suggested actions and guide students to recognize the need for scientists who specialize in identifying and stopping such disease outbreaks. Then, ask students to write down what they think they know about disease outbreaks.

mutation: a permanent change in the genetic material of a cell.

reasons. SARS was caus... change, or **mu...** virus. When a virus mutates, its shape changes. Then the antibodies that used to protect against it no longer work.

Get Started

In your group, share your ideas about what you think you know about disease outbreaks. Listen carefully to all the ideas presented. Consider what you know about

- how scientists find out about disease outbreaks, and
- why it is important to investigate outbreaks.

Think of several questions that might help you answer the question for this *Learning Set*. Develop two questions that might help you better understand disease outbreaks. Then share your two questions with your small group. Carefully consider each question and decide if it meets the criteria for a good question. With your group, refine the questions that do not meet the criteria. Choose the two most interesting questions to share with the class. Give your teacher the rest of the questions so they might be used later.

Get Started

20 min.

Students consider what they think they know about disease outbreaks and share their ideas in their groups. Students then collaborate in their groups to develop two questions about disease outbreaks that will help guide their investigation into how scientists identify and stop disease outbreaks.

△ Guide

Read the Get Started section together with the class and check to be sure that students understand the task:

1. Share what they think they know about disease outbreaks with their group.
2. Develop two questions that might help them better understand disease outbreaks.
3. Share their questions with their group.
4. Collaborate within their group to evaluate and revise their questions.
5. Select two questions from the group to share with the class.

Remind students of the criteria for the questions to be written. Check that students understand each of the criteria. Have students work individually to develop their questions and then share their questions with their group. Then, have groups create a chart listing each question shared by a group member in one column and criteria listed in the other column for each. As the group evaluates each question, have them place a check in the column for each of the criteria it meets. When all questions have been evaluated, have students select the two questions that best meet all of the criteria to be presented to the class. Have groups prepare to defend their choices when they are presented to the class.

NOTES

Update the *Project Board*

20 min.

Briefly review the Project Board *with the class. Add students' ideas and the questions developed during this activity to the* Project Board.

Update the *Project Board*

Share with the class your group's ideas about disease outbreaks. You can record these ideas in the *What do we think we know?* column. Next, share with the class your group's two questions. Your teacher will add your questions to the *Project Board*. Throughout this *Learning Set*, you will work to answer some of these questions.

△ Guide and Assess

Lead a discussion in which groups share their two most interesting questions with the class. The goal of this discussion is to engage the class in evaluating whether the questions meet the stated criteria. It also provides the basis for connecting the readings and investigations in *Learning Set 4* to questions the students have raised. Challenge students to support the questions they developed by describing how they meet the criteria.

TEACHER TALK

"What are the five criteria for good questions that we have used throughout the Unit?" *(Record criteria on the board as students answer.)*

As each group shares their questions, let's check to see if they meet the five criteria listed on the board. If you think a question does not meet the criteria, share your reasoning, and then allow the group to respond to your analysis.

Which questions do you think will best guide us to better understand disease outbreaks? Why?"

Assess students' ability to generate questions that meet the criteria and focus on how scientists find out about disease outbreaks and why it is important to investigate outbreaks.

Assessment Options

Targeted Concepts, Skills, and Nature of Science	How do I know if students got it?
Scientists often collaborate and share their findings. Sharing findings makes new information available and helps scientists refine their ideas and build on others' ideas. When another person's or group's idea is used, credit needs to be given.	**ASK:** When groups shared their questions and the class discussed whether they met the criteria, which of the five criteria was most often not met or only poorly met? **LISTEN:** Students should cite examples of valid comments relating to criteria for questions that were most often at issue in the class discussion. A common weakness in students' questions is a lack of reference to the collection and use of data.

Targeted Concepts, Skills, and Nature of Science	How do I know if students got it?
Scientists must keep clear, accurate, and descriptive records of what they do so they can share their work with others, consider what they did, why they did it, and what they want to do next.	**ASK:** How might the *Communicable-Disease Information Table* be useful to an investigation into disease outbreaks ? **LISTEN:** Students should cite examples of how the *Communicable-Disease Information Table* can provide information that could help identify and stop disease outbreaks. For example: • Information about symptoms can help identify the disease that is occurring in an outbreak. • Information about how the disease is transmitted is useful when trying to determine how to stop the spread of the disease. • Information about the type of germ that causes the disease is useful in determining the type of treatment to give for the disease.
Disease outbreaks are occurrences of more cases of a particular disease than expected in a given period of time or a given area.	**ASK:** According to the Centers for Disease Control, the average annual rate of measles infection between 2000 and 2007 was 0.2 cases per 100,000 population. San Diego, California has a population of about 1.2 million. In January and February of 2008, 11 cases of measles were reported in the entire city of San Diego. Why would this be considered a disease outbreak? **LISTEN:** Students' responses should make the point that an outbreak is the occurrence of more cases of a disease than expected in a given time period. The typical rate of measles infection expected in San Diego would be about 2.4 cases per year (0.2 cases/100,000 = X/1,200,000; X=2.4.) Eleven cases is more than 4 times the number of expected cases, therefore it is considered an outbreak.

Targeted Concepts, Skills, and Nature of Science	How do I know if students got it?
Mutation is a permanent change in genetic material. Mutation may result in changes in the traits of a germ that render existing antibodies ineffective.	**ASK:** What is a mutation? **LISTEN:** Students should state that a mutation is a permanent change in genetic material and connect a change in genetic material with a change in traits. **ASK:** How could a mutation in a microbe result in a disease outbreak? **LISTEN:** Students should make connections between a change in a microbe, a loss of protection against the disease provided by antibodies and an increased number of cases of the disease.

Teacher Reflection Questions

- Early adolescents and many adults have difficulty with proportional reasoning. The Centers for Disease Control reports annual infection rates in cases per 100,000 in the population. How could you help students distinguish between an outbreak and the number of cases of a disease that can be expected based on annual infection rates.
- What could you do to help students understand why mutation of a microbe can render antibodies for that microbe ineffective?
- How did you encourage students to listen and respond to the remarks of their peers during large and small groups discussions?
- Which questions raised by students will not be addressed by the activities in this *Learning Set?* How do you plan to address these questions?

NOTES

SECTION 4.2 INTRODUCTION

4.2 Read

How Can Disease Outbreaks Be Identified?

◀ 1 class period

A class period is considered to be one 40 to 50 minute class.

Overview

Students read a description of how scientists approached the task of identifying and containing several disease outbreaks. Students consider how incubation period influences disease spread and why quarantine can be an effective strategy to stop the spread of a disease. After examining a news release about an outbreak of avian flu, students consider ways in which avian flu spreads, the effects of incubation period, and possible methods of preventing its spread.

Targeted Concepts, Skills, and Nature of Science	Performance Expectations
Scientists often collaborate and share their findings. Sharing findings makes new information available and helps scientists refine their ideas and build on others' ideas. When another person's or group's idea is used, credit needs to be given.	Students share their ideas during a class discussion of their responses to the *Stop and Think* questions.
Scientists must keep clear, accurate, and descriptive records of what they do so they can share their work with others, consider what they did, why they did it, and what they want to do next.	Students make notes of key ideas encountered during their readings.
The incubation period of a disease is the time period between initial infection and the onset of symptoms. During this time, a person is contagious but may not realize that they have the disease.	Students describe how the absence of symptoms during the incubation period may result in a person unwittingly infecting others with the disease.
Quarantine is the isolation of infected people to prevent transmission of a disease to healthy people.	Students identify quarantine as a way of keeping infected persons from infecting others.

Homework Options

Reflection

- **Science Content:** Write a statement connecting the length of incubation period with the rate at which a disease can spread. Describe the reasoning behind your statement. *(The longer the incubation period, the faster a disease can spread. The disease will spread faster if it has a long incubation period because more people will be infected by the person with the disease before that person realizes they are sick.)*
- **Science and Human Society:** "At the height of the SARS outbreak, tens of thousands of people in Ontario were quarantined. Anyone who had visited certain hospitals during specific time periods was asked to observe quarantine. 1,700 high school students were quarantined after one student at the school became ill. Many health care workers had to abide by "work quarantine," which required them to travel directly from home to work without using public transit and without stopping at any other destination. At home, health care workers had to separate themselves from family members, wear masks when in contact with others in their household, and not have visitors." *(Source: Public Health Law and Ethics: Lessons from SARS and Quarantine, Nola M. Ries, Health Law Review, Vol. 13, No. 1 (2004)* What are some of the negative effects of imposing quarantine during a disease outbreak that must be weighed against the potential benefits of the quarantine? *(Quarantine seriously disrupts lives, isolates individuals from the outside world, and jeopardizes workers' livelihood unless appropriate compensation is available.)*

Preparation for 4.3

- **Science Content:** Suppose your class was experiencing an outbreak of the flu. What information would be helpful to you if you were trying to trace the outbreak back to its source? *(Answers will vary, but may include: identity of infected persons, length of incubation period, persons with whom the infected persons came into contact, and methods of transmission of flu.)*
- **Science in Historical Perspective:** From 1957-1958, almost 70,000 people in the United States died from the "Asiatic flu," a mutation of a virus in wild ducks in China. How do you think a virus that originated in wild ducks in China might have reached the United States? *(It was most likely carried by infected travelers from one country to the next. Airline travel in particular has made it possible for infected persons to cover great distances and infect persons in distant locations during the incubation period.)*

SECTION 4.2 IMPLEMENTATION

◀ 1 class period*

4.2 Read

How Can Disease Outbreaks Be Identified?

You have read about the contaminated spinach that killed a number of people in the United States in 2006. The spinach had been sprayed with water containing *E. coli*. The water used to irrigate the spinach fields had accidentally been contaminated by animal feces from a nearby farm.

It is important to stop outbreaks of serious diseases before they cause a lot of damage. Sometimes a disease can be **contained** by identifying the source (how the disease began) and eliminating it. Finding the source of a disease can be complicated. Epidemiologists had to make many discoveries to find the source of the *E. coli* outbreak you read about. The first discovery revealed that everybody who had the disease had eaten spinach. Scientists then discovered what fields the contaminated spinach came from. Finally, it became clear how the spinach had become contaminated, through animal feces in the water used for irrigation.

contained: keeping a disease from spreading.

quarantine: keeping infected people away from healthy people.

incubation period: the amount of time someone may be infected with the disease before he or she shows symptoms. During this time, an infected person is contagious.

All communicable diseases cause discomfort. Even mild diseases can cause serious problems, even death, especially in babies, elderly people, and people already sick with something else. To contain a communicable disease, it is important to keep infected people from infecting others. People who are contagious are kept in **quarantine**; they must stay in one place and are not allowed near healthy people. It is not always possible to quarantine every contagious person, especially when a disease has a long **incubation period**, the amount of time between exposure to a disease and when symptoms begin. During this time, a person is still contagious and can spread the disease to others, even though they show no symptoms. When incubation periods are long, many people will already be infected by the time an outbreak is discovered.

Outbreaks of disease in people can also be caused by disease in animals. Avian influenza is commonly called bird flu. Outbreaks occur worldwide in poultry from time to time. Some types of this flu virus can cause a severe and very contagious illness in birds and can be fatal to the birds. Some types of the virus can also infect people. People do not get bird flu from eating chicken. To catch bird flu, a person must be in close, daily contact with infected birds. Farmers who raise chickens are usually the people who become infected with bird flu.

4.2 Read

How Can Disease Outbreaks Be Identified?

10 min.

Students consider several examples of disease outbreaks and the ways in which scientists tried to identify and contain the outbreaks.

△ Guide and Assess

Before students begin reading, ask them to recall the earlier activity in which they studied gastrointestinal illnesses. Ask them to describe why Darryl's task was simpler than the task faced by an epidemiologist studying a disease outbreak. Elicit that Darryl already knew what all of the people that got sick had in common: they all ate food from his uncle's restaurant. Let the class know that as they read through this section, they should keep in mind the question *How can disease outbreaks be identified?* and make notes in order to keep track of important ideas.

*A class period is considered to be one 40 to 50 minute class.

Chickens can be carriers of avian influenza, or bird flu.

Read the news release from the Public Health Service for Wales. As you read the news release, think about how the disease spread. Then think about how the disease can be tracked and contained.

Human cases of avian influenza A/H7N2 in the United Kingdom

Following the confirmation on 25 May 2007 by Health Authorities of the United Kingdom, of influenza A/H7N2 virus infection in four individuals (two in Wales and two in northwest England) exposed to infected poultry at small-holding [small farm] Corwen Farm, Conwy, Wales, the National Public Health Service (NPHS) for Wales is continuing with the investigation of the incident and with the implementation [putting into action] of public health measures.

As of 30 May 2007, 256 individuals, exposed either to affected premises [buildings], infected poultry, or to another individual with confirmed or presumptive [assumed to have the flu] influenza A/H7N2 virus infection, were identified in the following settings: household (39), school (14), and workplace

Read through the first two paragraphs describing the *E.coli* outbreak together with the class. Emphasize that the epidemiologists had to make a number of discoveries in order to contain the outbreak. Ask students to identify the discoveries that had to be made in order to contain the outbreak. Assess student responses for the following information: epidemiologists had to discover that eating spinach was the common link between all of the cases, they had to discover the source of the contaminated spinach and they had to discover how the spinach became contaminated in order to eliminate the source of contamination.

(203), including, at least, 148 patients and staff at two hospitals. Seventy-nine (79) of the exposed individuals are no longer considered at risk as the 7-day incubation period has elapsed. As a precautionary measure, and in accordance with UK policy, the decision was made to offer oseltamivir [a medicine to treat this type of flu] to exposed individuals.

In addition to the two Wales residents with laboratory confirmed infection, 17 of the exposed individuals, including one healthcare worker, present [show] or have presented with influenza-like illness (fever above 38°C, aches and pains, cough/head cold, sore throat or conjunctivitis [eye infection]). None of them is seriously ill and they are receiving and have received oseltamivir treatment.

Although, according to the preliminary results of the epidemiological investigation [scientific study of the disease], limited human-to-human transmission cannot be ruled out, the public health risk is considered low.

From www.euro.who.int/flu/situation/20070601_1
(World Health Organization Regional Office for Europe)

Stop and Think

1. What are the different ways avian influenza (bird flu) can be spread?
2. Suppose that a person exposed to the bird flu does not develop any symptoms for six days after exposure. Describe why you cannot say for sure that the individual does not have the bird flu.
3. What are ways you can think of to keep from spreading bird flu?

What's the Point?

To track disease outbreaks, scientists need to identify the source of the disease or how it began. They must also keep track of who was exposed to the disease and who developed symptoms. This is particularly difficult if the disease has a long incubation period. To keep the disease contained, it is important to keep infected persons from infecting others.

Have students read the paragraph dealing with quarantine and incubation period individually. After students have finished reading, hold a brief class discussion of the relationship between incubation period and the spread of disease. Emphasize that during the incubation period, the infected person does not know that they are sick. Ask students how this could favor the rapid spread of a disease. You may also want to discuss some of the societal issues involved with imposing quarantine. Point out that in order for a quarantine to be effective, infected people have to be isolated before they exhibit symptoms of the disease. Ask students how they might react if they were forced to be isolated even though they did not feel sick and it hadn't been proven that they actually had a disease. Make a public list of the positive and negative aspects of imposing quarantine.

After students have read the introductory paragraph about avian influenza (bird flu), make clear that as they read the news release they should be looking for information about how the disease was spread and how it can be tracked and contained. Have students read the news release together in their groups. After groups have read the news release, have a discussion to develop a list of the facts of the outbreak. Ask students what additional information about the outbreak they might want to know if they were asked to contain the outbreak.

Stop and Think

15 min.

Stop and Think

1. What are the different ways avian influenza (bird flu) can be spread?
2. Suppose that a person exposed to the bird flu does not develop any symptoms for six days after exposure. Describe why you cannot say for sure that the individual does not have the bird flu.
3. What are ways you can think of to keep from spreading bird flu?

What's the Point?

To track disease outbreaks, scientists need to identify the source of the disease or how it began. They must also keep track of who was exposed

META NOTES

An excellent case that deals with the issues surrounding disease spread by an infected person (carrier) and quarantine, is the story of the complex case of Mary Mallon, aka "Typhoid Mary," who displayed no disease symptoms. Consider doing further research and presenting the class with details of Mary Mallon from either an internet search or a documentary video.

Get Going

Have students write down their answers to the *Stop and Think* questions. After students have finished answering the questions, hold a brief class discussion in which students share their responses.

Assess

Assess whether students' responses to the *Stop and Think* questions contain the following information:

1. Avian flu can be spread by close, daily contact with infected birds, and by contact with an infected person.
2. During the incubation period, the person does not have any symptoms, meaning that their body functions are all normal. If the person displays no symptoms, there are no observable differences between an infected person and an uninfected person.
3. Destroy infected birds and quarantine individuals that might have been exposed to the disease for at least the 7-day incubation period.

Assessment Options

Targeted Concepts, Skills, and Nature of Science	How do I know if students got it?
Scientists often collaborate and share their findings. Sharing findings makes new information available and helps scientists refine their ideas and build on others' ideas. When another person's or group's idea is used, credit needs to be given.	**ASK:** When the class discussed the answers to the *Stop and Think* questions, what answers given by other students had not occurred to you? **LISTEN:** Students' responses should refer to ideas that came up during the discussion.
Scientists must keep clear, accurate, and descriptive records of what they do so they can share their work with others consider what they did, why they did it, and what they want to do next.	**ASK:** Why is it important for an epidemiologist to keep detailed records when tracing the source of a disease outbreak? **LISTEN:** Students' responses should point out that in order to trace the disease to its source, epidemiologists need to keep a record of information such as: who got sick, when they got sick, who and/or what they came into contact with and when the contact occurred. Recording the information makes it possible for them to share their work and analysis with others, and identify ideas that need further investigation.
The incubation period of a disease is the time period between initial infection and the onset of symptoms. During this time, a person is contagious but may not realize that they have the disease.	**ASK:** Which disease is more likely to spread rapidly, one with a 3-day incubation period or one with an 8-12 day incubation period? Why? **LISTEN:** Students should identify the disease with the longer incubation period as more likely to spread rapidly. The longer the incubation period, the greater the number of people the infected person will come into contact with and infect before realizing that they are sick. The more people that are infected, the more people they, in turn, can infect and the more rapidly the disease spreads.

Targeted Concepts, Skills, and Nature of Science	How do I know if students got it?
The incubation period of a disease is the time period between initial infection and the onset of symptoms. During this time, a person is contagious but may not realize that they have the disease.	**ASK:** In order to contain a disease outbreak, for how long should people that were exposed to the disease be quarantined? **LISTEN:** People should be quarantined for at least the incubation period of the disease. If they then show no symptoms, they can be released from quarantine. If they start to show symptoms, they should remain in quarantine until it has been shown that they are not contagious (do not harbor any more of the microbes that cause the disease.)
Quarantine is the isolation of infected people to prevent transmission of a disease to healthy people.	**ASK:** What kinds of diseases can be contained by imposing quarantine? **LISTEN:** Disease caused by infectious agents that can be transmitted from one person to another.

Teacher Reflection Questions

- Students' interest and involvement may lag during long readings. How did you engage students during the reading and vary the method of reading to maintain high interest and involvement?
- What did you do to encourage students to consider both sides of the issues related to imposing quarantine?

SECTION 4.3 INTRODUCTION

4.3 Investigate

Track a Disease

◀ *$1\frac{1}{2}$ class periods*

A class period is considered to be one 40 to 50 minute class.

Overview

Students engage in two investigations in which they trace a disease outbreak to the sentinel case (initial carrier). The first investigation involves tracing a disease through a series of simple interactions by creating an interaction diagram and using it to determine the sentinel case. Guided by the teacher, the class collaborates to create an interaction diagram and to trace the outbreak to the sentinel case. In the second investigation, students apply what they have learned from the first investigation to trace a disease through a more complex set of interactions. Students collaborate in groups to develop interaction diagrams and evaluate one another's reasoning as they try to identify the sentinel case. Groups then share their diagrams and their reasoning with the class and consider why identifying the sentinel case was more difficult in this second investigation.

Targeted Concepts, Skills, and Nature of Science	Performance Expectations
Scientists often collaborate and share their findings. Sharing findings makes new information available and helps scientists refine their ideas and build on others' ideas. When another person's or group's idea is used, credit needs to be given.	Students share their interaction diagrams and their ideas about the identity of the sentinel case during a class discussion of how to identify the sentinel case in a disease outbreak. Students share the reasoning behind their determination of the sentinel case and supporting evidence for that reasoning from the interaction diagrams.
Scientists must keep clear, accurate, and descriptive records of what they do so they can share their work with others, consider what they did, why they did it, and what they want to do next.	Students create interaction diagrams that correctly show the sequence in which the interactions described in the narrative occurred.

Materials	
1 per classroom	Optional: projections of the interaction diagrams for the two investigations

Activity Setup and Preparation

For some parts of the investigation, students will be working in groups. Decide how you will assign students to groups. Also, decide how you would like students to keep a record of individual and group work. You may want to make projections of the interaction diagrams to use as a focal point for class discussions.

Homework Options

Reflection

- **Science Content:** Why is it possible that not every contact with an infected person will result in the spread of a communicable disease? *(The body's defenses may prevent the microbe from infecting the body, for example: the body may already have antibodies that can inactivate the microbe or it may be trapped and destroyed by saliva, mucus, or stomach acid.)*
- **Science in Historical Perspective:** Earlier you read about the work of Edward Jenner. How could Edward Jenner's discoveries help epidemiologists contain a disease outbreak? *(Edward Jenner discovered that vaccines help the body develop antibodies to certain diseases. If uninfected people are vaccinated against a disease, they will not get sick with the disease or be able to spread the disease to others if they come into contact with it.)*

Preparation for 4.4

- **Science Content:** Why would it be helpful for epidemiologists to know if a person has antibodies against a disease they are trying to trace? *(If the person has antibodies against the disease, they cannot spread the disease if they come into contact with an infected person. This limits possible routes of transmission the disease could take and enables the epidemiologist to focus time and effort on more likely sources of the disease.)*
- **Science Process:** If an infected person forgot or was unaware of an interaction they had, how would this affect the work of the epidemiologist trying to trace the source of the disease? *(If an unknown interaction occurred, the epidemiologist would lose track of the path of the infection and would be unable to trace those infections back to their source.)*

SECTION 4.3 IMPLEMENTATION

*◄ $1\frac{1}{2}$ class period**

4.3 Investigate

Track a Disease

When an outbreak happens, epidemiologists do two things: They make recommendations about how to keep the disease from spreading, and they try to trace where the disease began. If they can do both of these things, there is a good chance they can stop the outbreak and avoid further infection. It is important to understand the causes of an outbreak in order to stop it.

Like you did in your simulation of disease spread, epidemiologists trace the spread of an infection *backward* to determine where the infection began.They use a lot of different information to do that.

In the next two short investigations, you are going to trace the path of a disease just like an epidemiologist might. You will not be able to do everything an epidemiologist does, but these investigations will help you understand the kind of work epidemiologists do and might give you some ideas about how to prevent the spread of disease in your school.

Procedure: Tracking Disease Through Simple Interactions

Follow the interactions of the people who are described below to determine who made the other people sick.

Joe had breakfast with Peter and Mike. Joe then went to class with Sally. Sally and Peter went to the movies. The next day, Joe, Sally and Peter were sick.

1. To trace the interactions, sketch a diagram that shows how the disease spread through the group. Use each person's name and connect it with an arrow to the person with whom they interacted. One set of interactions has been done to get you started. Complete the interactions connecting Joe to Sally and Sally to Peter. Circle the names of the people who got sick.
2. The first person to get sick in a population is called the initial or sentinel case. Use the diagram you sketched to decide who was the sentinel case, the person who got the others sick.
3. What information did you use to figure this out? Did you make any mistakes along the way? If you did, what caused you to change your mind?

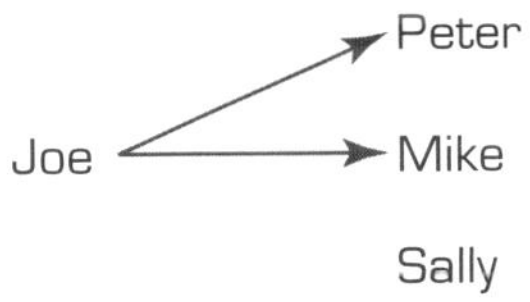

4.3 Investigate

Track a Disease

5 min.

Students investigate two disease outbreaks — one through a series of simple interactions and one through a more complex series of interactions.

△ Guide

Read the introduction together with students. Emphasize the two tasks of an epidemiologist: to make recommendations about how to keep the disease from spreading and to trace the disease to its source. Point out that this parallels the driving question for this section, *How do scientists identify and stop disease outbreaks?*

*A class period is considered to be one 40 to 50 minute class.

Remind students of the simulation of disease spread they carried out in *Learning Set 2.* Guide students to recall the logic of working backward to trace the source of the infection and the value of knowing who was not infected. Guide students to understand that the investigations they will carry out will simulate some, but not all, of the work an epidemiologist does.

the spre... *backward* to dete... e infection began. They ... ot of different information to do that.

In the next two short investigations, you are going to trace the path of a disease just like an epidemiologist might. You will not be able to do everything an epidemiologist does, but these investigations will help you understand the kind of work epidemiologists do and might give you some ideas about how to prevent the spread of disease in your school.

Procedure: Tracking Disease Through Simple Interactions

Follow the interactions of the people who are described below to determine who made the other people sick.

Joe had breakfast with Peter and Mike. Joe then went to class with Sally. Sally and Peter went to the movies. The next day, Joe, Sally and Peter were sick.

1. To trace the interactions, sketch a diagram that shows how the disease spread through the group. Use each person's name and connect it with an arrow to the person with whom they interacted. One set of interactions has been done to get you started. Complete the interactions connecting Joe to Sally and Sally to Peter. Circle the names of the people who got sick.
2. The first person to get sick in a population is called the initial or sentinel case. Use the diagram you sketched to decide who was the sentinel case, the person who got the others sick.
3. What information did you use to figure this out? Did you make any mistakes along the way? If you did, what caused you to change your mind?

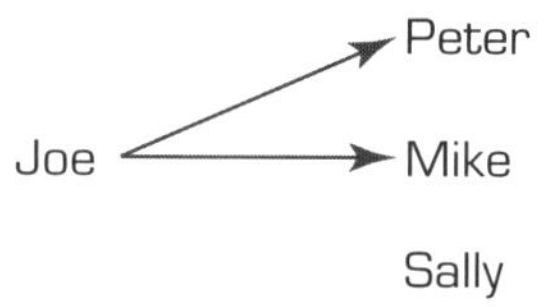

GF 137

GOOD FRIENDS and GERMS

Procedure: Tracking Disease Through Simple Interactions

10 min.

Students become epidemiologists by tracking a disease to find the sentinel case through reasoning.

△ Guide and Assess

Have students read through the entire procedure before starting work. Emphasize the importance of making detailed drawings that show the sequence in which interactions occurred. Check to be sure that students understand that arrows will connect people that interacted and that the names of people who got sick will be circled. Check that students understand what is meant by the sentinel case.

⬡ Get Going

Have students sketch their diagram of the interactions that occurred. Students should use the diagram to determine the sentinel case. Then, have students write down their answers to the questions in the procedure and prepare to share their decision about the identity of the sentinel case and the reasoning behind that decision with the class.

Communicate Your Results

Together with the class, sketch the interaction diagram and discuss how you figured out who was the sentinel case. Discuss what information you used to figure it out. Share with the class any mistakes you made along the way and what made you change your mind.

Procedure: Tracking Disease Through More Complex Interactions

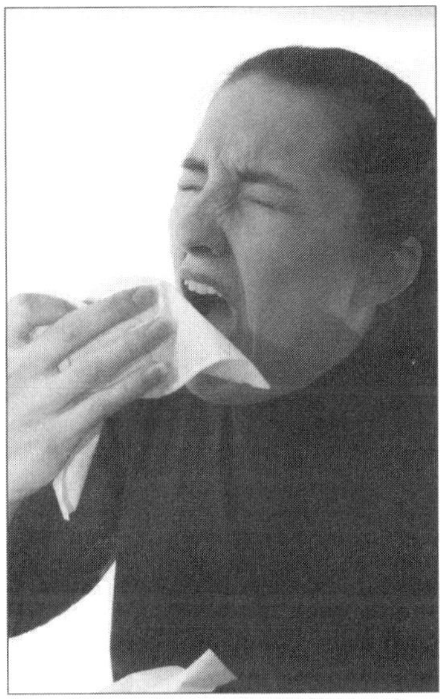

The interactions you just investigated were pretty simple. Usually, the interactions among people are more complicated. Remember the investigation you did at the beginning of the Unit when you shared liquid with others as a way of interacting with them? The interactions were hard to remember after they happened. You could remember them well only if you wrote them down as they were happening. In the real world, you interact with a lot of people. You interact directly with people. You shake hands. Others might sneeze or cough on you. You also interact indirectly with other people when you touch things that others have touched. When you are sick, you can spread germs either directly or indirectly.

The interactions among this next population are more complicated than the interactions in the first example. In this case, Kristin starts the interactions. The last interaction is from David to Kristin. The diagram of interactions is shown on the next page. Your job is to describe the interactions and to imagine how they might have happened. Kristin, Sue, and David all got the same illness within a few hours of each other. Kelly and Bill did not get sick.

Communicate Your Results

5 min.

Students share their findings about interaction sequences and use their reasoning to find the sentinel case.

△ Guide

Guide the class to create a sketch of an interaction diagram that correctly represents the sequence of interactions and discuss how it can be used to trace the disease to the sentinel case. Have the class share their ideas about the identity of the sentinel case and evaluate one another's statements to determine whether they are supported by the information in the interaction diagram.

- Joe, Sally, and Peter are sick.
- Since Joe and Peter had breakfast with Mike, and Mike didn't get sick, Joe, Peter and Mike were not infected at breakfast time.
- Therefore, Sally is the sentinel case—she infected Joe in class and Peter at the movies.

Procedure: Tracking Disease Through More Complex Interactions

15 min.

igured out who was the s... e. Discuss what ... sed to figure it out. Share with the class any mistakes you made along the way and what made you change your mind.

Procedure: Tracking Disease Through More Complex Interactions

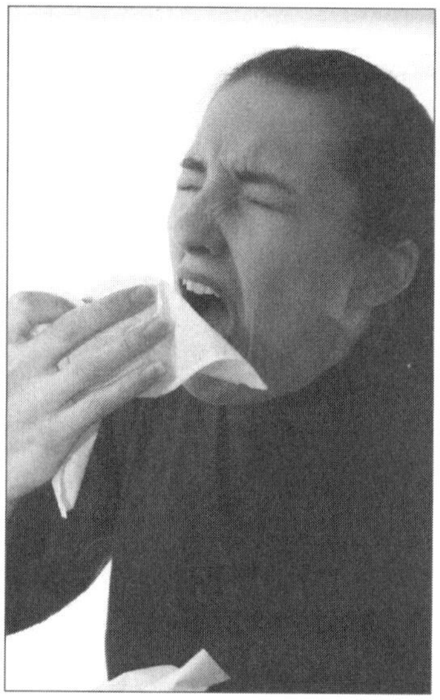

The interactions you just investigated were pretty simple. Usually, the interactions among people are more complicated. Remember the investigation you did at the beginning of the Unit when you shared liquid with others as a way of interacting with them? The interactions were hard to remember after they happened. You could remember them well only if you wrote them down as they were happening. In the real world, you interact with a lot of people. You interact directly with people. You shake hands. Others might sneeze or cough on you. You also interact indirectly with other people when you touch things that others have touched. When you are sick, you can spread germs either directly or indirectly.

The interactions among this next population are more complicated than the interactions in the first example. In this case, Kristin starts the interactions. The last interaction is from David to Kristin. The diagram of interactions is shown on the next page. Your job is to describe the interactions and to imagine how they might have happened. Kristin, Sue, and David all got the same illness within a few hours of each other. Kelly and Bill did not get sick.

GF 138

Project-Based Inquiry Science

△ Guide

Read through the first paragraph together with students, have students read it aloud, or have them read it independently. Guide students to recognize the challenges associated with keeping track of everyday interactions and the possibility of spreading disease both directly and indirectly through such interactions.

1. Write a set of sentences that describes all the interactions shown in the diagram. Use sentences from the exercise above to help you. Write your sentences as a simple story about everyday activities. Imagine what the people were doing and why they were interacting with each other.

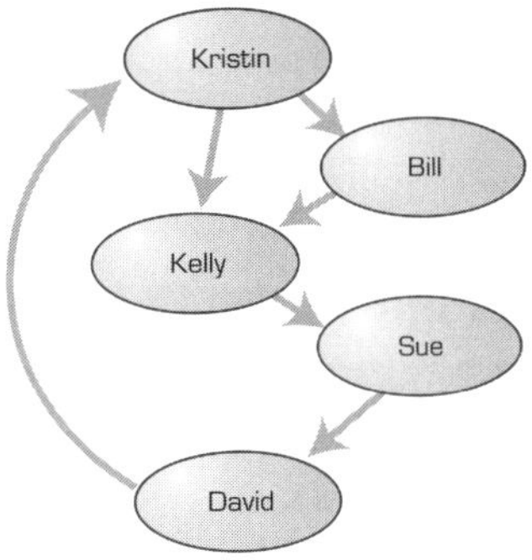

2. Now, try to figure out who the sentinel case is in this situation. Notice who got sick and who did not get sick. Make sure your answer can explain both.
   - Who was the sentinel case?
   - How do you know this?
   - Why do you suppose that Kelly or Bill did not get sick?

Communicate Your Results

This problem was more difficult to solve, and your group may have disagreed about the correct answer. Share the different answers, compare all the answers to the evidence provided, and figure out who was the sentinel case.

- Why is this the best answer? What makes it better than other answers your classmates came up with?
- What made this investigation harder than the first one?

What's the Point?

When an outbreak of a disease occurs, epidemiologists need to trace where the disease began and make recommendations to stop the spread of disease. Tracing where a disease began can be very tricky, especially when a disease has a long incubation period. Tracking where a disease came from requires tracing the spread of the disease backwards from those who are sick to identify what caused the disease. Epidemiologists must explain why some people got sick and why others did not.

Have students read through the next description of the interactions in the next paragraph. Check to make sure that students understand that their task is to write a description of how the interactions shown on the diagram might have occurred and then to trace the illness to the sentinel case. Remind students of the importance of knowing those who did not get the illness when working out the sequence in which infection occurred.

Get Going

Have students examine the interaction diagram and write their descriptions of how they imagine the interactions might have happened. Have students work on tracing the illness backward to the sentinel case and ask them to write down their answers to the questions in the procedure.

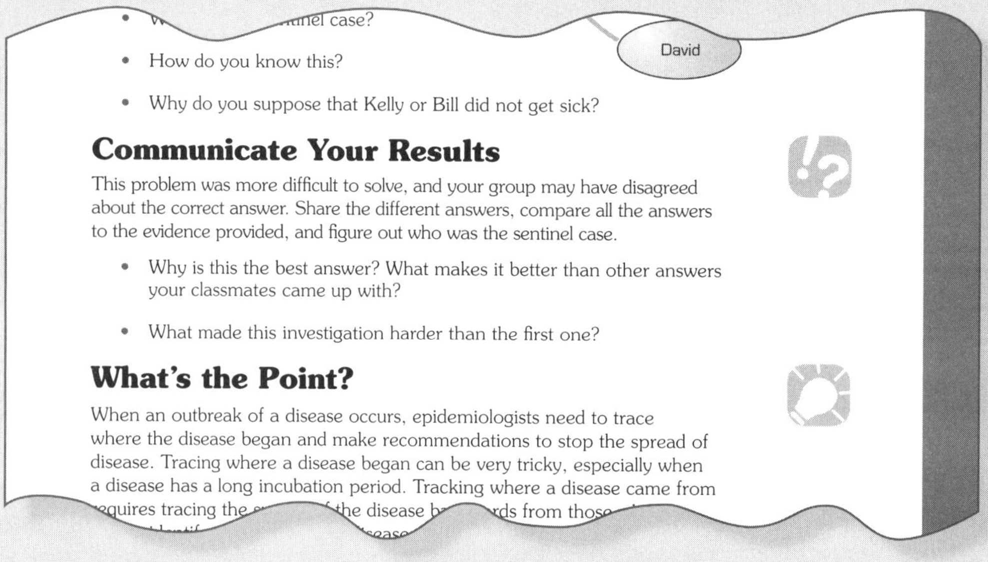

- How do you know this?
- Why do you suppose that Kelly or Bill did not get sick?

Communicate Your Results

This problem was more difficult to solve, and your group may have disagreed about the correct answer. Share the different answers, compare all the answers to the evidence provided, and figure out who was the sentinel case.

- Why is this the best answer? What makes it better than other answers your classmates came up with?
- What made this investigation harder than the first one?

What's the Point?

When an outbreak of a disease occurs, epidemiologists need to trace where the disease began and make recommendations to stop the spread of disease. Tracing where a disease began can be very tricky, especially when a disease has a long incubation period. Tracking where a disease came from

Communicate Your Results

15 min.

Students share their experiences with the previous interaction activity, then they share their conclusions of the sentinel case.

Guide

Have a few students briefly share their stories about how the interactions in the diagram might have happened. You may want to have the class vote on most creative story or funniest story.

Hold a class discussion in which the groups share who they decided was the sentinel case and the reasoning that led them to that decision. Guide students in comparing the answers and evidence provided. After groups have shared their ideas, have them write down their answers to the questions. Lead a class discussion in which students describe why they think a particular answer is the best and what made this investigation more difficult than the first one. Make a class list of students' ideas and supporting evidence.

- Since Bill and Kelly did not get the illness, they could not have been infected during their interaction with Kristin.
- Therefore, Kristen was uninfected at the time of her interactions with Kelly and Bill and is not the sentinel case.
- The only other interaction Kristen had was with David, therefore Kristen was infected by David.

- Since Sue interacted with Kelly and Kelly did not get sick, Sue was not infected at that time of that interaction.
- The only other person Sue interacted with was David, so David must have infected Sue.
- Therefore, the sentinel case must have been David.

Assessment Options

Targeted Concepts, Skills, and Nature of Science	How do I know if students got it?
Scientists often collaborate and share their findings. Sharing findings makes new information available and helps scientists refine their ideas and build on others' ideas. When another person's or group's idea is used, credit needs to be given.	**ASK:** When the class discussed the reasoning behind their decision about the sentinel case, what ideas suggested by other students prompted you to change your mind about your own ideas? **LISTEN:** Students' responses should refer to ideas that came up during the discussion that caused them to recognize faulty logic in their own ideas.
Scientists must keep clear, accurate, and descriptive records of what they do so they can share their work with others, consider what they did, why they did it, and what they want to do next.	**ASK:** Why was it more difficult to trace the disease in the second investigation? What information was most helpful in tracing the illness to the sentinel case? **LISTEN:** Students' responses should refer to the greater number of people and interactions in the second investigation. Students recognize that the identity of those who did not get sick after an interaction is a key factor in determining the identity of the sentinel case.

Teacher Reflection Questions

- In these investigations, students had to apply rules of logic when tracing the illness to the sentinel case. What logical inconsistencies did you observe when students shared their reasoning with the class? How might you address these weaknesses prior to carrying out this investigation with other classes?
- Writing logical expressions is a topic in many math curricula, how might you work collaboratively with your class' mathematics teacher to address weaknesses in logical thinking that you observe in your class?
- In the second investigation, students engaged in creative writing when they wrote descriptions of how they imagined the interactions in the diagram might have occurred. How might you work collaboratively with your class' English Language Arts teacher to prepare students for this task?

NOTES

SECTION 4.4 INTRODUCTION

4.4 Read

What Makes Disease Tracking Hard?

◀ *1 class period*

A class period is considered to be one 40 to 50 minute class.

Overview

Students read about several disease outbreaks that have become pandemics and how they have influenced history and the outcomes of wars. Students consider the roles of long incubation period and worldwide travel in causing a disease outbreak to become a pandemic. They also consider ways in which governments work together with scientists to identify disease outbreaks early enough for them to be contained.

Targeted Concepts, Skills, and Nature of Science	Performance Expectations
Scientists often collaborate and share their findings. Sharing findings makes new information available and helps scientists refine their ideas and build on others' ideas. When another person's or group's idea is used, credit needs to be given.	Students share their ideas during a class discussion of their responses to the *Stop and Think* questions.
Scientists must keep clear, accurate, and descriptive records of what they do so they can share their work with others, consider what they did, why they did it, and what they want to do next.	Students make notes of key ideas encountered during their readings.
An epidemic is a rapidly spreading outbreak of an infectious disease. If an epidemic spreads throughout the world it is considered a pandemic.	Students distinguish between an epidemic and a pandemic, and describe factors that contribute to an epidemic becoming a pandemic.

Materials	
Optional	projection of a timeline showing major pandemics.

Activity Setup and Preparation

For some parts of the investigation, students will be working in groups. Decide how you will assign students to groups. Decide how you would like students to keep a record of individual and group work.

Homework Options

Reflection

- **Science Content:** What allowed the influenza virus of 1918 to develop into a pandemic? *(World War I was taking place and soldiers traveling throughout the world carried the virus to places far from where the first infection occurred.)*
- **Science in Historical Perspective:** What role did a pandemic play in the defeat of the Mexican Aztecs by the Spanish Conquistadors? *(The Aztec population had no immunity to the smallpox virus carried by the Conquistadors. The large numbers of Aztecs that got sick or died from smallpox enabled the much smaller force of Conquistadors to defeat them in war.)*

Preparation for BBQ

- **Science Process:** Give two examples of techniques used by epidemiologists to contain an outbreak that could also prevent the spread of a communicable disease in your school. *(Answers will vary, but may include: immunization, voluntary quarantine of sick individuals, treatment of possibly infected persons with antibiotics in the case of disease caused by bacteria.)*
- **Science in Historical Perspective:** Do research online or at the library and make a timeline of major pandemics that have occurred throughout history. *(Answers will vary, but will typically include references to plague, cholera, influenza and smallpox.)*

SECTION 4.4 IMPLEMENTATION

◀ *1 class period**

4.4 Read

What Makes Disease Tracking Hard?

When you know everything about people's interactions, you can figure out where a disease started and how to keep it from spreading. You were able to figure out the sentinel case in the previous investigation by thinking about alternatives and which one explained the situation best. But you had to think about only a few alternatives.

When the population is larger and you cannot track every interaction, or when a disease has a long incubation period, it gets a lot harder. There are many more possibilities that need to be considered.

When you also add travel into the picture, it gets even harder to track diseases and stop them. The influenza virus of 1918 spread around the world. This was during World War I, and the virus was taken to distant places by soldiers. Scientists know how it was spread around the world, but they still do not know how it got started. When the SARS outbreak happened in 2003, people flying to and from China for business and vacations spread the virus around the world.

Soldiers may have unknowingly spread diseases to different parts of the world.

4.4 Read

What Makes Disease Tracking Hard?

25 min.

Students consider the factors that enable epidemics to become pandemics, how pandemics have influenced history, and how governments and scientists work together to identify disease outbreaks and contain them before they become pandemic.

△ Guide

Before students begin reading, ask them what factors other than interaction with an infected person, they think might affect the spread of a communicable disease. Make a class list of student ideas. Elicit factors such as indirect contact with an object the infected person touched, shared food or water supply, and immunization. Inform students that as you read this section you will edit the class list, adding interactions that contribute to the spread of disease and removing interactions that may not. This reading selection is long, so you may want to vary the reading method used throughout the selection to increase student interest and interaction.

*A class period is considered to be one 40 to 50 minute class.

When an **epidemic** spreads throughout the world, it is called a **pandemic**. Pandemics have ancient roots. Several documented pandemics affected the outcome of wars. In the 16th century (1500s), the Aztec population was beaten in war by the Conquistadors of Spain. The Aztecs lost even though they had more warriors and a bigger population. Unfortunately, the Spaniards brought smallpox with them to Mexico. The Aztecs were not immune to that disease, and many Aztecs got sick or died.

epidemic: a rapidly spreading outbreak of an infectious disease.

pandemic: an epidemic that has spread throughout the world.

Diseases spread very quickly in our world today. When people lived in rural places, like farms and small towns, they did not have contact with many people. Also, people didn't travel much outside their towns. Because there were few interactions outside of one's small town, diseases did not have a chance to spread outside of communities.

As cities have grown and more people have moved to them, diseases spread more easily than they used to. There are several reasons for this. In many large cities, in the USA and around the world, sanitation is still a problem. When there is dirt, waste, or sewage in the streets, a variety of diseases can live and spread to people. Also, when there is poor sanitation water sometimes becomes contaminated with diseases that spread to people who use the water. All these things make people sick. Then because of the number of people in a city, diseases can spread quickly because there are so many interactions among people.

Diseases spread more easily in crowded areas.

During the reading, stop at appropriate places and ask students to summarize important information and ask questions to assess comprehension. Guide students to recognize connections with previous *Learning Sets* by asking questions relating to earlier investigations. For example, when students read about the Aztec population contracting smallpox from Spanish Conquistadors in the 1500s, you might remind students of the work of Edward Jenner and ask them to recall details of what they learned, such as the first smallpox vaccine ever developed, and that Jenner made his discoveries in the late 1700s in England. Point out that

Finding ways to prevent the spread of disease is important. If diseases can be stopped before many people get sick, or if diseases cannot be spread from person to person, fewer people will get sick.

Scientists continually work to find ways to minimize outbreaks and prevent the spread of disease. Governments around the world have methods for reporting the incidence of disease outbreaks. Your doctor and school are required to report outbreaks of some communicable diseases to the health department where you live. When incidents of a disease are reported to a central place, scientists can identify outbreaks at an early stage and immediately get to work warning people about what they need to do so they won't catch it. For example, bacterial meningitis is highly contagious and very dangerous. When even one case occurs at a school or college, the school contacts everybody who works or goes to school in that place and tells everybody who has come in close contact with the sick person to be checked for infection. Those who are infected immediately get antibiotics. Even those who are not infected often get antibiotics to prevent them from getting the disease. The earlier this disease is treated with antibiotics, the easier it is to cure.

Stop and Think

1. Why does a long incubation period make it difficult to track a disease?
2. Why do diseases spread more quickly today than they did many years ago?
3. Doctors are required to report cases of some highly contagious diseases. Why do you think they need to do that?

What's the Point?

You have seen that tracking the interactions among small groups of people is sometimes difficult. Imagine how hard it would be to track the interactions of people when groups get very large. It gets even harder to track interactions when groups interact with each other. When it is hard to track interactions, it is also hard to track the spread of disease and to figure out how it got started. When a disease has an incubation period of more than a day or two, it becomes even harder to figure out how a disease got started, because it is practically impossible, when a population is large, to track who interacted with whom over a long period of time. Nonetheless, to stop a disease from spreading further, it is important to find out its source. Health officials need to find out who was the first, or sentinel, case and how that person got the disease.

at the time the Aztecs fought the Conquistadors, smallpox vaccine had not yet been discovered. Other examples include connections to: infection by indirect contact (*Learning Sets 1 and 2*); infection by eating infected food; prevention of disease spread by immunization and the use of antibiotics to control infection (*Learning Set 2*); as well as body defenses against disease including antibodies (*Learning Set 3*). Hold a brief discussion in which students identify how each of the actions taken when bacterial meningitis is detected help control the spread of the disease.

Stop and Think

25 min.

Assess students' understanding of the factors that make it difficult to identify and contain disease outbreaks using the Stop and Think *questions.*

...everybody who has ... contact with ... be checked for infection. Those who are infected immediately get antibiotics. Even those who are not infected often get antibiotics to prevent them from getting the disease. The earlier this disease is treated with antibiotics, the easier it is to cure.

Stop and Think

1. Why does a long incubation period make it difficult to track a disease?
2. Why do diseases spread more quickly today than they did many years ago?
3. Doctors are required to report cases of some highly contagious diseases. Why do you think they need to do that?

What's the Point?

You have seen that tracking the interactions among small groups of people is sometimes difficult. Imagine how hard it would be to track the interactions of people when groups get very large. It gets even harder to track interactions when groups interact with each other. When it is hard to track interactions, it is also hard to track the spread of disease and to figure out how it got started. When a disease has an incubation period of more than a day or two, ...

△ Guide and Assess

When students have completed the *Stop and Think* questions, hold a brief class discussion in which students share their answers.

Assess students' responses to the *Stop and Think* questions by listening for the following information in their responses:

1. A long incubation period makes it difficult to track a disease because there is more time during which interactions can occur before the person realizes they are sick. The larger number of interactions makes tracking the disease more difficult.
2. Diseases spread more quickly today than they did many years ago because population density has increased, which increases the number of interactions between people each day. Because travel is more rapid and wide-ranging, disease is more freely carried rapidly across greater distances.
3. Highly contagious diseases can spread very quickly. Early detection limits both the number of interactions that have to be analyzed to identify the source and the number of people that need to be quarantined. It also allows doctors to check those who came into contact with the infected person for infection. Treating those who are not infected can prevent them from getting the disease and treating those who are infected early enough can affect a cure.

Assessment Options

Targeted Concepts, Skills, and Nature of Science	How do I know if students got it?
Scientists often collaborate and share their findings. Sharing findings makes new information available and helps scientists refine their ideas and build on others' ideas. When another person's or group's idea is used, credit needs to be given.	**ASK:** When the class discussed the *Stop and Think* questions, how did the responses of other students differ from yours? **LISTEN:** Students' responses should refer to ideas that came up during the discussion that were different from their own ideas.
Scientists must keep clear, accurate, and descriptive records of what they do so they can share their work with others, consider what they did, why they did it, and what they want to do next.	**ASK:**What records of information did you use when making connections with previous *Learning Sets?* **LISTEN:** Students' responses should refer to the *Project Board*, the *Communicable-Disease Information Table*, interaction diagrams, and other information recorded during investigations in previous *Learning Sets*.
An epidemic is a rapidly spreading outbreak of an infectious disease. If an epidemic spreads throughout the world it is considered a pandemic.	**ASK:** Give two examples of pandemics that have influenced history. **LISTEN:** Students cite examples from their reading, such as the influenza pandemic of 1918, the SARS outbreak in 2003, and the 16th century smallpox pandemic that affected the Aztecs.

Teacher Reflection Questions

- Interdisciplinary/cross-curricular teaching can increase students' motivation for learning and their level of engagement. In contrast to learning skills in isolation, when students participate in interdisciplinary experiences they see the value of what they are learning and become more actively engaged (Resnick, 1989). What interdisciplinary/cross-curricular learning experiences centering on epidemics/pandemics could you plan with your class' social studies teacher?
- During long reading assignments, students' interest and involvement may decrease. How are the variations in reading methods you are choosing helping you get higher levels of interest and involvement? Is there any other way this could be handled?

NOTES

BACK TO THE BIG QUESTION INTRODUCTION

Learning Set 4

Back to the Big Question

◀ ***1 class period***

A class period is considered to be one 40 to 50 minute class.

Overview

Students collaborate in small groups to create explanations of how diseases spread through and across populations. Students then develop recommendations for helping populations avoid contagious diseases. Students conclude *Learning Set 4* by updating the *What are we learning?* and *What is our evidence?* columns of the *Project Board* and then move on to complete the *What does it mean for the challenge or question?* column.

Activity Setup and Preparation

Targeted Concepts, Skills, and Nature of Science	Performance Expectations
Scientists often collaborate and share their findings. Sharing findings makes new information available and helps scientists refine their ideas and build on others' ideas. When another person's or group's idea is used, credit needs to be given.	Students share their claims, and the explanations they developed using the *Create Your Explanation* page, with the class. Students share their ideas with the class as they consider what they think they have learned and what is their evidence when they fill out the class' *Project Board*.
Scientists must keep clear, accurate, and descriptive records of what they do so they can share their work with others, consider what they did, why they did it, and what they want to do next.	Students add information along with supporting evidence to the *Project Board*. Students record their claims, supporting evidence and science knowledge on *Create Your Explanation* pages.
Explanations are claims supported by evidence, accepted ideas, and facts.	Students develop explanations for how diseases spread through a population and across populations using *Create Your Explanation* pages

Materials	
1 per classroom	*Project Board*
4-5 per student	*Create Your Explanation* page

Display the class Project Board and make copies of the Create Your Explanation pages.

Homework Options

Reflection

- **Science Process:** Research and write about a communicable disease that is currently a health issue (e.g. Avian flu H5N1, West Nile virus). Describe how it spreads, how it is treated, and how you might prevent the spread of this disease. *(Answers will vary, but should address all of the issues listed in the question.)*
- **Nature of Science:** History often depends on scientific and technological developments. How might history have changed if Edward Jenner's discovery of how to immunize people against smallpox had taken place several hundred years earlier? *(Student responses should indicate that smallpox might not have been spread to the Aztecs and they might not have been defeated by the Spanish Conquistadors.)*

Preparation for ABQ

- **Nature of Science:** Some things, such as matters of ethics or morality, cannot be tested objectively in a scientific way. Science can sometimes identify the likely consequences of particular actions but cannot be used to establish that some action is either moral or immoral. Which of your recommendations for preventing the spread of disease involve ethical or moral issues? *(Answers will vary, but may include but are not limited to: issues relating to imposing quarantine, who should be immunized when supplies of vaccine are limited, and human testing of new vaccines.)*

BACK TO THE BIG QUESTION IMPLEMENTATION

Learning Set 4

Back to the Big Question

How can you prevent your good friends from getting sick?

The *Big Question* for this Unit is *How can you prevent your good friends from getting sick?* To help you answer this question, you investigated how scientists identify and stop disease outbreaks. You read about how tracking interactions among large groups of people is very difficult. You also experienced that it can be extremely difficult to find out who was the first person to get a disease.

Explain

Earlier, your explanations focused on how diseases are passed between individual people. In this *Learning Set*, you have focused on how diseases are passed around populations and between populations. Before going back to the *Big Question*, you will use that knowledge to explain two more things.

- How do diseases spread through populations?
- How do diseases spread across populations?

As you've done previously, use a separate *Create Your Explanation* page for your answer to each question. Use your answer as your claim. Then fill in evidence and science knowledge from this *Learning Set* that supports your claim. Then develop your explanation—a logical statement that ties together your claim and your evidence and science knowledge that makes your claim convincing. Be prepared to share your claims and explanations with the class.

Recommend

As you have experienced throughout this Unit, your claims and explanations about how people get sick are useful in developing recommendations about staying well. Use the claims and explanations you have just developed to come up with recommendations about helping populations avoid contagious diseases. Use a separate *Create Your Explanation* page for each of your recommendations, record evidence and science knowledge that supports

◂ *1 class period**

Learning Set 4

Back to the Big Question

5 min.

Students reflect on what they have come to understand in Learning Set 4 *about identifying and stopping disease outbreaks and the challenges associated with tracking interactions among large groups of people. Students create explanations of how diseases spread through and among populations and develop recommendations about helping populations avoid contagious diseases.*

△ Guide

Have students read the introduction and emphasize that the purpose of this activity is to relate what they have come to understand in *Learning Set 4* with the *Big Question: How can you prevent your good friends from getting sick?*

*A class period is considered to be one 40 to 50 minute class.

Hold a brief class discussion in which students identify ideas from *Learning Set 4* which they think will help them answer the *Big Question*. Make a class list of students' ideas. Point out that their next task will be to use these ideas to develop explanations and make recommendations.

Remind students that scientists do not pay much attention to claims unless the claims are backed up with evidence that can be confirmed and with a logical argument. Therefore, they will use *Create Your Explanation* pages to record their claims, evidence, and logical argument.

Explain

15 min.

Students create explanations for how diseases spread through and across populations.

> tracking ... among large groups ... very difficult. ... experienced ... it can be extremely difficult to find out who was the first person to get a disease.
>
> **Explain**
>
> Earlier, your explanations focused on how diseases are passed between individual people. In this *Learning Set*, you have focused on how diseases are passed around populations and between populations. Before going back to the *Big Question*, you will use that knowledge to explain two more things.
>
> - How do diseases spread through populations?
> - How do diseases spread across populations?
>
> As you've done previously, use a separate *Create Your Explanation* page for your answer to each question. Use your answer as your claim. Then fill in evidence and science knowledge from this *Learning Set* that supports your claim. Then develop your explanation—a logical statement that ties together your claim and your evidence and science knowledge that makes your claim convincing. Be prepared to share your claims and explanations with the class.
>
> **Recommend**
>
> ... you have ... throughout th... your claims ... tions

⬡ Get Going

Have groups use the *Create Your Explanation* pages as a framework for explaining how diseases spread through populations and across populations. Review the parts of a good explanation with the class: the claim, the evidence, the science knowledge, and a logical statement tying the claim to the evidence. Emphasize that students should use evidence gathered during all of the investigations in this *Learning Set*, because all provided information on how diseases spread.

☐ Assess

Check each group's explanation and ask them what claims they are making and what evidence they are using to support it. Some possible explanations are:

- The length of the incubation period affects the spread of disease because during the incubation period an infected person does not display symptoms and is unaware that they are infected. The longer a person is unaware that they are infected, the greater the number of interactions that may occur before they can be treated or isolated and the greater the number of new infections.
- The larger the group, the harder it is to trace a disease because of the greater number of interactions and possible connections between people through which the disease must be traced.

your ans... question. Use your a... ...ar claim. Then ... evidence and science knowledge from this *Learning Set* that supports your claim. Then develop your explanation—a logical statement that ties together your claim and your evidence and science knowledge that makes your claim convincing. Be prepared to share your claims and explanations with the class.

Recommend

As you have experienced throughout this Unit, your claims and explanations about how people get sick are useful in developing recommendations about staying well. Use the claims and explanations you have just developed to come up with recommendations about helping populations avoid contagious diseases. Use a separate *Create Your Explanation* page for each of your recommendations, record evidence and science knowledge that supports

GF 143 GOOD FRIENDS and GERMS

Recommend

15 min.

Students develop recommendations for helping populations avoid contagious diseases.

△ Guide

Have students read this section and emphasize that their task is to use their explanations about how diseases spread through and across populations to develop recommendations about how the public can be protected from contagious diseases. Have students evaluate their earlier recommendations and decide if they still think each is a good recommendation. Encourage students to consider ways in which they could restate the recommendations to make them more specific or more complete.

each, and develop an explanation for each. Make sure your recommendations answer the following question:

- How can the public be protected from contagious diseases?

Also, look back at your old recommendations, and see if any of those should be revised. Revise those that you now know more about. Be prepared to share your new recommendations with the class.

Update the *Project Board*

The *What are we learning?* column on the *Project Board* helps you pull together everything you learned. Remember always to include your evidence. You can then use what you have learned to answer the *Big Question*.

Each investigation you do is like a piece of a puzzle. You must fit together the pieces to help you address the challenge. Your *Big Question* was *How can you prevent your good friends from getting sick?* The last column, *What does it mean for the challenge or question?* is the place to record how learning about bacteria and viruses can help you answer the *Big Question*.

How can you prevent good friends from getting sick?

What do we think we know?	What do we need to investigate?	What are we learning?	What is our evidence?	What does it mean for the challenge or question?

⬡ Get Going

Have students make any necessary edits to their earlier recommendations to make them more specific and more complete. Once students have finished editing their earlier recommendations, have them develop their recommendation about how the public can be protected from contagious diseases. For each new recommendation, have students complete a new *Create Your Explanation* page using their recommendation as the claim and adding any evidence and science knowledge that supports it. Then have them develop an explanation statement connecting their recommendation to the evidence and science knowledge. Make clear to students that they should be prepared to present their recommendation to the class and be able to support it with evidence and science knowledge.

☐ Assess

Check to see that students have evaluated and made any necessary edits to their earlier recommendations. Ask students to describe the reasoning behind any changes they did or did not make. Check whether students' new recommendations and explanations are connected to evidence and science knowledge that supports it.

Update the *Project Board*

10 min.

Students update the Project Board *with what they have learned and any supporting evidence. Students then consider how this new information helps them answer the* Big Question *for the Unit.*

△ Guide

Lead a class discussion focusing on what groups think should be added to the *What are we learning?* and *What is our evidence?* columns of the *Project Board* now that they have completed *Learning Set 2.* During the discussion, guide students to identify what they have learned about how scientists identify and stop disease outbreaks and where that information should be recorded on the *Project Board.*

Guide students by asking:

- What information do epidemiologists need to trace a disease to its source?
- What strategies do epidemiologists use to stop disease outbreaks?
- What particular things that people do help spread disease?
- What do you recommend people do to avoid transmitting contagious diseases to others?
- What do you recommend populations do to combat contagious disease?

Lead a class discussion in which students share their ideas about how what they have learned will help them answer the *Big Question, How Can I Prevent My Good Friends From Getting Sick?* Record students' ideas in the What does it mean for the challenge or question? column.

Assessment Options

Targeted Concepts, Skills, and Nature of Science	How do I know if students got it?
Scientists often collaborate and share their findings. Sharing findings makes new information available and helps scientists refine their ideas and build on others' ideas. When another person's or group's idea is used, credit needs to be given.	**ASK:** Why was it important for you to revisit and evaluate recommendations you made in previous *Learning Sets?* **LISTEN:** *Students' responses should refer to the importance of refining their ideas based on new information.*
Scientists must keep clear, accurate, and descriptive records of what they do so they can share their work with others, consider what they did, why they did it, and what they want to do next.	**ASK:** What records of information did you use to find evidence and science knowledge to support your explanations and recommendations? **LISTEN:** Students' responses should refer to the *Project Board*, the *Communicable- Disease Information Table*, and other information recorded during investigations in previous *Learning Sets*.
Explanations are claims supported by evidence, accepted ideas, and facts.	**ASK:** What is the difference between an explanation and a claim? **LISTEN:** Students state that an explanation is a logical statement that connects a claim with evidence and science knowledge that supports it.

Teacher Reflection Questions

- What difficulties did students have developing explanations and recommendations? How did you address these? What ideas do you have for the future?
- Would modeling another explanation be beneficial for the students? What ideas do you have for modeling explanations?
- What difficulties, if any, did you have managing the class when they were updating the *Project Board?* What ideas do you have for managing *Project Board* updates?

NOTES

NOTES

Answer the Big Question

How can you prevent your good friends from getting sick?

◀ *5 class periods*

A class period is considered to be one 40 to 50 minute class.

Overview

Students synthesize and apply the understandings gained in this Unit to answering the *Big Question.* Student groups are assigned a specific disease and develop a comprehensive description of that disease. Groups then recommend ways of preventing the spread of that disease, providing supporting evidence and science knowledge on *Create Your Explanation* pages. Groups then create a poster that reports on their disease and makes a class presentation of their disease description and recommendations. The class then collaborates to update the *Project Board* with information from these presentations. The Unit concludes with the class developing a set of recommendations for making their classroom and school a healthier place and creating posters that can be displayed throughout the school.

Materials	
1 per classroom	*Project Board*
4-5 per student	*Create Your Explanation* page
1 per group	Poster board
1 per group	Markers or crayons for making poster

Activity Setup and Preparation

Display the class *Project Board*, make copies of the *Create Your Explanation* pages, and assemble materials for creating posters.

Homework Options

Reflection

- **Science Process:** What changes in the behavior of your fellow students do you hope that your recommendations will bring about? *(Students answers may include: washing hands more frequently, covering their mouth when coughing or sneezing, staying home when they are sick, and getting all of their immunizations before attending school.)*
- **Science Process:** Describe how you used both the Project Board and the Communicable-Disease Information Table in preparing the presentation about your disease to the class. *(Student should recognize that their descriptions utilized information from both sources and the importance of keeping records of what they have learned and supporting evidence.)*

NOTES

ANSWER THE BIG QUESTION IMPLEMENTATION

Answer the Big Question

How Can You Prevent Your Good Friends from Getting Sick?

It is now time to pull all your learning together. First, each group will focus on one communicable disease. They will describe the disease and how to prevent that disease from being spread to other people. Then they will make recommendations about what can be done to avoid transmitting the disease to others. When finished, each group will present its findings to the class in a *Solution Showcase*. Then you will make posters with recommendations for helping to keep the school healthy.

Describe Your Disease

Your first task is to describe enough about the disease you are assigned to make recommendations about how to prevent its spread. Use the questions on the next page to guide you in putting together a presentation about your disease. It is important to answer all of the questions. Some of them you already know the answers to. Some you might have to figure out based on what you have read and done. The explanations and recommendations you have been working on throughout the Unit will also help you answer the questions. You might also need to do some additional reading.

As you decide how much information to provide in answering each question, remember that your goal is to make recommendations about how to avoid spreading the disease to others. You need to answer each of the questions well enough so your classmates will trust statements you make about how to prevent the disease from being spread.

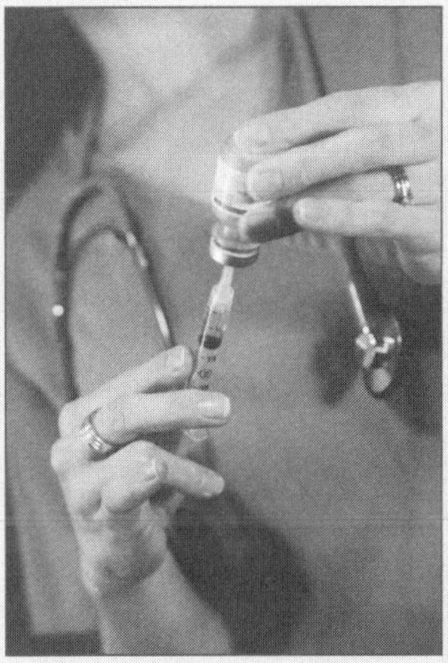

Vaccines help protect against diseases.

GF 145

GOOD FRIENDS and GERMS

△ Guide

Before students begin reading this section, ask them how they might go about accomplishing the task of describing a disease, including how it is spread and how it could be prevented. Guide students to recognize that the *Project Board* and *Communicable-Disease Information Table* are rich sources of information about diseases. Then, read through this section with students making sure they understand the overall task. Assign diseases to groups, let them know how much time they have and ask them to read through the entire section before they begin.

◂ *2 class periods**

Answer the Big Question

How can you prevent your good friends from getting sick?

5 min.

Students reflect on what they have learned throughout the Unit. Each group focuses on one communicable disease, describes the disease and how to prevent its spread. Groups then present their findings to the class in a Solution Showcase. Students collaborate in groups to make recommendations about how to avoid transmitting the disease and create posters of recommendations to keep the school healthy.

Students are given an overview of the tasks they will perform in this activity.

*A class period is considered to be one 40 to 50 minute class.

Describe Your Disease

20 min.

Students research a disease and develop detailed descriptions of the symptoms, causes, methods of transmission, epidemiology and treatment of the disease.

Symptoms of the Disease

- What are the symptoms of the disease?
- Which symptoms would help a healthcare worker diagnose the disease in a person? Which symptoms are unique to this disease?
- How sick do you get with this disease? How long are you sick? Can the disease kill a person?

Cause

- Is the infection bacterial or viral? Describe in detail what caused the disease.
- How does the bacteria or virus interact with the body to produce the disease?
- What cells or tissues does it infect, and how does it damage those cells or tissues?

Transmission

- How is this disease transmitted from person to person?
- What are the best ways to prevent this disease from spreading? How can people prevent this disease from being transmitted from one person to another?

Epidemiology

- Where did this disease first emerge?
- Is it spceific to a particual place or spread over a lager area? Describe the spread of this disease through a population or around the world.
- What human actions spread this disease?
- What human actions can prevent its spread?

Treatment

- How is this disease treated?
- Is there a vaccine available? If not, why is there no vaccine?

△ Guide

Have students read through this section and make sure that they understand what must be accomplished to describe their assigned disease. Emphasize the five major areas that their descriptions will address: Symptoms of the Disease, Cause, Transmission, Epidemiology and Treatment. Point out that the questions in each area will guide them in putting together a presentation on the disease. Make clear that they must prepare a class presentation on their disease that answers all questions well enough that their classmates

- Can antibiotics work against this disease? Explain why antibiotics can, or cannot, be used against this disease.
- How will this disease affect humans in the future? Will it be eradicated, or will it remain a disease that affects humans?

Plan Your Solution

Make Your Recommendations

Now, recommend ways of preventing the spread of your disease. Remember from other PBIS Units that the best recommendations are supported by evidence and science knowledge.

Try to make three types of recommendations:

- What do you recommend people do to avoid getting the disease?
- What do you recommend people do to avoid spreading the disease?
- What do you recommend populations do to avoid spreading the disease?

Use a *Create Your Explanation* page for each recommendation. Record your recommendation as a claim. Then provide evidence and science knowledge to support your claim. For each, try to include reasons why others should follow your recommendation. Your explanation should tie your claim, evidence, and science knowledge together and include the ways the disease is spread.

Communicate Your Solution

Solution Showcase

Make a poster that reports on your disease. Tell your class about the disease you are reporting about. Be sure to be especially clear about how the disease interacts with different body systems, how it is transmitted, the human actions that cause it to spread, and ways to prevent its spread. Your classmates will be particularly interested in understanding enough about your disease to know why the ways of preventing its spread will work. Make sure you answer the following questions in your presentation:

- What is the disease?
- What are its symptoms?
- What body systems does it affect?

will trust their recommendations. Guide students to recognize that they may have to do additional research and reading in order to be able to answer all of the questions about their disease. Assist students in identifying outside sources of information that may be useful to them.

⬡ Get Going

Have students gather information and prepare their presentations describing their assigned disease using the questions in the student text as a guide.

Assess

Groups' descriptions of the disease should address all 5 areas and answer all of the questions in each area. They should also be specific enough that a person reading the description would have confidence in recommendations made by the group.

> eradi... will it remain a dise... ects humans?
>
> **Plan Your Solution**
>
> **Make Your Recommendations**
>
> Now, recommend ways of preventing the spread of your disease. Remember from other PBIS Units that the best recommendations are supported by evidence and science knowledge.
>
> Try to make three types of recommendations:
>
> - What do you recommend people do to avoid getting the disease?
> - What do you recommend people do to avoid spreading the disease?
> - What do you recommend populations do to avoid spreading the disease?
>
> Use a *Create Your Explanation* page for each recommendation. Record your recommendation as a claim. Then provide evidence and science knowledge to support your claim. For each, try to include reasons why others should follow your recommendation. Your explanation should tie your claim, evidence, and science knowledge together and include the ways the disease is spread.
>
> ...mmuni... ...r So...

Plan Your Solution: Make Your Recommendations

15 min.

Students develop recommendations about what people can do to avoid getting and spreading a disease and what populations can do to avoid spreading a disease.

Guide

After groups have completed their disease descriptions, have students read this section and make sure they understand the task: to make three types of recommendations. Remind students of the characteristics of a good recommendation – that they be supported by evidence and science knowledge.

Get Going

Have groups work collaboratively using *Create Your Explanation* pages to develop their recommendations. Remind groups to record their recommendation as a claim and to use a separate *Create Your Explanation* page for each recommendation. Let groups know that they will be presenting their recommendations to the class.

Assess

As groups are developing their recommendations, assess what they are recording on the *Create Your Explanation* page. Pay particular attention to see if they have included opinions rather than science knowledge and if they are making logical connections using phrases such as: "I observed..., therefore,..." and "because,..."

NOTE: This is an opportune time to conclude the first session of this activity. You may want to assign groups to complete their disease descriptions for homework.

> recomm... claim. Then provide ... science know... support your claim. For each, try to include reasons why others should follow your recommendation. Your explanation should tie your claim, evidence, and science knowledge together and include the ways the disease is spread.
>
> **Communicate Your Solution**
>
> *Solution Showcase*
>
> Make a poster that reports on your disease. Tell your class about the disease you are reporting about. Be sure to be especially clear about how the disease interacts with different body systems, how it is transmitted, the human actions that cause it to spread, and ways to prevent its spread. Your classmates will be particularly interested in understanding enough about your disease to know why the ways of preventing its spread will work. Make sure you answer the following questions in your presentation:
>
> - What is the disease?
> - What are its symptoms?
> - What body systems does it affect?

Communicate Your Solution: Solution Showcase

2 class periods

Students make posters that report on their disease and make a class presentation of their findings and recommendations.

⬡ Get Going (1 class period)

Have groups work collaboratively to make a poster that reports on their disease. Emphasize that the poster should include answers to the questions listed in their text:

- What is the disease?
- What are its symptoms?
- What body systems does it affect?
- How is it treated?
- How does it spread to others?
- What are particular things people do that help spread this disease?
- How can the disease's spread be prevented?
- What do you recommend people do to avoid getting the disease?
- What do you recommend people do to avoid transmitting it to others?
- What do you recommend populations do to combat this disease?
- For each of your recommendations, explain why it would work?

Let groups know how long they will have for their *Solution Showcase* presentation and allow them time to prepare their presentation.

NOTE: This is an opportune time to conclude the second session of this activity. You may want to assign groups to complete their posters and prepare their presentations for homework.

△ Guide (1 class period)

As each group presents, encourage the other groups to ask questions about the disease and how to prevent its spread. You may want to model the kinds of questions students should be asking. These should be about how the evidence supports the recommendations and their underlying explanations. The questions listed in the text provide a starting point and will help students formulate their own questions. After all of the groups have made their presentations, lead a class discussion focused on how well each of the recommendations would work and identifying recommendations that many groups had in common. Guide students to recognize that some recommendations will help prevent the spread of a wide variety of diseases. Encourage students to ask questions about any of the recommendations that they do not understand or do not think will work.

NOTES

PBIS

- How does it spread to others?
- What particular things do people do that help spread this disease?
- How is it treated?
- How can its spread be prevented?
- What do you recommend people do to avoid getting the disease?
- What do you recommend people do to avoid transmitting it to others?
- What do you recommend populations do to combat this disease?
- For each one of your recommendations, explain why it would work.

As you listen to your classmates make their presentations, make sure you understand how to prevent the disease from spreading and how well each of their recommendations will work. If you do not understand those things, or if you do not think their recommendations will work, ask questions. The questions above might help you figure out what questions to ask.

Update the *Project Board*

The important thing to focus on in updating the *Project Board* this time is applying what you know to the *Big Question*. This goes in the last column. The *Big Question* is about how you can prevent your good friends from getting sick. What have you just learned from these presentations that can help you answer that question? In the last part of this Unit, you will be writing recommendations for making your classroom or school a healthier place. Update the *Project Board* with this in mind.

Make Your School a Healthier Place

Now that you know some things about how to prevent the spread of many different bacterial and viral illnesses, you can make recommendations about how to make your classroom or school a healthier place. You will certainly want to recommend things like sneezing into a tissue and washing your hands after you sneeze. But this Unit has a lot more than that in it about how to prevent the spread of diseases.

Your class's final product will be a set of recommendations for staying healthy and helping others stay healthy. Each recommendation will be supported by evidence from your investigations and science knowledge you learned.

Update the *Project Board*

15 min.

Students update the Project Board by applying what they know to answer the Big Question.

Make Your School a Healthier Place

1-2 class periods

△ Guide

Lead a class discussion in which you guide students to apply what they have learned to answering the *Big Question: How Can I Prevent My Good Friends From Getting Sick?* Record students' ideas in the What does it mean for the challenge or question? column. Inform students that in the last part of this Unit, they will be writing recommendations for making their classroom or school a healthier place.

Each group in the class will play a part in contributing to that list. It will be important to do several things in preparation for your group work:

- As a class, identify which diseases you will make recommendations about.
- As a class, create one recommendation together. Begin by identifying explanations and recommendations you've already made that are relevant to making a recommendation about this disease. From those explanations and recommendations, develop a more specific recommendation about how to prevent the spread of that disease in your school. Attach evidence from your investigations and science knowledge from your readings to support it, and develop an explanation statement to go with it. The explanations and recommendations you have developed during this Unit should help you with this.
- As a class, review the last column on the *Project Board*, the one that suggests answers to the *Big Question*.

Then each group will make recommendations about preventing the spread of disease in some place in the school—the classroom, the lunchroom, the bathroom, and so on. Use a Create Your Explanation page for each of your recommendations.

Make a poster that can be displayed in the school. Your poster should convince others in your school to follow your advice.

△ Guide

Lead a class discussion to identify the specific diseases that will be the focus of the class' recommendations. Guide the class to reach consensus on the list of diseases to address.

To model the process, guide the class in creating one recommendation together. Identify one recommendation from those already proposed and have the class work collaboratively to make it more specific to their school. Then have the class attach a list of evidence and science knowledge that supports the recommendation and develop an explanation statement to go with it. Next, have groups develop recommendations about how they could make their classroom or school a healthier place. You may want to suggest that recommendations be linked with a particular location in which they would be posted (e.g. the bathroom: wash your hands before leaving). Make sure that groups use a *Create Your Explanation* page for each of their recommendations. Have groups create posters of their recommendations that can be displayed in the school. Have each group present their poster. Encourage the class to ask questions and make suggestions about how the poster could be improved. Ask groups to update their recommendation posters and show them to you before making their final poster.

NOTES

NOTES

Black Line Masters

Good Friends and Germs Blackline Masters

* Number indicates Learning Set.section.sequence within section

Liquid Interactions Table

Name: ______________________ Date: ____________

Interaction	Name	Time of Interaction
1		
2		
3		
4		
5		
6		
7		
8		
9		
10		

Communicable Disease Information Table

Name: ______________________ **Date:** ____________

Name of disease	Cause (bacterial or viral)	How does it spread?	Which body systems does it affect?	What are the symptoms?	How is it treated? Is there a vaccine?	Other notes
Write the scientific name and the common name. People usually say that they have "the flu," which is a shorter way to say that they have "influenza."	Many diseases are caused by either a virus or a bacterium.	Some diseases are spread through blood, body fluids (like saliva), air, contaminated food, and animal bites.	Some diseases affect several systems. List the main one(s).	The symptoms are what help you to determine that you have a particular disease.	"Treat the symptoms" means doing something so the symptoms will not be as bad. Sometimes, you can treat a disease, sometimes you can only treat the symptoms.	Use this column to write anything you think is interesting or important to remember about a particular disease.
common cold	viral	Body fluids in the air or on surfaces—blowing your nose, touching your hands to other people or objects, sneezing or coughing.	respiratory	cough, body aches, congestion	It can't be cured. You can only relieve the symptoms to feel better, using medicine to reduce the fever and to stop the runny nose, and cough.	

Observing Cells

Name: ______________________________ **Date:** ______________

Animal Cells

Low power	Medium power	High power

Plant Cells Without Iodine Stain

Low power	Medium power	High power

Plant Cells with Iodine Stain

Low power	Medium power	High power

Communicable Disease Information Table

Name: ______________________ Date: ____________

Name of disease	Cause (bacterial or viral)	How does it spread?	Which body systems does it affect?	What are the symptoms?	How is it treated? Is there a vaccine?	Other notes
Write the scientific name and the common name. People usually say that they have "the flu," which is a shorter way to say that they have "influenza."	Many diseases are caused by either a virus or a bacterium.	Some diseases are spread through blood, body fluids (like saliva), air, contaminated food, and animal bites.	Some diseases affect several systems. List the main one(s).	The symptoms are what help you to determine that you have a particular disease.	"Treat the symptoms" means doing something so the symptoms will not be as bad. Sometimes, you can treat a disease, sometimes you can only treat the symptoms.	Use this column to write anything you think is interesting or important to remember about a particular disease.

My Breathing Rate Data

Name: ______________________________ **Date:** ______________

Trials	My number of breaths in 6 s without exercise	My number of breaths in 1 min without exercise (breaths in 6 s x 10)	My number of breaths in 6 s with exercise	My number of breaths in 1 min with exercise (breaths in 6 s x 10)
1		x 10 =		x 10 =
2		x 10 =		x 10 =
3		x 10 =		x 10 =
Average number of breaths per minute for all trials	Add together the number of breaths per minute for each trial and divide by 3			

Class Breathing Rate Data

Name: ______________________________ **Date:** ______________

	Resting breaths per minute	**Exercising breaths per minute**
1		
2		
3		
4		
5		
6		
7		
8		
9		
10		
11		
12		
13		
14		
15		
16		
17		
18		
19		
20		
21		
22		
23		
24		
25		
26		
27		
28		
29		
30		
Class average number of breaths per minute	Add together the number of breaths per minute for each student and divide by the number of students.	

My Heart Rate Data

Name: ____ **Date:** ________________

Trials	My number of heart beats in 6 s without exercise	My number of heart beats in 1 min without exercise (heartbeats in 6 s x 10)	My number of heart beats in 6 s with exercise	My number of heart beats in 1 min with exercise (heartbeats in 6 s x 10)
1		x 10 =		x 10 =
2		x 10 =		x 10 =
3		x 10 =		x 10 =
Averages				

Name: ______________________ **Date:** ______________

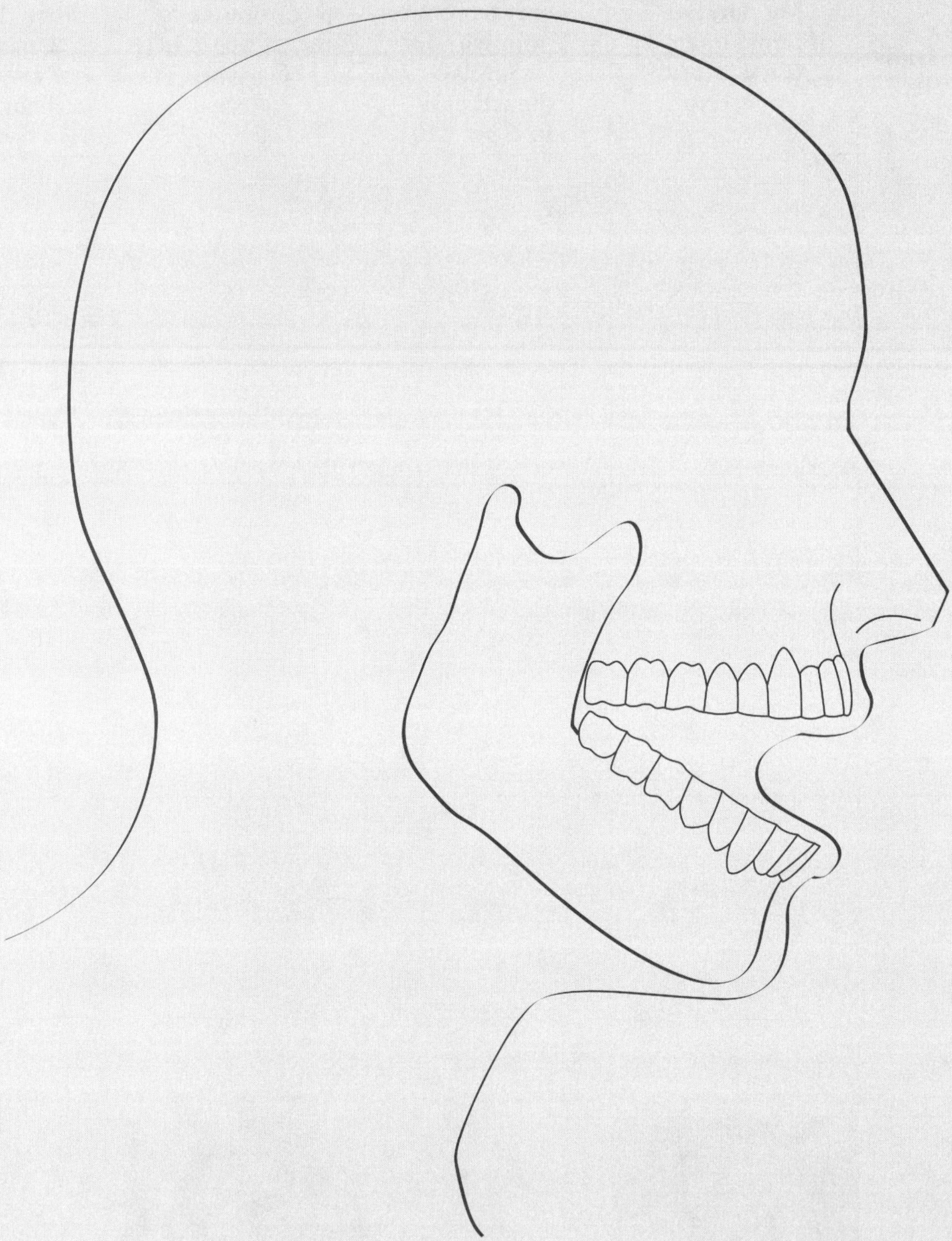

Create Your Explanation

Name: ______________________________ **Date:** ______________

Use this page to explain the lesson of your recent investigations.

Write a brief summary of the results from your investigation. You will use this summary to help you write your Explanation.

Claim – a statement of what you understand or a conclusion that you have reached from an investigation or a set of investigations.

Evidence – data collected during investigations and trends in that data.

Science knowledge – knowledge about how things work. You may have learned this through reading, talking to an expert, discussion, or other experiences.

Write your Explanation using the ***Claim***, ***Evidence*** and ***Science knowledge***.

NOTES

Credits

Credits

IT's ABOUT TIME®
HERFF JONES EDUCATION DIVISION

84 Business Park Drive, Armonk, NY 10504
Phone (914) 273-2233 Fax (914) 273-2227
www.its-about-time.com

Publishing Staff

President
Tom Laster

Director of Product Development
Barbara Zahm, Ph.D

Managing Editor
Maureen Grassi

Project Development Editor
Ruta Demery

Project Manager
Sarah V. Gruber

Assistant Editors, Student Edition
Gail Foreman
Susan Gibian
Nomi Schwartz

Assistant Editors, Teacher's Planning Guide
Danielle Bouchat-Friedman
Kelly Crowley
Edward Denecke
Heide M. Doss
Jake Gillis
Rhonda Gordon

Safety and Content Reviewers
Edward Robeck
Barbara Speziale

Creative Director
John Nordland

Production/Studio Manager
Robert Schwalb

Production
Sean Campbell

Illustrator
Dennis Falcon

Technical Art/Photo Research
Sean Campbell
Michael Hortens
Marie Killoran

Equipment Kit Developers
Dana Turner
Joseph DeMarco